TRANSLATIONS OF
MATHEMATICAL
MONOGRAPHS
VOLUME 64

M. M. Lavrent′ev
V. G. Romanov
S. P. Shishat·skiĭ

Ill-posed Problems of Mathematical Physics and Analysis

American Mathematical Society
Providence, Rhode Island

НЕКОРРЕКТНЫЕ ЗАДАЧИ МАТЕМАТИЧЕСКОЙ ФИЗИКИ И АНАЛИЗА

М. М. ЛАВРЕНТЬЕВ
В. Г. РОМАНОВ
С. П. ШИШАТСКИЙ

АКАДЕМИЯ НАУК СССР
СИБИРСКОЕ ОТДЕЛЕНИЕ
ВЫЧИСЛИТЕЛЬНЫЙ ЦЕНТР
ИЗДАТЕЛЬСТВО «НАУКА»
МОСКВА 1980

Translated from the Russian by J. R. Schulenberger
Translation edited by Lev J. Leifman

1980 *Mathematics Subject Classification* (1985 *Revision*). Primary 35R30; Secondary 45D05.

ABSTRACT. A number of applied problems connected with the interpretation of geophysical data and leading to mathematical problems which are ill-posed in the sense of Hadamard are considered. The exposition includes the basic concepts of ill-posed problems, problems of analytic continuation from continua and discrete sets, and analogous problems of continuation of solutions of elliptic and parabolic equations, the main ill-posed boundary value problems for partial differential equations, and results on the theory of Volterra equations of the first kind, in particular, on the theory of operator Volterra equations. A very broad presentation is given of modern results on the problem of uniqueness in integral geometry and on inverse problems for partial differential equations.

The monograph is of interest to specialists in applied mathematics, physics, and geophysics.

Library of Congress Cataloging-in-Publication Data

Lavrent'ev, M. M. (Mikhail Mikhaĭlovich)
Ill-posed problems of mathematical physics and analysis.
(Translation of mathematical monographs; v. 64)
Translation of: Nekorrektnye zadachi matematicheskoĭ fiziki i analiza.
Bibliography: p.
1. Boundary value problems—Improperly posed problems. 2. Numerical analysis—Improperly posed problems. 3. Mathematical physics. I. Romanov, V. G. (Vladimir Gavrilovich) II. Shishatskii, S. P. (Sergeĭ Petrovich) III. Title. IV. Series.
QC20.7.B6L3813 1986 530.1'55 86-3642
ISBN 0-8218-4517-9

10 9 8 7 6 5 4 3 2 02 01 00 99 98 97

Contents

Preface v

Introduction 1

CHAPTER I. Physical Formulations Leading to Ill-posed Problems 7
1. Continuation of static fields 7
2. Problems for the diffusion equation 10
3. Continuation of fields from discrete sets 11
4. Processing readings of physical instruments 13
5. Inverse problems of geophysics 15
6. Inverse problems of gravimetry 16
7. The inverse kinematic problem of seismology 19

CHAPTER II. Basic Concepts of the Theory of Ill-posed Problems 26
1. Problems well-posed in the Tikhonov sense 26
2. Regularization 32
3. Linear ill-posed problems 39

CHAPTER III. Analytic Continuation 48
1. Formulations of problems and classical results 48
2. Analytic continuation from continua 52
3. Analytic continuation from classes of sets including discrete sets 57
4. Recovery of solutions of elliptic and parabolic equations from their values on sets lying inside the domain of regularity 63

CHAPTER IV. Boundary Value Problems for Differential Equations 72
1. The noncharacteristic Cauchy problem for a parabolic equation. The Cauchy problem for an elliptic equation 73
2. A mixed problem for a parabolic equation with decreasing time 95
3. Cauchy problems with data on a segment of the time axis for degenerate parabolic and pseudoparabolic equations 106

4. Cauchy problems with data on a timelike surface for hyperbolic and ultrahyperbolic equations 123

CHAPTER V. Volterra Equations 155
1. Regularization of a Volterra equation of the first kind 155
2. Operator Volterra equations of the first kind 159

CHAPTER VI. Integral Geometry 168
1. The problem of finding a function from its spherical means 170
2. Problems of integral geometry on a family of manifolds which is invariant under a group of transformations of the space 175
3. Integral geometry in special classes of functions 183
4. Integral geometry "in the small" 190
5. The problem of integral geometry in plane curves and energy inequalities 202

CHAPTER VII. Multidimensional Inverse Problems for Linear Differential Equations 209
1. Examples of formulations of multidimensional inverse problems. Mathematical problems connected with investigating them 211
2. A general approach to investigating questions of uniqueness and stability of inverse problems 220
3. Inverse problems for hyperbolic equations of second order 227
4. Inverse problems for first-order hyperbolic systems 234
5. Inverse problems for parabolic equations of second order 251
   1. The connection between solutions of direct problems for equations of hyperbolic and parabolic types and inverse problems 251
   2. The method of descent in inverse problems 252
   3. Inverse problems for equations of parabolic type 254
6. An abstract inverse problem and questions of its being well-posed 255
   1. Reduction to the investigation of a two-parameter family of linear equations 257
   2. The method of linearization in investigating the inverse problem 261

Bibliography 265

Preface

The theory of ill-posed problems is a direction of mathematics which has developed intensively in the last two decades and is connected with the most varied applied problems: interpretation of readings of many physical instruments and of geophysical, geological, and astronomical observations, optimization of control, management and planning, synthesis of automatic systems, etc. Development of the theory of ill-posed problems was occasioned by the advent of modern computing technology.

Various areas of the theory of ill-posed problems can be included in traditional areas of mathematics such as function theory, functional analysis, differential equations, and linear algebra.

The concept of a well-posed problem is connected with investigations by the famous French mathematician Hadamard of various boundary value problems for the equations of mathematical physics. Hadamard expressed the opinion that boundary value problems whose solutions do not satisfy certain continuity conditions are not physically meaningful, and he presented examples of such problems.

It was subsequently found that Hadamard's opinion was erroneous. It turned out that many problems of mathematical physics which are ill-posed in the sense of Hadamard and, in particular, problems noted by Hadamard himself have real physical content. It also turned out that ill-posed problems arise in many other areas of mathematics which are connected with applications. Such a classical problem of mathematical analysis as the problem of differentiation is ill-posed if it is connected with processing experimental data.

In our country several monographs have been published which are devoted to the theory of ill-posed problems and their applications (see [113], [145], [152], [164], [184] and [266]). The present state of the theory of ill-posed problems is reflected most completely in the monograph of A. N. Tikhonov and V. Ya. Arsenin [266] and in the recently published monograph of V. K. Ivanov, V. V. Vasin and V. P. Tanana [113]. However, a number of important areas of the theory of ill-posed problems are absent or little reflected in the monograph literature noted. This refers, for example, to ill-posed problems for concrete types of differential equations, problems of analytic continuation, inverse problems for differential equations, and problems of integral geometry.

In the present monograph the authors have attempted to fill these gaps in the monograph literature on the theory of ill-posed problems, and also to give a new treatment of some of its areas.

The authors express their profound gratitude to A. L. Bukhgeĭm, who familiarized himself with the manuscript in detail and made a number of useful remarks.

Introduction

1. The concept of a well-posed (correct) problem of mathematical physics was formulated at the beginning of this century by the famous French mathematician Hadamard (see [306] and [307]). At the present time this concept is widely presented in textbooks on the equations of mathematical physics or partial differential equations.

A problem of mathematical physics or a boundary value problem for a partial differential equation is called well-posed if the following conditions are satisfied: 1) a solution of the problem exists; 2) the solution of the problem is unique; and 3) the solution of the problem depends continuously on the data of the problem (see, for example, [139]).

The conditions just formulated require refinement. Namely, both the solution and the data of the problem are considered as elements of some function space, and the conditions for a problem to be well-posed are formulated as follows.

A. A solution of the problem exists for all data belonging to some closed subspace in a normed linear space of the type C^k, L_p, H^l_p, W^l_p, and belongs to a space of the same type. The subspace is most often either the entire space or a part of the space on which a finite collection of linear functionals vanishes.

B. The solution of the problem is unique in some analogous space.

C. To infinitesimal variations of the data of the problem in the data space there correspond infinitesimal variations of the solution in the solution space (see [75], [199], and [248]).

Having formulated the concept of a well-posed problem, Hadamard presented an example of an ill-posed problem for a differential equation which in his opinion did not correspond to any real physical formulation. This example was the Cauchy problem for the Laplace equation.

It can be shown that the solution of the Cauchy problem for the Laplace equation is unique, i.e., this problem satisfies the second condition for being well-posed. Without dwelling on the question of existence, we shall show that in the Cauchy problem for the Laplace equation continuous dependence of the solution on the data does not hold for any pair of the spaces mentioned.

We restrict ourselves to the simplest version of the problem. Let

$$u = u(x,y), \qquad \Delta u = 0, \quad y > 0, \tag{1}$$

$$u(x,0) = 0, \qquad \frac{\partial}{\partial y} u(x,0) = \alpha \sin nx, \quad x \in [0,\pi]. \tag{2}$$

The Cauchy problem (1), (2) has the solution

$$u(x,y) = \frac{\alpha}{n} \sinh ny \cdot \sin nx.$$

We consider the problem of finding the solution u for some fixed $y > 0$ on the basis of the Cauchy data at $y = 0$. It is obvious that for any pair of the function spaces mentioned above and any $\varepsilon > 0$, $c > 0$, and $y > 0$, it is possible to choose α and n such that

$$\|\alpha \sin nx\| < \varepsilon, \qquad \left\| \frac{\alpha}{n} \sinh ny \cdot \sin nx \right\| > c.$$

This example, just as the concept of a well-posed problem itself, is well known.

An analogous situation with continuous dependence of the solution on the data occurs also in the Cauchy problem for the heat equation with reverse time:

$$u = u(x,y), \qquad \frac{\partial u}{\partial t} = -\frac{\partial^2 u}{\partial x^2}, \quad t > 0, \tag{1'}$$

$$u(x,0) = f(x). \tag{2'}$$

The opinion was expressed that the Cauchy problem for the Laplace equation and for elliptic equations in general is ill-posed, since the Cauchy problem is natural for the description of processes developing in time, while elliptic equations describe steady-state processes and physical fields. The fact that the Cauchy problem for the heat equation with reverse time is ill-posed was associated with the second law of thermodynamics.

2. It turned out that the opinion of Hadamard regarding the Cauchy problem for the Laplace equation and a number of other problems of this same type was erroneous. Problems ill-posed in the classical sense were encountered long ago in the mathematical description of physical phenomena. A number of classical results in mathematical analysis and differential equations of the last century and the beginning of this one can be attributed to the theory of ill-posed problems. However, the systematic study of questions in the theory of ill-posed problems began comparatively recently.

In discussing ill-posed problems with representatives of the natural sciences, mathematicians repeatedly asserted that consideration of ill-posed problems is devoid of meaning, referring here to the opinion and example of Hadamard. In the thirties and forties, geophysicists discovered a connection between a problem equivalent to the Cauchy problem for the Laplace equation and certain questions of the interpretation of gravitational and magnetic anomalies. Expert mathematicians evaded discussing these problems, and attempts to carry out calculations without a theoretical foundation were ineffectual. A trial-and-error method widely used by practicing geophysicists also remained without theoretical basis.

In recent years it was established that the set of phenomena in physics whose mathematical description is connected with ill-posed problems is not of the same type as the set of phenomena associated with problems which are well-posed in the classical sense. In the first chapter of this book we present a series of examples of such phenomena and problems: a number of other examples are given in the monographs [93], [164], and [266]. However, all these examples comprise only a small part of the various published formulations.

3. The method of integral equations is one of the main methods of investigating boundary value problems for the equations of mathematical physics. Problems well-posed in the classical sense reduce to Fredholm integral equations of the second kind or to singular integral equations. The latter, after some transformations, usually also reduce to equations of the second kind. Ill-posed problems frequently reduce relatively simply to integral equations of the first kind. In particular, this goes for the two ill-posed problems mentioned above. The integral equations to which the Cauchy problem for the Laplace equation and the Cauchy problem for the heat equation with reverse time reduce have the respective forms

$$\frac{y}{\pi}\int_{-\infty}^{\infty}\frac{\varphi(\xi)}{y^2+(x-\xi)^2}\,d\xi = f(x), \tag{3}$$

$$\frac{1}{2\sqrt{\pi t}}\int_{-\infty}^{\infty}\exp\left[-\frac{(x-\xi)^2}{4t}\right]\varphi(\xi)\,d\xi = f(x). \tag{3$'$}$$

The concept of an operator equation of the first kind is a natural generalization of the concept of an integral equation of the first kind. An *operator equation of the first kind* is defined to be an equation of the form

$$Ax = f, \tag{4}$$

where x and f are elements of some spaces X and F; A is a compact (linear) operator acting from X to F.

It is known that the operator inverse to a compact operator (in an infinite-dimensional space), if it exists, is not continuous, and the problem of solving equation (4) is thus ill-posed. It is obvious that the problem of solving (4) may be well-posed if it is considered for some other pair of spaces X_1, F_1. Here if $X_1 = X$, then in order that the problem of solving (4) be well-posed it is necessary that F_1 be a subspace of F, while if $F_1 = F$, then it is necessary that X_1 be an extension of X.

In discussing questions related to ill-posed problems the opinion was expressed that in problems of the type of the Cauchy problem for the Laplace equation it suffices to choose an appropriate pair of function spaces, and the problem for this pair will be well-posed. Indeed, in the Cauchy problem for the Laplace equation, in the Cauchy problem for the heat equation with reverse time, and also in many other problems of similar type, it actually is comparatively easy to choose pairs of spaces in which the problems become well-posed.

However, this approach to ill-posed problems leaves aside an aspect which is very important from the point of view of applications. The fact of the matter is that if an equation of the type (4) is considered in connection with the mathematical modelling of a real physical phenomenon, then the right side of the equation is often obtained on the basis of readings of physical instruments. Since the instruments possess errors, in such cases we cannot assume that the right side of (4) is given with absolute accuracy. We may suppose only that we are given an element f_ε satisfying the inequality

$$\|f - f_\varepsilon\| \le \varepsilon, \tag{5}$$

where the number ε is determined by the accuracy of the instruments. Here the norm of the space in which we know an estimate of the error of the right side cannot be prescribed arbitrarily: it is dictated by the organization of the system of measurements. As a rule, this is either the norm in the space C (we know an estimate of the maximal error of the measurements) or the norm in L_2 (we know the mean square error).

Although it presents additional technical difficulties, an organization of the system of measurements is possible where the error is small together with its derivative—the norm in the space C^1 or $W_2^{(1)}$. An organization of measurements where the error is small together with its second derivative (the spaces C^2 and $W_2^{(2)}$) either cannot be accomplished or presents considerable technical difficulties.

Let us consider the integral equations (3) and (3′). It is obvious that if solutions of these equations exist, then the right sides of the equations have derivatives of all orders. Moreover, the right sides are analytic functions in some complex neighborhood of the real axis, which imposes a limitation on the order of growth of the norms of the derivatives.

Let F be a normed function space such that the problem of solving (3) is well-posed for a pair of spaces consisting of any ordinary space of solutions, for example, C or L_2, and the space F of right sides. From what has been said above it follows that derivatives of all orders must participate in the definition of the norm in F, and this is unacceptable from the viewpoint of real systems of measurement.

An analogous situation occurs also in problems not connected with physical measurements, for example, in problems of synthesis and control [266].

Let us consider a classical problem of mathematical analysis—the problem of differentiaton. If a function is given in the form of a formula, a composition of elementary functions, or integrals depending on a parameter, then the derivative is found by the classical rules for differentiation also in the form of a formula. If, however, the function is obtained on the basis of experimental data in the form of a table or graph, the problem of finding derivatives becomes more complicated. As already noted, the error can most often be considered small in the spaces C or L_2, and it is desirable to obtain a derivative with an error small in these same spaces. In this case the problem of differentiation is a problem which is ill-posed in the classical sense.

4. Principles for approaching formulations of ill-posed problems which are natural from the viewpoint of applications were first stated in the work of A. N. Tikhonov in 1943 [257] which was related to the justification of the trial-and-error method in the interpretation of data of geophysical measurements, and attracted the attention of many researchers, especially geophysicists. In many works, monographs, and even textbooks on geophysics the authors refer to this work in presenting certain techniques of interpretation. Intensive development of the theory of ill-posed problems began in the mid-fifties of our century. It was connected with the advent of computers and the start of their broad application (see [134], [135], [140]–[146], [160], [161], [249], [277], [298], [301], [302], [305], [311]–[315], [319], and [322]–[324]). The concept of regularization introduced by A. N. Tikhonov in a series of works in the sixties ([258]–[268]) played a significant role in the development of the theory of ill-posed problems.

5. The theory of ill-posed problems received further substantial development in the works of a number of authors (see [24]–[28], [30]–[33], [62], [69]–[72], [90]–[94], [97], [98], [104], [105]–[114], [132], [150]–[155], [164], [166], [167], [174], [178]–[186], [203], [204], [252]–[255], [276], [279], [282], [283], [297], and [327]).

A number of directions can be distinguished in the work on ill-posed problems. In the present book the authors have attempted to present achievements in some of these directions.

The first chapter is devoted to an exposition of applied problems leading to ill-posed problems. As already noted, the range of such problems is very wide. The authors have limited themselves to presenting a number of examples connected with the processing and interpretation of data of physical measurements.

The basic postulates and some results in the general theory of ill-posed problems are presented in the second chapter. This direction can also be considered to be an area of functional analysis.

Problems of analytic continuation are considered in the third chapter. This direction actually starts from classical uniqueness theorems obtained in the last century. Estimates of conditional stability for problems of analytic continuation were first obtained by Carleman [299]. Full consideration of a problem of analytic continuation as an ill-posed problem was first given in [143].

The main results known at the present time on the theory of ill-posed boundary value problems for differential equations are presented in the fourth chapter.

Results on Volterra integral equations of the first kind and on Volterra operator equations are presented in the fifth chapter. Since the beginning of our century, Volterra equations and operators have attracted the attention of many researchers. However, consideration of problems of solving Volterra equations as ill-posed problems began not long ago, in [59], [66], and [150].

Results on a number of formulations of ill-posed problems of integral geometry are presented in the sixth chapter.

The seventh chapter contains some results from a vast area of mathematical physics—the theory of inverse problems for differential equations.

CHAPTER I

Physical Formulations Leading to Ill-posed Problems

§1. Continuation of static fields

1. As already noted in the Introduction, some problems of interpretation of gravitational and magnetic fields connected with the search for mineral resources lead to ill-posed problems equivalent to the Cauchy problem for the Laplace equation.

If the earth were a spherically homogeneous ball, the intensity of the gravitational field at the surface would be constant. Inhomogeneities of the terrain and of the distribution of the density of matter cause the intensity of the gravitational field on the surface of the earth to deviate from the mean value. Percentagewise these derivations are small but can be well recorded by physical instruments—gravimeters.([1]) Data from gravitational observations are used in prospecting for deposits of mineral resources.

Figures 1 and 2 show in cross section typical geologic structures of portions of the earth's crust and the data of gravitational measurements—the vertical components of anomalies of the intensity of the gravitational field. These anomalies are caused by the fact that the density of the base and of intrusions([2]) are usually notably higher than the density of sediments.([3]) The problem of interpreting gravitational data is to draw conclusions on the basis of the gravitational measurements regarding the topography of the base and the distribution and form of intrusions. Deposits of mineral resources are often associated with characteristic forms of the topography of the base and

([1])The average intensity of the gravitational field of the earth is 1 kg=980 gal. Typical anomalies have an intensity of 0.5–10 mgal. The accuracy of readings of gravimeters is 0.02–0.05 mgal.

([2])Geological bodies formed after introduction of molten magma into the surrounding rock.

([3])The density of sediments is 2–4 g/cm^3. The density of the base and of intrusions is 4–6 g/cm^3.

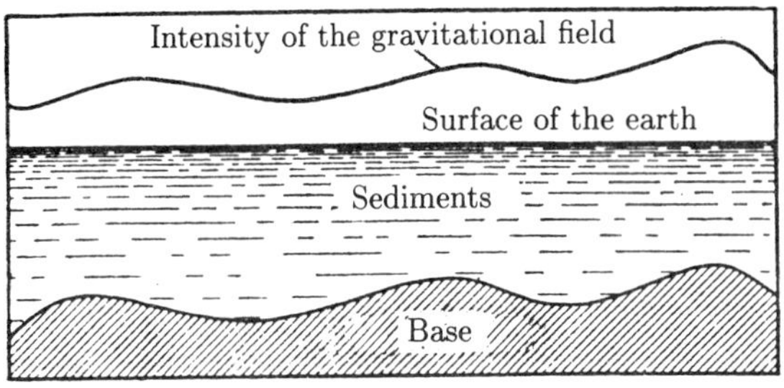

Figure 1

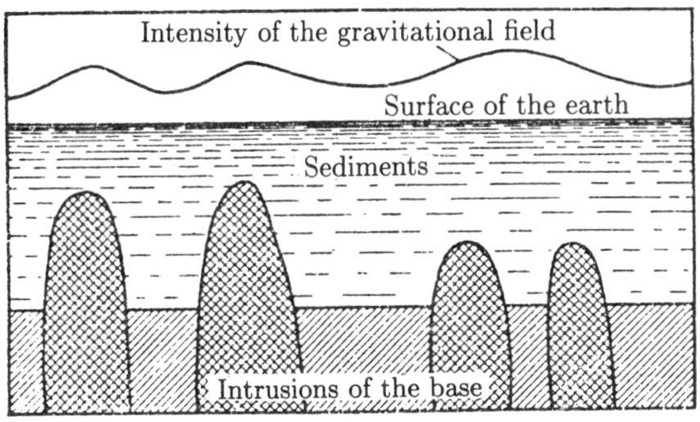

Figure 2

intrusions; for example, in oil-bearing regions oil deposits are frequently associated with uplifts of the topography of the base—domes, while ore deposits are associated with intrusions.

We shall consider in more detail the problem of determining the site of intrusions from the data of gravitational measurements. In the case where the distance between geological bodies—intrusions—is greater than the distance from the bodies to the surface of the earth the positions of the bodies correspond to local maxima of the anomalies (see the left part of Figure 2). In the case where the distance between the bodies is less than the distance from the bodies to the surface, one local maximum corresponds to two bodies (see the right part of Figure 2). In prospecting for mineral resources, geophysical measurements and their interpretation are a preparatory step. The main step in finding deposits hidden by sediments is drilling prospecting wells and analyzing the drilling data.

If we decide on the basis of the form of the anomaly shown on the right side of Figure 2 that it is caused by a single body, then we naturally choose a site for drilling in the center of the anomaly, and the hole then passes between the bodies of interest to us. This situation was repeatedly encountered in the practice of geological prospecting.

Geophysicists suggested the following: on the basis of data of gravitational measurements on the surface of the earth, calculate the anomalous gravitational field at some depth below the surface. In the case where at that depth the anomaly preserves the form of a single local maximum, it is possible with greater reliability to draw the conclusion that the anomaly is caused by a single body. In the case where, after recalculation, two local maxima appear, it is possible to conclude that there are two bodies and to choose the drilling site accordingly.

We now consider the mathematical formulation corresponding to the problem stated above of recalculating the gravitational field. For simplicity we restrict ourselves to the case of two variables. We note that this restriction frequently corresponds to real geophysical situations, i.e., there are geological structures extended in a particular direction, while geophysical measurements are carried out on a course perpendicular to this direction.

As is known, away from masses generating the gravitational field, the potential of the field and the intensity components satisfy the Laplace equation. Thus, suppose that a line on the surface of the earth coincides with the x-axis; let $u(x, y)$ be the vertical component of the intensity of the gravitational field generated by several bodies lying below the level $y = -H, H > 0$. Then the function $u(x, y)$ is a solution of the Laplace equation $\Delta u = 0$ which is regular in the half-plane $y > -H$. The values of $u(x, y)$ are given on the x-axis, i.e., there is given a function $f(x)$ such that $f(x) = u(x, 0)$. It is required to determine the values of the function

$$\varphi(x) = u(x, -h), \quad 0 < h < H.$$

Solving this problem by means of Poisson's formula reduces to the integral equation of the first kind

$$\int_{-\infty}^{\infty} \frac{h\varphi(\xi)}{h^2 + (x - \xi)^2}\, d\xi = f(x). \tag{1.1}$$

The problem we have formulated is equivalent to a Cauchy problem for the Laplace equation. For this problem the example of Hadamard has the form: if

$$f(x) = \alpha \sin nx,$$

then

$$\varphi(x) = \alpha e^{nh} \sin nx.$$

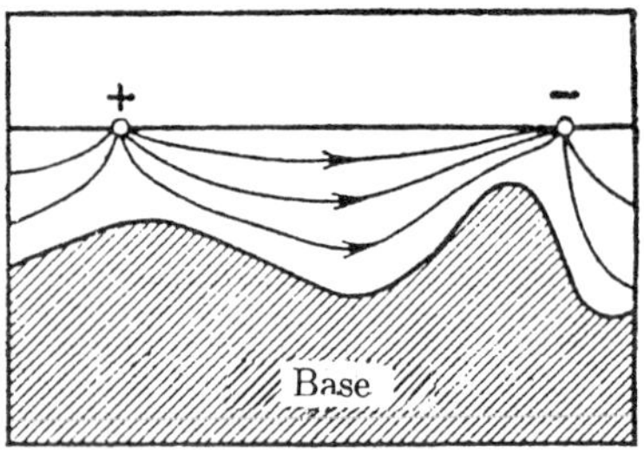

Figure 3

An entirely analogous formulation arises in interpreting anomalies of a constant magnetic field, since the potential and the intensity components of this field away from magnetic masses also satisfy the Laplace equation.

2. Geophysical prospecting by electrical means is one of the important methods of searching for mineral resources. In electrical prospecting with a constant current, electrodes are applied at two points on the earth's surface (Figure 3), and a constant current is started. The difference of potentials is measured on the surface. As in any geophysical prospecting, it is required to draw conclusions regarding the internal structure of the region on the basis of the measurements. In the case where the layer of sediments is a homogeneous conducting medium and the resistance of the base is much higher, the current lines will flow about the contour of the base. Hence, to determine the contour of the base it suffices to find the current lines in the layer of sediments. In a homogeneous medium the potential of a constant current satisfies the Laplace equation. The normal derivative of the potential on the earth's surface is equal to zero, and the potential itself is measured. We thus arrive at a Cauchy problem for the Laplace equation.

§2. Problems for the diffusion equation

We consider the equation

$$\partial u/\partial t = \partial^2 u/\partial x^2, \quad x \in [0, l], \ t \in [0, T], \tag{1.2}$$

describing the process of variation of temperature in a homogeneous rod and the process of diffusion—variations in the concentration of a substance in a solution. The classical problem for equation (1.2) is the mixed problem

$$u(x, 0) = f(x), \qquad u(0, t) = u(l, t) = 0. \tag{1.3}$$

The physical meaning of problem (1.2), (1.3) consists in the following: at some initial time we know the temperature distribution in the rod or the concentration of a substance in a cylindrical vessel. It is required to determine the temperature distribution or the concentration of the substance at later times. The boundary conditions mean that a constant temperature or concentration is maintained at the end points in our process.

We now consider an ill-posed problem for (1.2):

$$u(x,T) = f(x), \qquad u(0,t) = u(l,t) = 0. \tag{1.3$'$}$$

It is obvious that problem (1.3′) for equation (1.2) is equivalent to problem (1.3) for the equation

$$\partial u/\partial t = -\partial^2 u/\partial x^2. \tag{1.2$'$}$$

This problem also has an altogether definite physical meaning. We know (on the basis of physical measurements) the temperature distribution or concentration at some time. We are interested in the history of the process: what was the temperature distribution or concentration at previous times.

The following ill-posed problem for equation (1.2) also has a physical meaning:

$$u(0,t) = f_1(t), \quad u'_x(0,t) = f_2(t). \tag{1.4}$$

We suppose that we are unable to measure the concentration of substance in the entire cylinder but can only measure the concentration and flow of substance at one of the ends. The following example shows that problem (1.4) is ill-posed:

$$f_1(t) = \sqrt{2/n}\sin nt, \qquad f_2(t) = \cos nt + \sin nt,$$
$$u(x,t) = \sqrt{2/n}\sin(nt + \sqrt{n/2x})\exp\sqrt{n/2x}.$$

§3. Continuation of fields from discrete sets

1. In the problem of interpreting data of gravitational measurements discussed in §1, geophysicists actually have at their disposal the values of the component of the gravitational field not on the entire surface of the earth nor on a segment but only at some finite set of points. The following problem thus arises: it is required to determine the function $\varphi(x)$ in (1.1) if the values of the right side of (1.1) are known at a finite set of points, i.e., we know $f_k = f(x_k)$, $k = 1,\dots,n$. A solution of this problem is not unique, but it is obvious that if the points $\{x_k\}$ fill some segment $[a,b]$ with sufficient density, it is possible to extend the function $f(x)$ to the entire segment by interpolation. There thus arises an additional error on the right side of (1.1).

In applied problems, in particular in the problem of interpreting gravimetric data, measurement of the value of a function at a point requires certain expenditures. In connection with this there arises the problem of rational choice of the system of points of measurement. It is clearly not expedient to choose too fine a grid for which the errors of the simplest interpolation are considerably less than the errors of the measuring instruments. Resolution of the following question is of practical interest: to what accuracy is it possible to determine a solution of (1.1) if we know the numbers

$$\{f_k^\varepsilon\}, |f_k^\varepsilon - f(x_k)| \leq \varepsilon, \quad k = 1, \dots, n.$$

An analogous formulation is meaningful also in the case where the set of points $\{x_k\}$ is infinite. If the set $\{x_k\}$ has a limit point, a solution of (1.1) is uniquely determined by the sequence of numbers $f_k = f(x_k)$, $k = 1, \dots, \infty$.

Uniqueness of the solution of equation (1.1) on the basis of the data $\{f_k\}$ follows from a familiar theorem in the theory of analytic functions: if an analytic function vanishes on a set having a limit point in the interior of the domain of regularity, then it is identically equal to zero.

2. We note that many physical fields are described by analytic functions of the spatial variables.

The intensity components of a variable electromagnetic field in a medium with constant properties satisfy the Helmholtz equation

$$\Delta u + k^2 u = 0. \tag{1.5}$$

Solutions of (1.5) with a constant coefficient k are analytic functions, and the problem of determining the structure of the electromagnetic field from the data of measurements on a discrete set thus reduces to the problem of recovering an analytic function (of one or several variables) from its values on this set.

3. In short-term (up to three days) weather forecasting, the system of equations of gas dynamics is solved. The initial conditions are given mainly from the data of measurements made at meteorological stations. The network of meteorological stations is very nonuniform, and the location of the stations depends on the region, its closeness to population centers, and communications.

The system of equations of gas dynamics is presently solved with considerable simplifications. Under certain simplifications, equations of parabolic type are obtained from the system. The following mathematical formulation is thus related to a problem of weather forecasting which is very important in practice.

Let $u(x, y, t)$ be a solution of the diffusion equation

$$\partial u / \partial t = \Delta u. \tag{1.6}$$

It is required to determine the function u for $t > 0$ from the data

$$f_k(t) = u(x_k, y_k, t), \quad k = 1, \dots, n; \ t > 0.$$

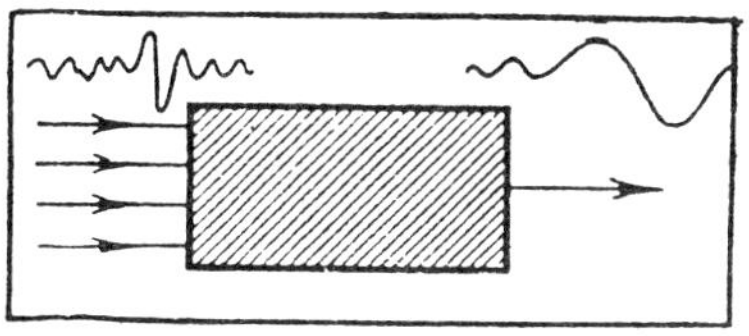

Figure 4

§4. Processing readings of physical instruments

1. The actions of many instruments recording time-dependent physical fields are described by the following scheme (Figure 4). A signal $\varphi(t)$ arrives at the input of the instrument, and a function $f(t)$ is recorded at the output. In the simplest case the functions $\varphi(t), f(t)$ are connected by the relation

$$\int_0^t g(t-\tau)\varphi(\tau)\,d\tau = f(t). \tag{1.7}$$

The function $g(t)$ in this case is called the *impulse transition function* of the instrument. Theoretically $g(t)$ is the function registered by the instrument in the case where the generalized function $\delta(t)$ (the Dirac delta function) arrives at the input of the instrument. In practice, a sufficiently short impulse is fed to the input to obtain the function $g(t)$ characterizing the operation of the instrument.

Thus, the problem of interpreting readings of an instrument, i.e., determining the form of the arriving signal, reduces to the solution of the integral equation of the first kind (1.7).

The connection between the signal $\varphi(t)$ arriving at the input of the instrument and the function $f(t)$ recorded at the output may also be more complex. In the case of a "linear" instrument this connection has the form

$$\int_0^t K(t,\tau)\varphi(\tau)\,d\tau = f(t). \tag{1.8}$$

A nonlinear connection between the functions $\varphi(t)$ and $f(t)$ is also possible, for example,

$$\int_0^t K(t,\tau,\varphi(\tau))\,d\tau = f(t). \tag{1.9}$$

In particular, instruments recording variable electromagnetic fields, regimes of pressures and stresses in a continuous medium, and seismographs recording oscillation of the surface of the earth function according to this scheme.

The Fourier and Laplace transforms can be used to solve the simplest equation (1.7). Let $\tilde{g}(\lambda), \tilde{\varphi}(\lambda)$, and $\tilde{f}(\lambda)$ be the Fourier transforms of the functions $g(t), \varphi(t)$, and $f(t)$ of (1.7):

$$\tilde{g}(\lambda) = \int_0^\infty e^{i\lambda t} g(t)\,dt, \quad \tilde{\varphi}(\lambda) = \int_0^\infty e^{i\lambda t}\varphi(t)\,dt, \quad \tilde{f}(\lambda) = \int_0^\infty e^{i\lambda t} f(t)\,dt.$$

Then by the convolution theorem we have $\tilde{g}(\lambda)\tilde{\varphi}(\lambda) = \tilde{f}(\lambda)$, whence, inverting the Fourier transform, we obtain for the solution of (1.7) the formula

$$\varphi(t) = \frac{1}{2\pi}\int_{-\infty}^{\infty} e^{-i\lambda t}\tilde{\varphi}(\lambda)\,d\lambda = \frac{1}{2\pi}\int_{-\infty}^{\infty} e^{-i\lambda t}\frac{\tilde{f}(\lambda)}{\tilde{g}(\lambda)}\,d\lambda. \tag{1.10}$$

Formula (1.10) is unstable, since the function $\tilde{g}(\lambda)$ (the Fourier transform of the impulse transition function of the instrument for real instruments) tends to zero as $\lambda \to \infty$, and thus arbitrarily small noise in the determination of $\tilde{f}(\lambda)$ for sufficiently large λ may lead to large variations in the solution $\varphi(t)$.

In the case where the function $g(t)$ is constant the problem of solving (1.7) is the problem of differentiation.

2. Let us consider the problem of processing photometric images. This problem has become very important in recent years in connection with photographing regions of the earth from space. We imagine an ideal photographic instrument with an infinitely narrow diaphragm (Figure 5). In photographing by means of this instrument, the illuminated points on the surface being photographed are projected onto points of a photographic plate.

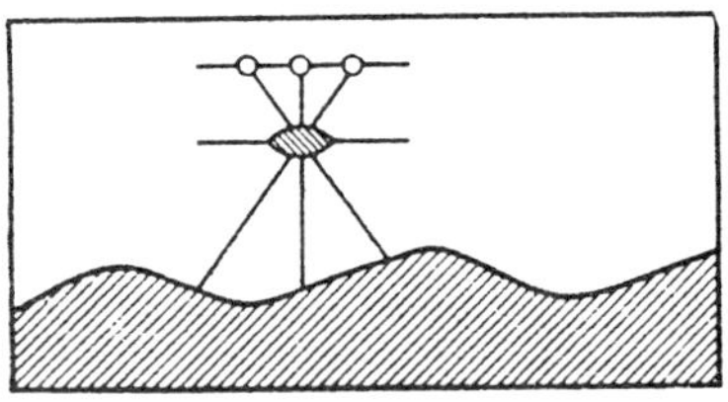

Figure 5

In real photographic instruments a luminous point is projected into a spot with variable illuminance depending on the direction and intensity of the ray. Thus, in solving the problem of recovering the true luminance of the surface being photographed we arrive at the integral equation

$$\int_S K(x,y,\xi,\eta)\varphi(\xi,\eta)\,d\xi\,d\eta = f(x,y). \tag{1.11}$$

Here $f(x, y)$ is the illuminance of the photographic plate; $\varphi(x, y)$ is the luminance of the surface being photographed; $K(x, y, \xi, \eta)$ is the illuminance of a point (x, y) on the plate by a point of unit luminosity (ξ, η).

§5. Inverse problems of geophysics

The problem of interpreting gravitational data discussed in §1 is one of a broad circle of problems called inverse problems of geophysics.

Geophysics is concerned with the study of connections of various physical fields with the structure of our planet [54]. Traditional fields in geophysics are the gravitational and constant magnetic fields, seismic fields (regimes of oscillation of the elements of the earth's surface), electric and electromagnetic fields, and thermal fields. Fields connected with radioactive radiations have also come to be studied in recent years. These fields may be of natural origin such as oscillations caused by earthquakes and telluric currents, or they may be specially induced as in seismic prospecting and geophysical prospecting by electrical means. Problems of mathematical physics arising in the interpretation of geophysical data are divided into direct and inverse problems.

Direct problems consist in the following: the sources of the physical field, the laws of interaction of the field with the medium written in the form of equations, and also the distribution of physical characteristics of the medium are known. It is required to determine the corresponding physical field.

In inverse problems, known physical laws and data of field measurements are presumed. It is required to determine the sources of the field or elements of the distribution of the characteristics of the medium.

The laws of interaction of traditional geophysical fields with the medium were described long ago. These are Newton's law of universal gravitation, Faraday's, Ohm's, and Hooke's laws, Huygens' principle, Maxwell's equations for electromagnetic fields, and the equations of the dynamic theory of elasticity, the Lamé equations.

In the first half of our century, fundamental existence and uniqueness theorems were proved for solutions of boundary value problems for differential equations and systems of equations corresponding to the main formulations of direct problems of geophysics.

Solution of boundary value problems corresponding to direct problems of geophysics may be considered as mathematical modelling of geophysical processes. Numerical solution of these problems plays an essential role in the correct understanding and estimation of the effectiveness of the results of geophysical observations, in designing geophysical instruments, and in planning experiments.

Numerical solution of direct problems of geophysics may present considerable difficulty. Before the advent of computers, only the simplest versions of models of the medium were accessible for calculations.

The most important purpose of geophysical research is determination of the earth's internal structure, i.e., the distribution within the earth of physical characteristics—density, magnetization, magnetic susceptibility, electrical conductivity, elasticity, and temperature. In connection with this it can be stated that the inverse problems are the main mathematical problems in geophysics.

The theory of inverse problems both in geophysics as well as in differential equations in general is much less developed than the theory of direct problems. It is obvious that many versions of inverse problems can be associated with each direct problem.

The simplest formulations of inverse problems are well-posed in the classical sense; these are certain one-dimensional problems where it is assumed that the properties of the medium depend on a single variable, for example the depth, and geometric inverse problems—problems of determining the shape of a surface on the basis of fields of times of arrival of reflected waves. Under less restrictive assumptions regarding the unknown medium, inverse problems are, as a rule, ill-posed.

§6. Inverse problems of gravimetry

In §1 it was shown how the problem of interpreting data of gravimetric measurements leads to an ill-posed Cauchy problem for the Laplace equation. In general, the problem of interpreting gravimetric data is an inverse problem of geophysics. We shall limit ourselves to considering the local problem, i.e., we assume that measurements are carried out on a bounded portion of the earth's surface, and the anomalies measured are caused by inhomogeneities situated in a bounded volume directly abutting the surface. We assume that the portion of the earth's surface being considered is part of a plane.

Thus, let S be part of the (x, y)-plane, and let D be the part of three-dimensional space (x, y, z) abutting S; let $\rho(x, y, z)$ be the deviation of the density of matter in the region D from the mean value; let $u(x, y, z)$ be the potential of the anomalous gravitational field. Then

$$u(x, y, z) = \gamma \int\int\int_D \frac{1}{R}\rho(\xi, \eta, \varsigma)\, d\xi\, d\eta\, d\varsigma,$$

where

$$R = [(x-\xi)^2 + (y-\eta)^2 + (z-\varsigma)^2]^{1/2};$$

γ is the gravitational constant.

The usually measured vertical component of the intensity of the anomalous field $H(x,y)$ is given by

$$H(x,y) = \frac{\partial}{\partial z} u(x,y,0) = \gamma \int\int\int_D \frac{\varsigma}{R^3} \rho \, d\xi \, d\eta \, d\varsigma. \tag{1.12}$$

Equality (1.12) may be considered an integral equation of the first kind for determining the unknown function $\rho(x,y,z)$. A solution of equation (1.12) is not unique in rather large classes of functions of three variables (for example, in classes of functions having finitely many continuous derivatives). This is natural, since the left side of (1.12) contains a function of two variables, while a function of three variables is to be determined.

In inverse problems of gravimetry, density distributions ρ which are determined by giving a function of two variables are considered. In practical interpretation the hypothesis is often adopted that the function $\rho(x,y,z)$ in some domain D_1 (a part of D) assumes a constant and known value, while in the remainder of D it is equal to zero:

$$\rho(x,y,z) = \rho_0, \quad (x,y,z) \in D_1,$$
$$\rho(x,y,z) = 0, \quad (x,y,z) \notin D_1.$$

Regarding the domain D_1, one of the following assumptions is additionally made.

1. The domain D_1 is starlike relative to some point. In this case (1.12) assumes the form

$$H(x,y) = \gamma \rho_0 \int_0^{2\pi} \int_0^{\pi} \int_0^{\varphi(\alpha,\theta)} \frac{1}{R^3} r^2 \sin\theta \, dr \, d\theta \, d\alpha, \tag{1.13}$$

where r, θ, α are polar coordinates.

In gravimetry this formulation of the inverse problem is called the problem of determining the form of an intrusion or ore deposit.

2. The domain D_1 is bounded by a surface $z = \varphi(x,y)$. In this case (1.12) takes the form

$$H(x,y) = \rho_0 \gamma \int_S \int_{z_0}^{\varphi(\xi,\eta)} \frac{1}{R^3} \, d\varsigma \, d\xi \, d\eta. \tag{1.14}$$

(It is assumed that D_1 is a cylindrical domain, and S is the base of D_1.)

Equations (1.13) and (1.14) for the functions $\varphi(\alpha,\theta)$ and $\varphi(x,y)$ are nonlinear equations of the first kind of Urysohn type.

It can be shown that for these equations there are examples analogous to the Hadamard example. We limit ourselves to presenting an example for a two-dimensional version of (1.14):

$$H(x) = \rho_0 \int_a^b \int_{y_0}^{y_0+\varphi(\xi)} \frac{\eta}{(x-\xi)^2 + \eta^2} \, d\eta \, d\xi = \frac{\rho_0}{2} \int_a^b \ln \frac{(x-\xi)^2 + \varphi^2(\xi)}{(x-\xi)^2 + y_0^2} \, d\xi. \tag{1.15}$$

If the function $y_0 + \varphi(x)$ is bounded away from zero:

$$y_0 + \varphi(x) \le y_0 + \varphi_0 < 0,$$

then the function $H(x)$ has derivatives of all orders. We set

$$z(\xi) = x - \xi + iy_0, \qquad w(\xi) = x - \xi + i\varphi(\xi).$$

Then

$$\frac{1}{2}\ln\frac{(x-\xi)^2 + \varphi^2(\xi)}{(x-\xi)^2 + y_0^2} = \operatorname{Re}\ln w - \operatorname{Re}\ln z,$$

whence

$$\frac{d^n H(x)}{dx^n} = \rho_0(-1)^{n-1}(n-1)!\int_a^b \left[[\operatorname{Re}\frac{1}{w^n} - \operatorname{Re}\frac{1}{z^n}\right] d\xi. \tag{1.16}$$

We consider the functions

$$\varphi_1(x) = \begin{cases} \varphi_0, & a + 2kh \le x < a + (2k+1)h, \\ 0, & a + (2k+1)h \le x < a + (2k+2)h, \end{cases}$$

$$\varphi_2(x) = \begin{cases} 0, & a + 2kh \le x < a + (2k+1)h, \\ \varphi_0, & a + (2k+1)h \le x < a + (2k+2)h, \end{cases}$$

$$k = 0, \dots, m-1, \qquad h = (b-a)/2m.$$

We denote the corresponding components of the intensity by $H_1(x)$ and $H_2(x)$. From (1.15) and (1.16) we obtain

$$|H_1(x) - H_2(x)| = \frac{\rho_0}{2}\left|\int_a^b \ln\frac{(x-\xi)^2 + \varphi_1^2(\xi)}{(x-\xi)^2 + \varphi_2^2(\xi)}\, d\xi\right| \tag{1.17}$$

$$\le \frac{\rho_0}{2}\sum_{k=0}^{m-1}\int_{a+2kh}^{a+(2k+1)h}\left|\ln\frac{(x-\xi)^2 + \varphi_0^2}{(x-\xi-h)^2 + \varphi_0^2}\right| d\xi,$$

$$\left|\frac{d^n H_1(x)}{dx^n} - \frac{d^n H_2(x)}{dx^n}\right|$$

$$\le \rho_0(n-1)!\sum_{k=0}^{m-1}\int_{a+2kh}^{a+(2k+1)h} |\operatorname{Re} w^{-n}(\xi) - \operatorname{Re} w^{-n}(\xi+h)|\, d\xi.$$

It is easy to see that as $m \to \infty$ the right sides in (1.17) have order $h = (b-a)/2m$, and hence for any finite n and $\varepsilon > 0$, for sufficiently large m we have the inequalities

$$|H_1(x) - H_2(x)| < \varepsilon, \qquad \left|\frac{d^k H_1(x)}{dx^k} - \frac{d^k H_2(x)}{dx^k}\right| < \varepsilon, \quad k = 1, \dots, n.$$

The modulus of the difference of the functions $\varphi_1(x)$ and $\varphi_2(x)$ is equal to

$$|\varphi_1(x) - \varphi_2(x)| = \varphi_0.$$

The example we have presented is based on the fact that the external gravitational field changes so little under a sufficiently small shift of the mass. For (1.12) it is possible to present an analogous example based on the fact that the field of a ball in which the distribution of the density is spherically homogeneous is determined only by the mass of the ball and the position of its center.

§7. The inverse kinematic problem of seismology

Apparently, the inverse kinematic problem of seismology was the first inverse problem considered for partial differential equations. It arose in connection with the attempt to study the earth's internal structure from observations on the earth's surface of the propagation of seismic wave fronts generated by earthquakes. It is known that in earthquakes, waves of two types arise: longitudinal and transverse waves. Longitudinal waves, which are compression and rarefaction waves, are connected with the oscillation of particles in the direction of propagation of the wave front; transverse waves are connected with the oscillation of particles in a direction orthogonal to the wave front and characterize the resistance of the elastic substance to shear [58].

Longitudinal and transverse waves propagate in the elastic body of the earth according to the laws of geometric seismology which are altogether analogous to the laws of propagation of a light ray. The trajectories, which are orthogonal to the wave fronts, longitudinal and transverse respectively, are here called *seismic rays*, by analogy with a light ray.

The propagation speeds of longitudinal and transverse seismic waves (in geophysics they are customarily denoted by v_P and v_S respectively) are connected with the elastic Lamé parameters λ, μ, and the density of matter ρ by the formulas

$$v_P = \sqrt{(\lambda + 2\mu)/\rho}, \qquad v_S = \sqrt{\mu/\rho}, \tag{1.18}$$

from which it is evident that $v_P > v_S$, i.e., longitudinal waves propagate faster than transverse waves.

From the point of view of mechanics, the Lamé parameters λ, μ, and the density completely characterize an elastic substance. Of course, the speeds of seismic waves v_P and v_S cannot completely characterize the elastic substance of the earth, but if they are known, then, as formulas (1.18) show, they provide two relations between the three parameters λ, μ, ρ and thus contain considerable information regarding the substance of the earth. Therefore, one of the most important problems of seismology consists in finding the propagation speeds of longitudinal and transverse seismic waves.

In the body of the earth the speeds v_P and v_S are not constant but vary from point to point. In connection with this, seismic rays are not straight lines. It is known from variational calculus (see, for example, [246] and [292]) that a ray joining any pair of points x^0, x is an extremal of the functional

$$J(L) = \int_{L(x^0,x)} \frac{1}{v(x)}|dx|. \tag{1.19}$$

Here $L(x^0, x)$ is an arbitrary sufficiently smooth curve joining the pair of points x^0, x; $|dx|$ is the element of its length in the Euclidean metric and $v(x)$ is the propagation speed of the wave—a longitudinal or transverse wave depending on the ray in question. Below it is irrelevant for us which speed is understood by $v(x)$. From a physical point of view it may simply be said that $v(x)$ is the transmission speed of signals in the medium.

It is evident from (1.19) that $J(L)$ gives the time a signal takes to transverse the curve $L(x^0, x)$. Indeed, the ratio $|dx|/v(x)$ is the elementary transit time of the signal along the curve. Thus, a seismic ray is a curve $L(x^0, x)$ on which the transit time of the signal is a minimum. Actually, for rather complex media where the function $v(x)$ differs strongly from a constant, the pair of points x^0, x can be joined by several rays (or even an uncountable set of rays), each of which possesses the property that the functional (1.19) assumes a minimal value on it as compared with all other values corresponding to curves sufficiently near the ray in question. Thus, seismic rays are local extrema of the functional (1.19). It is known that, in n-dimensional space, extrema of the functional (1.19) are integral curves of the system of Euler equations, which consists of n ordinary differential equations of second order. These integral curves are subject to an additional condition: they pass through the points x^0 and x.

We denote the ray joining the points x^0 and x by $\Gamma(x^0, x)$. Then the transit time $\tau(x^0, x)$ of the signal along this ray is calculated by the formula

$$\tau(x^0, x) = \int_{\Gamma(x^0,x)} \frac{|dx|}{v(x)}. \tag{1.20}$$

The surface defined by the equality $\tau(x^0, x) = \text{const}$ for fixed x^0 is the wave front from a point source of perturbations concentrated at the point x^0. Let $\nu(x^0, x)$ be the unit vector tangent to the ray $\Gamma(x^0, x)$ constructed at the point x and directed to the side of increasing τ. From (1.20) it then follows that

$$\nabla_x \tau(x^0, x) = \frac{1}{v(x)} \nu(x^0, x). \tag{1.21}$$

Here $\nabla_x \tau$ denotes the gradient of the function τ computed with respect to the variable point x. A consequence of equality (1.21), which expresses the condition of orthogonality of the fronts and rays, is the eikonal equation

$$|\nabla_x \tau(x^0, x)|^2 = \frac{1}{v^2(x)}. \tag{1.22}$$

The characteristics of this nonlinear first-order equation are precisely the rays.

The classical problem for equation (1.22) is the problem of constructing the function $\tau(x^0, x)$ for a given function $v(x)$. This construction is realized by means of the Hamilton-Jacobi method in which the characteristics of equation (1.22) and the function $\tau(x^0, x)$, i.e. the rays and fronts of the wave process and hence its kinematics, are constructed simultaneously. In seismology the problem of constructing rays and fronts has received the name of the direct kinematic problem. As already mentioned, the inverse problem, where the function $v(x)$ is unknown while the law of motion of wave fronts along the boundary of the region in which the wave process occurs is known, is of major practical importance. We shall present a precise mathematical formulation of it or, more precisely, of one of its possible versions.

The inverse kinematic problem is posed as follows. Suppose that in a region D of the space $R^n, n > 1$, bounded by a surface S, a wave process occurs which is generated by a point source of oscillations applied at the point x^0, and suppose the function $\tau(x^0, x)$ is known for all possible $x^0 \in S$ and $x \in S$. It is required to find the transmission speed $v(x)$ of signals for $x \in D$.

This formulation is a mathematical idealization of the real physical situation. Actually, observations carried out at seismological stations make it possible to compute the transit time of a signal from the source of the perturbation (an earthquake or artificial explosion) to the seismological station. By possessing a large collection of such data from sources lying in various regions of the terrestrial ball to seismological stations also scattered over the surface of the entire ball, it is possible to pose the question of finding the transmission speed of signals within the earth. Here, as is frequently the case in applications, it is convenient to give up the discrete model of the positions of seismological stations and sources of earthquakes and pass to a continuous analogue of it—a mathematical model in which each point of the surface can be a source of perturbations and each such point can also be a recorder of these perturbations.

The inverse kinematic problem of seismology was first considered in 1905–1907 by G. Herglotz and E. Wiechert, assuming spherical symmetry of the earth. The mathematical expression of this assumption is that $v = v(r), r = |x|$ (the center of the coordinate system is assumed to coincide with the center of the earth). The assumption of spherical symmetry was based on the experimentally observed fact that the transit times of seismic waves between a source and a receiver situated at a fixed distance from one another are almost independent of the region where they are located. It was therefore altogether natural to suppose that the speed also does not depend on the angular spherical coordinates of the point. In the case $v = v(r)$ the seismic ray $\Gamma(x^0, x)$ is a plane curve and lies in the plane passing through the points x^0, x and the center of the earth. The problem thus becomes a plane problem. It suffices to consider an arbitrary cross section of the earth by the plane of a great circle and points x^0 and x situated only on the circumference of this circle. Herglotz and Wiechert showed that in the case of the one-dimensional model $v = v(r)$ it is possible to fix the point x^0 and allow the point x to run through the set of points on the circumference of the circle. If here the function

$$m(r) = r/v(r) \tag{1.23}$$

increases monotonically with increasing r and is continuously differentiable, then it is uniquely determined by prescribing the function $\tau = \tau(\Delta)$, called the *hodograph* of the wave. Here Δ is the angular distance between the points x^0 and x; $\tau(\Delta)$ is the transit time of the signal between these points. It was found that local recovery of the function $v(r)$ is also possible; namely, if the function $\tau(\Delta)$ is known for $\Delta \in [0, \Delta_0], \Delta_0 > 0$, then $v(r)$ can be found in some layer $r_0 \leq r \leq R$ where R is the radius of the earth, and r_0 is determined from Δ_0; here $r_0(\Delta_0)$ decreases monotonically with increasing Δ_0, and $r_0(\Delta_0) \to R$ as $\Delta_0 \to 0$.

Solution of the one-dimensional inverse kinematic problem reduces to two operations: 1) calculation, from the hodograph $\tau(\Delta)$, of the derivative in terms of which $m(r)$ is found at some point $r = \rho$; 2) evaluation of an integral depending on a parameter and determining ρ as a function of Δ. Solution of the problem turns out to be continuous from $C^1 \to C$. In practice the function $\tau(\Delta)$ is given discretely at a sequence of points; therefore, the inverse kinematic problem is an ill-posed problem. The character of its ill-posed behavior is similar to that of the problem of differentiation of a tabulated function.

Solution of the inverse kinematic problem in the one-dimensional formulation greatly influenced the development of views on the structure of the earth (see [96]). The first calculations carried out on the basis of data on transit times of seismic waves revealed characteristic elements of the earth's structure: the earth's crust, mantle, and core. The earth's crust is a layer of relatively small thickness; its average thickness is of the order of 30 km. The mantle differs from the crust in having considerably greater values for the speeds of seismic waves; a rather abrupt increase in them occurs at the crust-mantle boundary. The thickness of the mantle is of the order of 3000 km. The material of the core differs substantially in its physical properties from the material of the mantle. Transverse waves do not pass through it, while they do propagate in the mantle. The lack of transverse waves is characteristic for a fluid. This provides a basis for suspecting that the material of the earth's core is in a liquid state.

At present there are several dozen velocity models of the earth constructed by different authors. They do not differ greatly from one another, although they are constructed on the basis of different seismological material.

Further progress in the inverse kinematic problem was related to giving up the condition of monotonicity of the function $m(r)$ in the one-dimensional model, and to the study of multidimensional models.

Later geophysical investigations revealed that the assumption of monotonicity of the function $m(r)$ is, generally speaking, not justified. Without this assumption, however, the inverse kinematic problem does not have a unique solution. The nature of the nonunicity of the solution of the problem was studied by M. L. Gerver and V. M. Markushevich (see [87]–[89]).

The first result for a multidimensional inverse kinematic problem was obtained by M. M. Lavrent′ev and V. G. Romanov [158]. It was formulated in two-dimensional space x_1, x_2 for a region constituting the half-plane $x_2 \geq 0$, and consisted in the following. Suppose that the transmission speed $v(x)$ of signals can be represented in the form

$$v(x) = v_0(x) + v_1(x), \tag{1.24}$$

where the function $v_0(x)$ is known, and $v_1(x)$ is small as compared with $v_0(x)$. Up to a small quantity of higher order than $v_1^2(x)$, the transit time of a signal between two points x^0 and x belonging to the boundary of the half-plane $x_2 \geq 0$ can then be represented in the form

$$\tau(x^0, x) = \tau_0(x^0, x) + \tau_1(x^0, x).$$

Here $\tau_0(x^0, x)$ is the transit time of the signal between the points x^0 and x in a medium with speed $v = v_0(x)$, and $\tau_1(x^0, x)$ is found from the formula

$$\tau_1(x^0, x) = -\int_{\Gamma_0(x^0,x)} \frac{v_1(x)}{v_0^2(x)} |dx|, \tag{1.25}$$

in which $\Gamma_0(x^0, x)$ is the ray joining the points x^0 and x in the medium with speed $v = v_0(x)$. For the case

$$v_0(x) = ax_2 + b, \quad a > 0, \ b > 0, \tag{1.26}$$

it was shown that for known a and b the continuous function $v_1(x)$ is uniquely determined in the region $x_2 \geq 0$ by giving the function $\tau(x^0, x)$ for arbitrary pairs of points x^0, x of the line $x_2 = 0$.

The result obtained thus referred to a linearized formulation of the inverse kinematic problem. It was here found that the problem of determining v_1 on the basis of τ_1 from (1.25) was ill-posed. The character of its ill-posed behavior is the same as in the Cauchy problem for the Laplace equation. For a transmission speed of signals defined by (1.26) the rays $\Gamma(x^0, x)$ are arcs of circles passing through the points x^0 and x and having center on the line $x_2 = -b/a$. The problem of solving equation (1.25) thus reduces to the problem of integral geometry of finding a function from its mean values over all possible circles with center on the line $x_2 = -b/a$.

We shall present an example like Hadamard's for this problem of integral geometry. Thus, let

$$v(x_1, r) = \frac{1}{2\pi r} \int_{\Gamma(x_1,r)} u(\xi_1, \xi_2)\,|d\xi|, \quad \xi = (\xi_1, \xi_2), \tag{1.27}$$

where $\Gamma(x_1, r)$ is a circle of radius r with center at the point $(x_1, 0)$. It is required to determine the function $u(x_1, x_2)$ from the function $v(x_1, r)$. Relation (1.27) is a linear operator equation. A solution of (1.27) is not unique, since in the case where the function $u(x_1, x_2)$ is odd in the variable x_2,

$$u(x_1, -x_2) = -u(x_1, x_2),$$

the function $v(x_1, r)$ is equal to zero. In Courant's book [139] there is a proof of uniqueness of a solution of (1.27) in a class of functions even in the variable x_2 or in a class of functions equal to zero in the lower half-plane. The uniqueness theorem for (1.27) has local character in the following sense: if $v(x_1, r)$ is known only for all circles $\Gamma(x_1, r)$ lying inside some circle $\Gamma(x_1^0, r^0)$, then $u(x_1, x_2)$ is uniquely determined inside $\Gamma(x_1^0, r^0)$.

Let

$$u(x_1, x_2) = a \sin(nx_1)\cosh(nx_2).$$

Then

$$v(x_1, r) = a \sin(nx_1)$$

by virtue of the fact that $u(x_1, x_2)$ is a solution of the Laplace equation and satisfies the mean value theorem.

We note that the problem of solving (1.27) is equivalent to an ill-posed nonhyperbolic Cauchy problem for the wave equation:

$$\frac{\partial^2 u}{\partial t^2} = \Delta u, \quad u = u(x, y, t), \ u(x, 0, t) = f_1(x, t), \ u_y(x, 0, t) = f_2(x, t).$$

Subsequently Romanov [210] (see also [217]) considered a linearized version of the inverse problem for a three-dimensional ball in the more general case where linearization is carried out relative to an arbitrary smooth function

$v_0(r)$ such that the function $m_0(r) = r/v_0(r)$ increases monotonically with increasing r. In this case the problem of solving (1.25) decomposes into a one-parameter family of plane problems in each of the cross sections of a great circle of the earth.

In the nonlinear formulation, the multidimensional inverse kinematic problem of seismology was investigated by Yu. E. Anikonov [12], [15], [22], R. G. Mukhometov and V. G. Romanov [188]–[191], [226], G. Ya. Beĭl′kin [38], and I. N. Bernshteĭn and M. L. Gerver [46].

In [3], [4], and [23], numerical algorithms are given for solving the multidimensional inverse kinematic problem of seismology.

CHAPTER II

Basic Concepts of the Theory of Ill-posed Problems

§1. Problems well-posed in the Tikhonov sense

1. Many ill-posed problems can be reduced relatively easily to operator equations of the first kind. We consider a more general nonlinear equation than equation (4) mentioned in the Introduction.

Let X and F be complete metric spaces, and let $A(\cdot)$ be a continuous operator (mapping) acting from X to F (defined on all of X).

The equation

$$A(x) = f, \tag{2.1}$$

where $x \in X$ and $f \in F$, is called an *operator equation of the first kind* if the operator A is compact and no ball in the space X is compact.([1])

We shall formulate the concept of a classical well-posed problem (in the Hadamard sense) for equation (2.1).

DEFINITION 1. The problem of solving (2.1) is called (*classically*) *well-posed* if:

1) for any $f \in F$ there exists a solution $x \in X$ of (2.1),

2) the solution of (2.1) is unique, and

3) the solution x depends continuously on the right side f, i.e., to infinitesimal variations in f (in the metric of the space F) there correspond infinitesimal variations of the solution x (in X).

We call the problem of solving (2.1) *uniformly well-posed* if in place of condition 3 the following stronger condition is satisfied:

3′) the solution x depends on the right side f in a uniformly continuous manner.([2])

For classical problems of mathematical physics and the operator equations corresponding to them the fact that the conditions for being well-posed are

([1])Other concepts of an operator equation of the first kind can also be found in the literature (see [113] and [266]).

([2])In the monograph [266] a solution satisfying condition 3′) is called *stable*, while a problem satisfying conditions 1), 2), and 3′) is called *well-posed*.

satisfied is established by theorems of existence, uniqueness, and continuous dependence of the solution on the data.

The existence of a continuous operator B from F to X inverse to the operator A is necessary and sufficient for the problem of solving (2.1) to be well-posed.

It is easy to see that the problem of solving an operator equation of the first kind cannot be well-posed. We shall show that an operator inverse to a compact operator is not continuous. Let $X_0 \subset X$ be some ball. It follows from the compactness of $A(\cdot)$ that the set $F_0 = A(X_0)$ (the image of the ball X_0) is compact.

Since X_0 is not compact there exists a sequence $\{x_k\}, x_k \in X_0,\ k = 1, 2, \ldots$, such that

(a) the sequence $\{x_k\}$ does not converge to any element of the set X_0, and

(b) the sequence $f_k = A(x_k)$ converges: $\lim_{k\to\infty} f_k = f_0$.

This implies that the operator inverse to $A(\cdot)$ is not continuous at the element f_0.

DEFINITION 2. Suppose that in the space X a subspace $Y \subset X$ is additionally distinguished. The problem of solving (2.1) is called *well-posed in the Tikhonov sense* if

1) it is known a priori that a solution x exists and belongs to Y: $x \in Y$;

2) the solution is unique; and

3) to infinitesimal variations of the right side f not taking the solution x out of the set Y there correspond infinitesimal variations in the solution x.

The subspace Y is called a *set of well-posedness.*

We shall indicate the differences in the concepts of problems well-posed in the Tikhonov sense from problems well-posed in the classical sense.

Existence theorems are not proved in a theoretical investigation of whether a problem of mathematical physics is well-posed in the Tikhonov sense. The existence of a solution and its membership in a set of well-posedness are postulated in the very formulation of the problem. One here proceeds from additional "physical" considerations (examples of such considerations will be presented below).

Investigations of uniqueness questions in formulations well-posed in the Tikhonov sense do not differ from investigations of uniqueness in formulations well-posed in the classical sense; uniqueness is established by uniqueness theorems.

Although for problems well-posed in the Tikhonov sense it suffices to establish uniqueness in the subspace Y, uniqueness theorems are usually proved for all of X.

We denote by $Y_A \subset F$ the image of the set Y under the mapping realized by the operator A, and by A_Y the restriction of A to the set Y. For the problem of solving (2.1) to be well-posed in the Tikhonov sense it is necessary and sufficient that there exist an operator B_Y, continuous on Y_A, which is inverse to A_Y.

Suppose now that the operator B_Y (inverse to A_Y) on the set Y_A is uniformly continuous. We denote by $\gamma(\tau)$ the modulus of continuity of the operator B_Y on Y_A. The function $\gamma(\tau)$ characterizes the stability of the solution of equation (2.1). In the theory of ill-posed problems another characteristic of the stability of the solution of (2.1) is often considered, namely, the function

$$\omega(\varepsilon) = \sup_{x_j \in Y} \rho_X(x_1, x_2), \qquad \rho_F(A(x_1), A(x_2)) \leq \varepsilon, \tag{2.2}$$

where ρ_X and ρ_F are the distances in the spaces X and F. It is easy to see that if one of the functions $\gamma(\tau)$ or $\omega(\varepsilon)$ is continuous, then the other function is also continuous, and these functions are mutually inverse.

2. That the third condition of Tikhonov well-posedness is satisfied often follows directly from the second condition—the uniqueness theorem.

In the case where the set Y of well-posedness is compact we have the following result.

THEOREM 1 (TIKHONOV). *Suppose that the solution of equation* (2.1) *is unique, the set Y is compact, and $Y_A \subset F$ is the image of Y under the mapping realized by the operator A. Then on the set Y_A the operator B inverse to A is uniformly continuous.*

PROOF. We suppose that the theorem is false. Then there exists an $\varepsilon_0 > 0$ such that for any $\delta > 0$ there exist $x_1, x_2 \in Y$ such that

$$\rho_X(x_1, x_2) > \varepsilon_0, \qquad \rho_F(A(x_1), A(x_2)) < \delta.$$

Let $\{\delta_k\}$ be a sequence tending to zero as $k \to \infty$, and let $\{x_{k1}\}$ and $\{x_{k2}\}$ be sequences of elements of Y such that

$$\rho_X(x_{k1}, x_{k2}) > \varepsilon_0, \qquad \rho_F(A(x_{k1}), A(x_{k2})) < \delta_k. \tag{2.3}$$

Because of the compactness of Y the sequences $\{x_{kj}\}$ have convergent subsequences. It may obviously be assumed that these subsequences coincide with the sequences themselves.

Let

$$x_1 = \lim_{k \to \infty} x_{k1}, \qquad x_2 = \lim_{k \to \infty} x_{k2}.$$

Then by (2.3) and the continuity of the operator A

$$\rho_x(x_1, x_2) \geq \varepsilon_0, \qquad \rho_F(A(x_1), A(x_2)) = 0, \text{ i.e. } A(x_1) = A(x_2),$$

which contradicts the uniqueness of the solution of (2.1).

3. We return to considering the physical formulations presented in Chapter I. In the problem of interpreting gravimetric data and the problem of determining (calculating) the anomalous gravitational field below a level of observations (see §1), the existence of a solution is determined by the very formulation of the problem, the correctness of the readings of the instruments, and our ideas regarding the geologic structure of the region in question. Indeed, gravimetric instruments record real anomalies of the gravitational field. If the geologic objects generating the anomalous field lie below the level at which we wish to determine the field by solving a problem ill-posed in the classical sense, the anomalous field exists at this level and is connected with the field measured on the surface of the earth by relation (1.1).

Indeed, as is known, the excess density (specific gravity) of geologic objects is usually within the limits 1–3 g/cm^3 (granite, basalt). Geologic bodies containing iron ores can have excess densities 3–5 g/cm^3. The highest excess density in the interior of the earth may occur in gold deposits—up to 17.5 g/cm^3. Thus, excess densities of geologic objects are bounded by an altogether definite known quantity.

By analyzing formula (1.15), it can be shown that away from the surface separating the body with excess density the intensity of the anomalous field is bounded by some definite constant and satisfies a Hölder condition also with a definite constant. This circumstance makes natural the assumption that the desired solution belongs to a compact set of well-posedness—a set of functions uniformly bounded and satisfying a Hölder condition with a fixed constant.

In the case where the level at which we wish to determine the anomalous field is taken "with margin," i.e., the geologic objects are located at some distance from this level, the vertical component of the intensity of the anomalous field is an analytic function, and its derivatives of all orders are bounded by particular constants.

In the problem of determining the history of a process of variation in temperature or concentration of a substance (see §2) the existence of a solution is guaranteed by the setting of the physical experiment or by the correctness of our idea of the nature of the process. The desired function characterizing the state of the object (the temperature or the concentration of the substance) during the course of the entire process is bounded above and below by altogether definite constants. In the process the temperature of a solid body cannot be above the melting temperature nor below absolute zero (in practice it may be assumed that it is not below a higher level). The concentration (that is, the ratio of the mass of the component of the solution to the total mass) is obviously no less than zero and no greater than one.

If the process began at time $t = 0$ the above conditions guarantee that the solution of problem (2.1), (2.2) for any $t > 0$ belongs to a compact set of well-posedness.

In problems of processing readings of physical instruments discussed in §4 of Chapter I, hypotheses are usually adopted regarding the frequency composition of the fields in question. It is assumed that the spectrum of the field (the Fourier transform) is concentrated mainly on some fixed segment, while away from this segment it is negligibly small. One attempts to choose the characteristics of the instrument in correspondence with this. The assumption that the field being recorded is bounded in norm and that the spectrum of the field is equal to zero outside a fixed segment constitutes the assumption that the desired solution belongs to a compact set of well-posedness.

The concept of a problem well-posed in the Tikhonov sense was suggested with the purpose of justifying the trial-and-error method widely used in interpreting geophysical data (solution of inverse problems). The essence of the trial-and-error method consists in the following. On the basis of his or her ideas regarding the geologic structure of a territory being investigated, the geophysicist constructs some physical model. For example, in interpreting gravimetric data a model of the following type is often taken: in a homogeneous layer of sediments there is an isolated intrusion—a geologic body with higher density. The shape of the body is an ellipsoid. In seismological prospecting the following model is often adopted: beneath a homogeneous layer of sediments there is a base, and the instruments record waves reflected from the base.

The direct problem corresponding to the model is then solved, and the solutions are compared with data from measurements. On the basis of this comparison, a new model is chosen so that the data of the solution of the direct problem in the new model will be closer to the experimental data.

At the basis of the trial-and-error method is the assumption that the medium under study does not have too complex a structure. It is assumed that the physical characteristics of the medium (the density, the propagation speed of seismic waves, etc.) are functions which are either piecewise constant or piecewise smooth, the number of surfaces or lines of discontinuity is bounded, and the surfaces themselves are piecewise smooth. Moreover, the values of the functions are within altogether definite limits. These assumptions are hypotheses that the desired solutions belong to particular compact sets of well-posedness.

4. The concept of a quasisolution introduced by V. K. Ivanov in [106] is connected with the concept of problems well-posed in the Tikhonov sense.

Below we shall consider the case where the right side of (2.1) is given approximately with some error. The right side of (2.1) without error we denote by f_T, and the corresponding exact solution we denote by x_T: $A(x_T) = f_T$. The right side of (2.1) with error we denote by f_ε, assuming that $\rho_F(f_T, f_\varepsilon) \leq \varepsilon$.

Suppose the problem of solving (2.1) is well-posed in the Tikhonov sense. A *quasisolution* of equation (2.1) is an element $x \in Y$ minimizing the functional $\rho_F(A(y), f)$ on the set Y:

$$\rho_F(A(\bar{x}), f) = \inf_{y \in Y} \rho_F(A(y), f).$$

If the set Y is compact, then a quasisolution of (2.1) exists for any right side f.

In the case where the right side of (2.1) is given without error the quasisolution obviously coincides with the exact solution of (2.1): $\bar{x} = x_T$.

We now consider a quasisolution of (2.1) corresponding to a right side of (2.1) given with error. Thus, suppose that

$$\rho_F(f_T, f_\varepsilon) \leq \varepsilon, \qquad \rho_F(A(\bar{x}), f_\varepsilon) = \inf_{y \in Y} \rho_F(A(y), f_\varepsilon). \tag{2.4}$$

From the definition of a quasisolution and the triangle inequality we obtain

$$\rho_F(A(\bar{x}), f_T) \leq 2\varepsilon. \tag{2.5}$$

It follows from (2.5) that a quasisolution depends continuously on the right side of (2.1). In the case where there is uniformly continuous dependence of the solution on the right side on the set of well-posedness, it follows from (2.5) that

$$\rho_X(\bar{x}, x_T) \leq \omega(2\varepsilon), \tag{2.6}$$

where $\omega(\varepsilon)$ is the function in (2.2), i.e., a quasisolution is an approximate solution of (2.1) with an accuracy guaranteed by inequality (2.6).

5. The definition of the concept of a quasisolution contains no indication of how to actually find it. Generally speaking, the problem of constructing a quasisolution is the problem of finding the minimum of the residual functional on a set. A description of many methods of finding extrema of functionals on sets is given in the literature (see, for example, [113], [121], and [266]). Convergence theorems have been proved and estimates of efficiency obtained under various assumptions regarding the sets and functionals. The method of enumeration is the most general method of minimizing functionals on sets. We present a description of this method in application to the problem of solving equation (2.1) which is well-posed in the Tikhonov sense with a compact set of well-posedness.

As is known from functional analysis, for any $\delta > 0$ in a compact set Y there exists a finite δ-net: $\{y_k\}$, $k = 1, \dots, n(\delta)$.(3)

Since it is assumed that we are given the set Y and the operator A, it is natural to suppose that for any $\delta > 0$ we can determine both the elements $y_{k\delta}$ and their images:

$$\varphi_{k\delta} = A(y_{k\delta}).$$

We denote by p the index of the element $\varphi_{k\delta}$ for which

$$\rho_F(\varphi_{p\delta}, f_\varepsilon) = \inf_{k=1,\dots,n(\delta)} \rho_F(\varphi_{k\delta}, f_\varepsilon),$$

by q the index of the element $y_{k\delta}$ for which

$$\rho_X(x_T, y_{q\delta}) \leq \delta,$$

and by $\alpha(\tau)$ the modulus of continuity of the operator A on the set Y.

By (2.4) and the triangle inequality,

$$\rho_F(\varphi_{p\delta}, f_T) \leq \rho_F(\varphi_{p\delta}, f_\varepsilon)+\varepsilon \leq \rho_F(\varphi_{q\delta}, f_\varepsilon)+\varepsilon \leq \rho_F(\varphi_{q\delta}, f_T)+2\varepsilon \leq \alpha(\delta)+2\varepsilon.$$

Hence,

$$\rho_X(y_{p\delta}, x_T) \leq \omega(\alpha(\delta) + 2\varepsilon).$$

§2. Regularization

1. We consider the operator equation

$$A(x) = f. \tag{2.7}$$

As in the preceding section, it is assumed that the solution of (2.7) is unique.

DEFINITION 1. An operator $R(f, \alpha)$ (depending on a scalar parameter α) is called a *regularizing operator* for equation (2.7) *in a neighborhood of an element* $f_T = A(x_T)$ if the following two conditions hold:

1) there exist numbers α_0 and δ_0 such that the operator $R(f, \alpha)$ is defined for any α and f satisfying the conditions(4)

$$0 < \alpha < \alpha_0, \qquad \rho(f, f_T) < \delta_0;$$

2) there exists a function $\alpha(\delta)$ such that for any $\varepsilon > 0$ there is a number $\delta(\varepsilon) < \delta_0$ such that if $\rho(f, f_T) < \delta(\varepsilon)$, then

$$\rho(x_\alpha, x_T) < \varepsilon,$$

(3)For any $x \in Y$ there exists a p such that $\rho_{\bar{X}}(y_{p\delta}, x) \leq \delta$.

(4)In the notation for distance in this section we omit indices affixed to the spaces X and F.

where $x_\alpha = R(f, \alpha(\delta))$. The scalar parameter α is called the *regularization parameter*.

DEFINITION 2. An operator $R(f, \alpha)$ is called a *regularizing operator* for equation (2.7) *on a set* $F_1 \subset F$ if it is a regularizing operator in a neighborhood of each element of F_1. We note that $F_1 \subset X_A$, where X_A is the range of the operator A.

Suppose that we have somehow managed to construct a regularizing operator for (2.7) in a neighborhood of a point f_T. We consider the problem of finding an approximate solution of (2.7) on the basis of an element f_δ (the right side of (2.7)) with error $\rho(f_T, f_\delta) < \delta$.

From the definition of a regularizing operator it follows that for $\delta < \delta_0$ it is possible to take as an approximate solution the value of the regularizing operator at the element f_δ for a value of the parameter α consistent with the error estimate δ.

In solving a concrete equation (2.7) connected, for example, with some problem of physics there are two possible cases.

1. We are given the number ε, the maximum error we can allow in a solution of the equation. If a regularizing operator has been constructed, the problem then consists in finding $\delta(\varepsilon)$ (the required accuracy on the right side of (2.7)) and the regularization parameter $\alpha(\delta)$.

2. We are given the number δ, the error estimate for the right side of (2.7). In this case the problem consists in determining the value of the regularization parameter with corresponding error in the solution as small as possible.

The functions $\alpha(\delta)$ and $\delta(\varepsilon)$ appearing in the definition of a regularizing operator depend on f_T. Since in the formulation of the problem of finding an approximate solution on the basis of data with error the element f_T is obviously not assumed to be known, additional information is needed for the choice of a regularization parameter consistent with the parameters ε and δ.

DEFINITION 3. An operator $R(f, \alpha)$ is called *uniformly regularizing* for equation (2.7) *on a set* F_1 if the following two conditions hold:

1) there exist numbers α_0 and δ_0 such that the operator $R(f, \alpha)$ is defined for any α, f, and f_T satisfying the conditions

$$0 < \alpha < \alpha_0, \qquad \rho_F(f, f_T) < \delta_0, \quad f_T \in F_1;$$

2) there exists a function $\alpha(\delta)$ such that for any $\varepsilon > 0$ there is a $\delta(\varepsilon) < \delta_0$ such that if $\rho_F(f, f_T) < \delta(\varepsilon)$, $f_T \in F_1$, then

$$\rho_X(x_T, x_\alpha) < \varepsilon,$$

where $x_\alpha = R(f, \alpha(\delta))$.

If the problem of solving (2.7) is well-posed in the Tikhonov sense and F_1 contains Y_A (the image of the set Y of well-posedness), then a uniformly regularizing operator on F_1 makes it possible to construct approximate solutions with guaranteed accuracy on the basis of data with error.

2. We shall indicate some sufficient conditions on an operator depending on a parameter in order that it be regularizing for equation (2.7).

THEOREM 1. *Suppose an operator $G(\cdot, \alpha)$ from F to X satisfies the following conditions:*

1) *the operator $G(\cdot, \alpha)$ is defined for all α and f with $0 < \alpha < \alpha_0$ and $\rho(f_T, f) < \delta_0$, and is continuous at the element f_T;*

2) $\lim_{\alpha \to 0} G(f_T, \alpha) = x_T$.

Then the operator $G(\cdot, \alpha)$ is regularizing for equation (2.7) in a neighborhood of the element f_T.

If conditions 1) and 2) are satisfied for any $f_T \in F_1$, the operator $G(\cdot, \alpha)$ is regularizing for (2.7) on the set F_1.

If for any $\alpha, 0 < \alpha < \alpha_0$, the operator $G(\cdot, \alpha)$ is uniformly continuous on F_1 and $G(A(x), \alpha)$ converges uniformly to x on X_1 as $\alpha \to 0$, then the operator $G(\cdot, \alpha)$ is uniformly regularizing for (2.7) on F_1 ($F_1 = A(X_1)$).

PROOF. By the triangle inequality

$$\rho(G(f, \alpha) x_T) \leq \rho(x_{T\alpha}, G(f, \alpha)) + \rho(x_{T\alpha}, x_T),$$

where $x_{T\alpha} = G(f_T, \alpha)$. We set $\rho(f, f_T) = \delta$. From the continuity of $G(\cdot, \alpha)$ for $\alpha > 0$ it follows that

$$\lim_{\delta \to 0} \beta(\delta, \alpha, f_T) = 0. \tag{2.8}$$

It may obviously be assumed that the function β depends monotonically on δ.

We consider any arbitrarily small number $\varepsilon > 0$. Because of the second property of $G(\cdot, \alpha)$ there exists an $\alpha(\varepsilon)$ such that for any $\alpha < \alpha(\varepsilon)$ we have

$$\rho(x_{T\alpha}, x_T) < \frac{\varepsilon}{2}.$$

By (2.8) for any $\alpha > 0$ there exists $\delta(\alpha)$ such that for $\delta < \delta(\alpha)$

$$\rho(x_{T\alpha}, G(f, \alpha)) < \frac{\varepsilon}{2}$$

and hence

$$\rho(x_T, G(f, \alpha)) < \varepsilon.$$

The second assertion of the theorem obviously follows from the first.

Suppose now that $G(\cdot, \alpha)$ is uniformly continuous and the convergence of $G(A(x), \alpha)$ to x is uniform on X_1. It may then be assumed that the function β on the right side of (2.8) does not depend on $f_T \in F_1$:

$$\beta(\delta, \alpha, f_T) = \beta(\delta, \alpha),$$

and the inequalities

$$\rho(x_T, x_{T\alpha}) < \frac{\varepsilon}{2}, \qquad \rho(x_{T\alpha}, G(f, \alpha)) < \frac{\varepsilon}{2}$$

hold for $\alpha < \alpha(\varepsilon)$ for any $x_T \in X_1$ and $\rho(f, f_T) < \delta(\alpha)$; this means the third assertion of the theorem is valid.

3. It is obvious that for a single equation of type (2.7) there exists an infinite set of regularizing operators (RO). In concrete cases the choice of regularizing operator for the purpose of constructing an approximate solution on the basis of data with error depends on various factors such as the availability of a priori information regarding the solution, error estimates, and simplicity from the point of view of constructing numerical algorithms. In this section we shall present some of the most general ways of constructing RO based on variational principles.(5)

Let $\varphi(x)$ be a nonnegative functional defined on a dense set Z in X.

DEFINITION 4. The functional $\varphi(x)$ is called a *stabilizing functional* if for any number $a > 0$ the set of $x \subset Z$ for which $\varphi(x) \leq a$ is compact.

We denote by $Q(f_0, \alpha)$ the ball in the space F of radius α:

$$Q(f_0, \alpha) = \{f : \rho(f, f_0) \leq \alpha\}.$$

Let $\tilde{x}_\alpha$ be an element of Z satisfying the relation

$$\varphi(\tilde{x}_\alpha) \leq q \inf_{A(y) \in Q(f_0, \alpha)} \varphi(y),$$

where $q \geq 1$ is a constant. It is obvious that for f_0 and α such that

$$\rho(f_0, Z_A) \leq \alpha_0, \quad \alpha \leq \alpha_0$$

(Z_A is the image of the set Z under the mapping $A(\cdot)$) an element $\tilde{x}_\alpha$ exists.

We denote by $\tilde{R}(\cdot, \alpha)$ the operator from F to X which assigns to the element f_0 the element $\tilde{x}_\alpha$ defined above:

$$\tilde{R}(f_0, \alpha) = \tilde{x}_\alpha.$$

THEOREM 2. *The operator $\tilde{R}(f, \alpha)$ is regularizing for equation (2.7) on the set Z_A.*

PROOF. Let f_T be an arbitrary element of the set Z_A. It was noted above that condition 1) in the definition of a regularizing operator is satisfied, and $\alpha_0 = \delta_0$.

(5)Here we follow mainly the exposition given in [266].

We now suppose that condition 2) for the operator $\tilde{R}(f,\alpha)$ is not satisfied. This means that for some $\varepsilon > 0$ there exist sequences f_k, α_k, and $\tilde{x}_{\alpha k} = \tilde{R}(f_k, \alpha_k)$ with $\lim_{k\to\infty} \alpha_k = 0$ such that

$$\rho(\tilde{x}_{\alpha k}, x_T) \geq \varepsilon. \tag{2.9}$$

From the definition of $\tilde{x}_{\alpha k}$ it follows that

$$\varphi(\tilde{x}_{\alpha k}) \leq q\varphi(x_T),$$

whence we see that the sequence $\tilde{x}_{\alpha k}$ belongs to a compact set.

Let $\tilde{x}$ be the limit of a convergent subsequence of $\tilde{x}_{\alpha k}$. From the uniqueness of the solution of (2.7) and the inequality

$$\rho(A(x_{\alpha k}), f_T) \leq \alpha_k$$

we have $\tilde{x} = x_T$, which contradicts (2.9).

4. We denote by T_δ the set of strictly increasing, continuous functions defined on the segment $[0, \delta]$.

DEFINITION 5. Let $\gamma(\tau)$ be a function in T_∞; let $\varphi(x)$ be a stabilizing functional. The functional

$$\psi(x, f, \alpha) = \gamma[\rho(A(x), f)] + \alpha\varphi(x)$$

is called a *smoothing functional* for equation (2.7). We denote by $\bar{x}_\alpha$ an element of X satisfying the relation

$$\psi(\bar{x}_\alpha, f, \alpha) \leq q \inf_{x\in Z} \psi(x, f, \alpha), \quad q \geq 1,$$

and by $\overline{R}(\cdot, \alpha)$ the operator assigning to the element f the element $\bar{x}_\alpha$, i.e., $\overline{R}(f, \alpha) = \bar{x}_\alpha$.

THEOREM 3. *For any $\varepsilon > 0$ and any functions $\beta_1(\delta)$ and $\beta_2(\delta)$ in T_{δ_0} satisfying the relations*

$$\beta_2(0) = 0, \qquad \frac{\gamma(\delta)}{\beta_1(\delta)} \leq \beta_2(\delta),$$

there exists $\delta_1 = \delta_1(\varepsilon, \beta_1, \beta_2)$ such that for any $f \in F, \delta$ and α satisfying the inequalities

$$\rho(f, f_T) \leq \delta \leq \delta_1, \qquad \frac{\gamma(\delta)}{\beta_1(\delta)} \leq \alpha \leq \beta_2(\delta), \tag{2.10}$$

the inequality

$$\rho(\bar{x}_\alpha, x_T) \leq \varepsilon$$

holds, where $\bar{x}_\alpha = \overline{R}(f, \alpha)$.

PROOF. By the definition of the operator $\overline{R}(f, \alpha)$ we have

$$\psi(\bar{x}_\alpha, f, \alpha) \leq q\psi(x_T, f, \alpha),$$

whence we obtain

$$\begin{aligned}\alpha\phi(\bar{x}_\alpha) &\leq \psi(\bar{x}_\alpha, f, \alpha) \leq q\psi(x_T, f, \alpha)\\ &= q\{\gamma[\rho(A(x_T), f)] + \alpha\varphi(x_T)\} \leq q[\gamma(\delta) + \alpha\varphi(x_T)].\end{aligned}$$

It follows from (2.10) that

$$\frac{\gamma(\delta)}{\alpha} \leq \beta_1(\delta) \leq \beta_1(\delta_0)$$

and hence

$$\varphi(\bar{x}_\alpha) \leq q[\beta_1(\delta_0) + \varphi(x_T)] = a_0.$$

Therefore, the elements $\bar{x}_\alpha$ and x_T belong to the compact set of elements of X satisfying the relation $\varphi(x) \leq a_0$. We denote this compact set by X_0, and the image of X_0 under the mapping A by F_0. By Theorem 1 of the preceding section the operator A_0 (the restriction of A to X_0) has a uniformly continuous inverse operator on the set F_0. Hence, for any $\varepsilon > 0$ there exists $\omega(\varepsilon)$ such that for any $x_1, x_2 \in X_0$ satisfying the relation

$$\rho(A(x_1), A(x_2)) \leq \omega(\varepsilon),$$

we have the inequality $\rho(x_1, x_2) \leq \varepsilon$.

We now estimate $\rho(A(\bar{x}_\alpha), f)$:

$$\begin{aligned}\gamma[\rho(A(\bar{x}_\alpha), f)] &\leq \psi(\bar{x}_\alpha, f, \alpha) \leq q\psi(x_T, f, \alpha)\\ &= q\gamma(\rho(f_T, f) + \alpha\varphi(x_T)) \leq q\gamma(\delta) + qa_0\alpha \leq q\gamma(\delta) + qa_0\beta_2(\delta),\end{aligned}$$

whence we obtain

$$\rho(A(\bar{x}_\alpha), f) \leq \gamma^{-1}[q\gamma(\delta) + qa_0\beta_2(\delta)] = \beta_3(\delta), \tag{2.11}$$

where γ^{-1} is the function inverse to γ.

It is obvious that $\beta_3(\delta) \in T_{\delta_0}$. From (2.11) and the triangle inequality we have

$$\rho(A(\bar{x}_\alpha), f_T) \leq \delta + \beta_3(\delta) = \beta_4(\delta).$$

We denote by β_4^{-1} the function inverse to $\beta_4(\delta)$, and we set $\delta_1 = \beta_4^{-1}(\omega(\varepsilon))$. Then

$$\rho(A(\bar{x}_\alpha), A(x_T)) \leq \omega(\varepsilon),$$

and, since $\bar{x}_\alpha, x_T \in X_0$, it follows that $\rho(\bar{x}_\alpha, x_T) \leq \varepsilon$, as required.

COROLLARY. *The operator* $\overline{R}(f, \alpha)$ *is a regularizing operator for* (2.7) *on the set* A.

5. The concepts of stabilizing and smoothing functionals formulated above are generalizations of concepts presented in [266], where a stabilizing functional was additionally assumed to be continuous. Regularizing operators corresponding to these functionals are defined by the formulas

$$\tilde{R}_1(f,\alpha)=\tilde{x}_{1\alpha},\qquad \varphi(\tilde{x}_{1\alpha})=\inf_{A(y)\in Q(f,\alpha)}\varphi(y),$$

$$\bar{R}_1(f,\alpha)=\bar{x}_{1\alpha},\qquad \psi(\bar{x}_{1\alpha},f,\alpha)=\inf_{x\in Z}(\psi(x,f,\alpha)).$$

6. Construction of stabilizing functionals for concrete spaces usually occasions no difficulty. The efficiency of regularization by means of stabilizing and smoothing functionals naturally depends on the coordination of the functionals with the equation, a priori information regarding the solution, and the complexity of the minimization problem.

In the case where the problem of solving (2.7) is well-posed in the Tikhonov sense the operator assigning to the right side f a quasisolution $\bar{x}$ (see §1) is obviously regularizing on the image of the set Y_A of well-posedness.

We note that the set of well-posedness is often given as a collection of elements of the space X for which some stabilizing functional does not exceed a fixed constant: $Y=\{x\colon \varphi(x)\le\alpha\}$. In this case the problems of finding a quasisolution and of constructing regularizing algorithms $\tilde{R}_1(\cdot,\alpha)$ are problems of finding conditional extrema. If the method of Lagrange is applicable to these problems, then their solutions are stationary elements for the smoothing functional $\psi(x,f,\alpha)$. We shall present sufficient conditions for a quasisolution and the value of the regularizing algorithm $\tilde{R}_1(f,\alpha)$ to coincide with the value of the regularizing algorithm $\bar{R}_1(f,\bar{\alpha})$ for some value $\bar{\alpha}$, i.e., with the element minimizing the smoothing functional for $\alpha=\bar{\alpha}_0$.

Let X and F be Banach spaces, and suppose that the set on which the stabilizing functional $\varphi(x)$ is defined is a linear subspace of X.

THEOREM 4. *Suppose the stabilizing function $\varphi(x)$ is strictly concave*([6]) *and Gâteaux differentiable, and that the operator A and the functional $\gamma(\tau)$ are such that the functional $\gamma(\|A(x)-f\|)$ for any f is strictly concave and Gâteaux differentiable. Then the following statements are true*:

1) *for any set of well-posedness $Y_a=\{x\colon \varphi(x)\le a\}$ and any $f\in F$ there exists an α_1 such that a quasisolution $\bar{x}$ of* (2.7) *minimizes the functional*

$$\psi(x,f,\alpha_1)=\gamma(\|Ax-f\|)+\alpha_1\varphi(x),$$

i.e.,

$$\psi(\bar{x},f,\alpha_1)=\inf_{x\in Z}\psi(x,f,\alpha_1);$$

2) *for any δ and f there exist α_2 and $\tilde{x}_{\alpha_2}\in Z$ such that*

$$\inf_{x\in Q(f,\delta)}\varphi(x)=\varphi(\tilde{x}_{\alpha_2}),\qquad \inf_{x\in Z}\psi(x,f,\alpha_2)=\psi(\tilde{x}_{\alpha_2},f,\alpha_2).$$

([6])A functional $\varphi(x)$ is called *concave* if the functional $-\varphi(x)$ is convex.

PROOF. It follows from known properties of convex functionals that the quasisolution $\bar{x}$, the element $\tilde{x}_{\alpha_2}$, and the element $\tilde{x}_\alpha$ minimizing the functional $\psi(x, f, \alpha)$ for fixed α are unique:

$$\psi(x, f, \alpha) = \inf_{x \in Z} \psi(x, f, \alpha).$$

Further, by the principle of Lagrange, the elements $\bar{x}$ and $\bar{x}_{\alpha_2}$ on which the functionals $\gamma(\|A(x) - f\|)$ and $\varphi(x)$ achieve conditional extrema for some α_1 and α_2 are stationary elements of the functionals $\psi(x, f, \alpha_1)$ and $\psi(x, f, \alpha_2)$. The assertion of the theorem follows from what has been said above and the fact that for fixed α the functional $\psi(x, f, \alpha)$ has a unique stationary element x_α.

§3. Linear ill-posed problems

The theory of linear ill-posed problems is presented rather completely in the monograph [113]. We shall present some aspects of this theory.

1. Let X and F be Banach spaces, and let A be a continuous linear operator acting from X to F. In this section all spaces are real.

We consider the linear operator equation

$$Ax = f. \tag{2.12}$$

We assume that the solution of (2.12) is unique, i.e., an operator A^{-1} inverse to A exists and is defined on $X_A \subset F$ (the image of the space X); we assume that X_A is dense in F. In this case for problem (2.12) to be well-posed it is necessary and sufficient that the operator A^{-1} be bounded.

As noted earlier, if the operator A is compact the inverse operator is not continuous, and the problem of solving (2.12) is ill-posed.

In some cases it is natural to consider linear ill-posed problems as problems of determining the values of unbounded operators. The problem of differentiation and problems of determining the values of linear differential operators in general, for example, are such problems. The concepts and results pertaining to equation (2.12) carry over to these problems as well (see [113] and [179]).

2. For linear equations in linear spaces, Theorem 1 of §1 extends to a larger class of sets.

THEOREM 1. *Suppose that the solution of equation* (2.12) *is unique and the set Y is the algebraic sum $Y = Y' + Y''$, where Y' is a compact set and Y'' is a finite-dimensional subspace of the space X. Then on the set Y_A, the image of Y in F under the mapping A, the operator A^{-1} is uniformly continuous.*

PROOF. Suppose not. Then there exist sequences $x'_{jk} \in Y'$, $x''_{jk} \in Y''$, $\delta_{k,j} = 1, 2,\ k = 1, \ldots, \infty$, and a number $\varepsilon > 0$ such that

$$\|A(x'_{1k} + x''_{1k} - x'_{2k} - x''_{2k})\| \leq \delta_k, \qquad \|x'_{1k} + x''_{1k} - x'_{2k} - x''_{2k}\| \geq \varepsilon,$$
$$\lim_{k\to\infty} \delta_k = 0.$$

We denote by $\tilde{Y}$ the set of elements x which can be represented in the form $x = x_1 - x_2$, $x_j \in Y'$. It is obvious that the set $\tilde{Y}$ is compact. Thus, there exist sequences $\tilde{x}_k \in \tilde{Y}$ and $x_k \in Y''$ such that

$$\|A(\tilde{x}_k + x_k)\| \leq \delta_k, \qquad \|\tilde{x}_k + x_k\| \geq \varepsilon. \tag{2.13}$$

If the sequence $\|x_k\|$ is bounded, then it is possible to extract a convergent subsequence from the sequence of elements $\tilde{x}_k + x_k$, and for the limit of this subsequence $\tilde{x}$ we obtain the relations $\|A\tilde{x}\| = 0$ and $\|\tilde{x}\| \geq \varepsilon$, which contradict the uniqueness of the solution of (2.12).

Suppose now that the sequence $\|x_k\|$ is not bounded. Equation (2.12) in the subspace Y'' is equivalent to solving a finite system of linear algebraic equations, whence it follows that $\|Ax\| \geq q\|x\|$, for $x \in Y''$, where $q > 0$ is some constant. Hence, the sequence $\|Ax_k\|$ is not bounded, while the sequence $\|A\tilde{x}_k\|$ is bounded by virtue of the boundedness of A. This contradicts the first of the inequalities (2.13).

3. Functionals connected with the linear structure of Banach spaces are distinguished among possible stabilizing functionals for these spaces. Let $\tilde{X}$ be a Banach subspace which is compactly imbedded and dense in X. In this case the functional

$$\tilde{\varphi}(x) = \tilde{\gamma}(\|x\|_{\tilde{X}}),$$

where $\tilde{\gamma}(t)$, a function in T_∞, is obviously stabilizing.

If the spaces F and $\tilde{X}$ and the functions $\tilde{\gamma}(t), \gamma(t) \in T_\infty$ are strictly convex(7) then the functionals

$$\tilde{\varphi}(x) = \tilde{\gamma}(\|x\|_{\tilde{X}}), \quad \bar{\varphi}(x, f) = \bar{\gamma}(\|Ax - f\|_F), \quad \psi(x, f, \alpha) = \bar{\varphi}(x, f) + \alpha\tilde{\varphi}(x) \tag{2.14}$$

are also strictly convex, and hence the assertions of §2 regarding the connection between a quasisolution and variational methods of regularizing equation (2.12) are applicable to them.

Indeed, let t_1 and t_2 be positive numbers with $t_1 + t_2 = 1$. If $x_1 \neq \lambda x_2$, then

$$\tilde{\varphi}(t_1x_1 + t_2x_2) = \tilde{\gamma}(\|t_1x_1 + t_2x_2\|)$$
$$< \tilde{\gamma}(t_1\|x_1\| + t_2\|x_2\|) \leq t_1\tilde{\varphi}(x_1) + t_2\tilde{\varphi}(x_2).$$

(7)A Banach space is called *strictly convex* if for any $x_1, x_2 \in X$ the equality $\|x_1+x_2\| = \|x_1\| + \|x_2\|$ implies $x_1 = \lambda x_2$.

If $x_1 = \lambda x_2$ but $x_1 \neq x_2$, then

$$\tilde{\varphi}(t_1x_1 + t_2x_2) = \tilde{\gamma}(t_1\|x_1\| + t_2\|x_2\|) < t_1\tilde{\varphi}(x_1) + t_2\tilde{\varphi}(x_2),$$

which means that the functional $\tilde{\varphi}(x)$ is strictly convex. Strict convexity of $\bar{\varphi}(x)$ is proved similarly.

4. If the problem of solving (2.12) is well-posed in the Tikhonov sense and the set of well-posedness is

$$Y_a = \{x\colon \|x\|_{\tilde{X}} \leq a\},$$

then the function characterizing conditional stability depends on two variables:

$$\omega(\varepsilon, a) = \sup \|x_1 - x_2\|, \quad x_j \in Y_a, \quad \|A(x_1 - x_2)\| \leq \varepsilon.$$

It is easy to see that

$$\tilde{\omega}(\varepsilon, a) \leq \omega(\varepsilon, a) \leq \tilde{\omega}(\varepsilon, 2a), \qquad \tilde{\omega}(\varepsilon, a) = \sup \|x\|, \quad x \in Y_a, \ \|Ax\| \leq \varepsilon.$$

The function $\tilde{\omega}(\varepsilon, a)$ has the form

$$\tilde{\omega}(\varepsilon, a) = a\tilde{\omega}(\varepsilon/a).$$

5. In the theory of ill-posed problems, the subspaces $\tilde{X}$ are often given as follows:

$$\tilde{X} = \{x\colon x = Bx_1, \ x_1 \in X_1\},$$

where B is a compact operator from X_1 to X.

In the case where X_1, X, and F are additionally Hilbert spaces, an element minimizing the smoothing functional can be determined explicitly, as will be shown below.

Thus, we consider a smoothing functional of the type (2.14) for equation (2.12), setting $\gamma(t) = \tilde{\gamma}(t) = t^2$:

$$\psi(x, \alpha) = (Ax - f, \ Ax - f)_F + \alpha(B^{-1}x, \ B^{-1}x)_{X_1}.$$

The problem of minimizing $\psi(x, \alpha)$ is obviously equivalent to the problem of minimizing the functional

$$\tilde{\psi}(x_1, \alpha) = \tilde{\psi}(B^{-1}x, \alpha) = (ABx_1 - f, \ ABx_1 - f)_F + \alpha(x_1, x_1)_{X_1}.$$

An element x_1 at which $\tilde{\psi}(x_1, \alpha)$ achieves a minimum is a solution of the Euler equation

$$(AB)^*(AB\bar{x}_1 - f) + \alpha\bar{x}_1 = 0$$

and hence

$$\begin{aligned} \bar{x}_1 &= [\alpha E + (AB)^*AB]^{-1}(AB)^*f, \\ \bar{x} &= B\bar{x}_1 = B[\alpha E + (AB)^*AB]^{-1}(AB)^*f. \end{aligned} \tag{2.15}$$

We note that because of the connection established in Theorem 4 of §2 between a quasisolution and regularizing operators for the set of well-posedness, defined by the relation

$$Y_a = \{x\colon x = Bx_1, \|x_1\| \le a\},$$

a quasisolution can be constructed according to formula (2.15) for a particular value of the parameter $\alpha = \bar{\alpha}$. The number $\bar{\alpha}$ is found from the relations

$$\|[\bar{\alpha}E + (AB)^*AB]^{-1}(AB)^*f\| = a \qquad \text{for } \|(AB)^{-1}f\| > a,$$
$$\bar{\alpha} = 0 \qquad \text{for } \|(AB)^{-1}f\| \le a$$

(in the case where the operator $(AB)^{-1}$ is not defined at the element f we set $\|(AB)^{-1}f\| = \infty$).

6. DEFINITION 1. A family of linear operators $\{R_\alpha\}, 0 < \alpha < \alpha_0$, acting from F to X is said to *regularize* equation (2.12) on the set Z if

1) for any α, $0 < \alpha < \alpha_0$, the operator R_α is defined on all of F and is bounded, and

2) $\lim_{\alpha\to 0} R_\alpha Ax = x$ for any $x \in Z$.

If convergence in 2) is uniform on Z, then R is called *uniformly regularizing.*

The set Z is called the *regularization set* (*uniform regularization set*) of equation (2.12) for the family $\{R_\alpha\}$. We note that a regularization set is always a linear subspace of X.

If the operator A^{-1} is not bounded and $Z = X$, then $\|R_\alpha\|$ as a function of α, $0 < \alpha < \alpha_0$, is not bounded, and $\{R_\alpha\}$ is not a uniformly regularizing family. Indeed, the boundedness of $\|R_{\tilde{\alpha}}\|$ would imply boundedness of $\|A^{-1}\|$. Further, since $R_\alpha \ne A^{-1}$, for some $\tilde{X}$ we have $R_\alpha A\tilde{x} \ne \tilde{x}$, and hence the quantity

$$\|R_\alpha A(\lambda\tilde{x}) - \lambda\tilde{x}\| = |\lambda|\,\|R_\alpha A\tilde{x} - \tilde{x}\|$$

can be arbitrarily large.

It follows from Theorem 1 of §2 that regularizing families are a special case of the regularizing operators defined in §2.

It is sometimes more convenient to use regularizing families depending on an integral parameter.

DEFINITION 2. A family of linear operators $\{R_n\}$, $n = 1, \dots, \infty$, acting from F to X is called *regularizing* for equation (2.12) on a set Z if

1) for any $n > 0$ the operator R_n is defined on all of F and is bounded, and

2) $\lim_{n\to\infty} R_n Ax = x$ for any $x \in Z$.

We suppose that we know a regularizing family $\{R_\alpha\}$ for (2.12), and we consider the problem of determining an approximate solution of (2.12) on the basis of approximate data.

Thus, let

$$Ax_T = f_T \tag{2.12'}$$

and suppose we know an element f_ε with $\|f_T - f_\varepsilon\| \le \varepsilon$. We denote by $x_{\alpha\varepsilon}$ the result of applying R_α to the approximate right side of (2.12): $x_{\alpha\varepsilon} = R_\alpha f_\varepsilon$, and we estimate the norm of the difference $x_{\alpha\varepsilon} - x_T$:

$$\begin{aligned}\|x_{\alpha\varepsilon} - x_T\| &\le \|x_{\alpha\varepsilon} - R_\alpha A x_T\| + \|x_T - R_\alpha A x_T\| \\ &\le \|R_\alpha\|\varepsilon + \|x_T - R_\alpha A x_T\|.\end{aligned} \tag{2.16}$$

If $x_T \in Z$, then the second term on the right side of (2.16) tends to zero as $\alpha \to 0$, and by Theorem 1 of §2 for any $x_T \in Z$ there exists a function $\alpha(\varepsilon)$ such that the right side of (2.16) for $\alpha = \alpha(\varepsilon)$ tends to zero as $\varepsilon \to 0$. If R_α is a uniformly regularizing family, then

$$\begin{aligned}&\|x_T - R_\alpha A x_T\| \le \beta(\alpha), \quad x_T \in Z,\\ &\beta(\alpha) = \sup_{x\in Z} \|R_\alpha A x - x\|, \qquad \lim_{\alpha\to 0} \beta(\alpha) = 0,\end{aligned}$$

and hence

$$\|x_{\alpha\varepsilon} - x_T\| \le \|R_\alpha\|\varepsilon + \beta(\alpha).$$

In this case for the appropriate choice of α we have a guaranteed estimate of the accuracy of the approximate solution $x_{\alpha c}$. In particular, we may set $\alpha = \bar{\alpha}(\varepsilon)$, where $\bar{\alpha}$ is defined by the relation

$$\|R_{\bar{\alpha}}\|\varepsilon + \beta(\bar{\alpha}) = \inf_\alpha[\|R_\alpha\|\varepsilon + \beta(\alpha)].$$

7. We shall present some examples of regularizing families for equations with compact operators in Hilbert spaces.

Reduction to equations of the second kind. Suppose that the operator A is compact, selfadjoint, and positive. Together with (2.12) we consider the family of equations

$$\alpha x_\alpha + A x_\alpha = f. \tag{2.12''}$$

As is known from functional analysis, under our assumptions a solution of (2.12″) for $\alpha > 0$ always exists, is unique, and depends continuously on the right side, i.e., the problem of solving (2.12″) is well-posed.

We denote by R_α the operator $(\alpha E + A)^{-1}$. Then $x_\alpha = R_\alpha f$.

We shall show that the family $\{R_\alpha\}$ is regularizing for equation (2.12) on the entire space X. Since A is a positive operator, it follows that

$$\|(\alpha E + A)x\| \ge \alpha\|x\|,$$

and hence

$$\|R_\alpha x\| \le \|x\|/\alpha, \qquad \|R_\alpha\| \le 1/\alpha.$$

We now consider the difference

$$R_\alpha Ax - x = \alpha(\alpha E + A)^{-1}x.$$

As is known from functional analysis, the operator A has a complete orthonormal system of eigenfunctions $\{\varphi_k\}$ and eigenvalues $\{\lambda_k\}, \lambda_k \geq \lambda_{k+1} > 0$. We set $\xi_k = (x, \varphi_k)$. Then

$$x = \sum_{k=1}^{\infty} \xi_k \varphi_k, \qquad R_\alpha Ax - x = -\alpha \sum_{k=1}^{\infty} \frac{\xi_k}{\alpha + \lambda_k} \varphi_k,$$

$$\|R_\alpha Ax - x\|^2 = \alpha^2 \sum_{k=1}^{\infty} \frac{\xi_k^2}{(\alpha + \lambda_k)^2}$$

$$= \alpha^2 \sum_{k=1}^{n} \frac{\xi_k^2}{(\alpha + \lambda_k)^2} + \alpha^2 \sum_{k=n+1}^{\infty} \frac{\xi_k^2}{(\alpha + \lambda_k)^2} \leq \frac{\alpha^2}{\lambda_n^2}\|x\|^2 + \sum_{k=n+1}^{\infty} \xi_k^2. \quad (2.17)$$

Because of the convergence of $\sum_{k=1}^{\infty} \xi_k^2$ the second term on the right side of (2.17) is arbitrarily small for sufficiently large n. The first term is obviously arbitrarily small for sufficiently small α and fixed n, whence it follows that

$$\lim_{\alpha \to \infty} \|R_\alpha Ax - x\| = 0.$$

If the operator A is not positive and selfadjoint, equation (2.12) can be reduced to an equation with a positive selfadjoint operator by applying to both sides of (2.12) the operator A^* adjoint to A:

$$A_1 x = f_1, \qquad A_1 = A^* A, \qquad f_1 = A^* f. \qquad (2.12''')$$

We shall show that any solution of (2.12‴) is also a solution of (2.12), i.e., these equations are equivalent. For this it suffices to show that a solution of (2.12‴) is unique. Let $\bar{x}$ be a solution of the homogeneous equation (2.12‴): $A_1\bar{x} = 0$. Then $(A_1\bar{x}, \bar{x}) = (A\bar{x}, A\bar{x}) = 0$, whence by the uniqueness of the solution of (2.12) it follows that $x = 0$. For (2.12) with an arbitrary compact operator the regularizing family constructed according to the method outlined above has the form

$$R_\alpha = (\alpha E + A^* A)^{-1} A^*.$$

Expansion in eigenfunctions. Suppose that the operator A is compact and selfadjoint, and let $\{\varphi_k\}$ be the eigenfunctions and $\{\lambda_k\}$ the eigenvalues, $|\lambda_{k+1}| \leq |\lambda_k|$, $k = 1, \ldots, \infty$. We denote by $\overline{R}_n$ the operator defined by

$$\overline{R}_n f = \sum_{k=1}^{n} \frac{1}{\lambda_k} f_k \varphi_k, \qquad f_k = (f, \varphi_k). \qquad (2.18)$$

It is obvious that the operators $\overline{R}_n$ are continuous, and that $\|\overline{R}_n\| = 1/|\lambda_n|$ and

$$\lim_{n\to\infty} \overline{R}_n Ax = \lim_{n\to\infty} \sum_{k=1}^{n} \xi_k \varphi_k = x, \qquad \xi_k = (x, \varphi_k).$$

Thus, the family of operators $\{\overline{R}_n\}$ defined by (2.18) is a regularizing family depending on an integral parameter.

The method of successive approximations. Suppose that in (2.12) the operator A is positive and selfadjoint, and let $\{\varphi_k\}$ and $\{\lambda_k\}$ be the sequences of eigenfunctions and eigenvalues of the operator A, $\|A\| = \lambda_1 \leq 1$. We consider the sequence $\{x_k\}$:

$$x_{k+1} = x_k - Ax_k + f, \qquad x_0 = f. \tag{2.19}$$

We shall show that $\lim_{n\to\infty} x_n = x$. From (2.19) it follows that

$$x_n = \sum_{k=0}^{n} (E - A)^k Ax, \qquad x - x_n = (E - A)^{n+1} x.$$

Expanding the elements x and x_n in terms of the basis $\{\varphi_k\}$, we obtain

$$x - x_n = \sum_{k=1}^{\infty} (1 - \lambda_k)^{n+1} \xi_k \varphi_k, \qquad \xi_k = (x, \varphi_k),$$

$$\begin{aligned} \|x - x_n\|^2 &= \sum_{k=1}^{m} (1 - \lambda_k)^{2(n+1)} \xi_k^2 + \sum_{k=m+1}^{\infty} (1 - \lambda_k)^{2(n+1)} \xi_k^2 \\ &\leq (1 - \lambda_m)^{2(n+1)} \|x\|^2 + \sum_{k=m+1}^{\infty} \xi_k^2. \end{aligned} \tag{2.20}$$

In analogy to (2.17), the second term in (2.20) is arbitrarily small for sufficiently large m, while the first term is arbitrarily small for sufficiently large n and fixed m.

The regularizing family corresponding to the above method of successive approximations has the form

$$\tilde{R}_n = \sum_{0}^{n} (E - A)^k = [(E - A)^{n+1} - E](E - A)^{-1}.$$

Obviously, $\|\tilde{R}_n\| = n + 1$.

8. Regularizing families $\{R_\alpha\}$ can give criteria for the existence of a solution of equation (2.12).

THEOREM 2. *Suppose $\{R_\alpha\}$ is a regularizing family for equation (2.12) on the entire space X, $X = F$, and the operators R_α commute with A ($R_\alpha A = AR_\alpha$). If for some f_0 the limit $\lim_{\alpha\to 0} R_\alpha f_0 = x_0$ exists, then this limit is a solution of (2.12) with right side f_0, i.e., $Ax_0 = f_0$.*

PROOF. Because of the continuity of A,

$$\lim_{\alpha\to 0} AR_\alpha f_0 = Ax_0,$$

whence

$$Ax_0 = \lim_{\alpha\to 0} R_\alpha A f_0 = f_0.$$

9. Suppose the problem of solving (2.12) is well-posed in the Tikhonov sense, Y is the set of well-posedness, and $\{R_\alpha\}$ is a family of operators uniformly regularizing (2.12) on Y. As noted above, the error of an approximate solution obtained by applying the operator R_α can be estimated by the inequality

$$\|R_\alpha f_\varepsilon - x_T\| \le \|R_\alpha\|\varepsilon + \beta(\alpha), \qquad \beta(\alpha) = \sup_{x\in Y} \|R_\alpha Ax - x\|. \tag{2.21}$$

For a choice of the regularization parameter $\alpha = \alpha^*$ which is optimal from the point of view of the estimate (2.21) the error can be estimated by the inequality

$$\|R_{\alpha^*} f_\varepsilon - x_T\| \le \gamma(\varepsilon), \qquad \gamma(\varepsilon) = \|R_{\alpha^*}\|\varepsilon + \beta(\alpha^*) = \inf[\|R_\alpha\|\varepsilon + \beta(\alpha)].$$

On the other hand, conditional stability of the problem of solving (2.12) can be estimated by the function

$$\omega(\varepsilon) = \sup \|x_1 - x_2\|, \quad x_j \in Y, \ \|A(x_1 - x_2)\| \le \varepsilon.$$

Obviously, $\omega(\varepsilon) \le 2\gamma(\varepsilon)$.

In works on the theory of ill-posed problems, various relations between the functions $\omega(\varepsilon)$ and $\gamma(\varepsilon)$ have been presented both for particular problems and for general equations under additional conditions on the set of well-posedness and on the regularizing families. We present one such relation.

THEOREM 3. *Let X and F be Hilbert spaces, and suppose that the set of well-posedness is defined as follows:*

$$Y = \{x \colon x = By, \ y \in X, \ \|y\| \le 1\},$$

where B is a positive selfadjoint operator from X to X, and

$$R_\alpha = B[\alpha E + (AB)^* AB]^{-1}(AB)^*.$$

Then

$$\gamma(\varepsilon) \le \frac{2a}{\tilde{\alpha}(\varepsilon)}\varepsilon = 2\omega\left[\frac{\sqrt{\tilde{\alpha}(\varepsilon)}}{2}\right], \tag{2.22}$$

where $a = \|B\|^2\,\|A\|$; $\tilde{\alpha}(\varepsilon)$ *is a root of the equation* $a\varepsilon/\alpha = \omega(\sqrt{\alpha}/2)$.

PROOF. We set

$$z_\alpha = R_\alpha f_T - x_T = (R_\alpha A - E)x_T.$$

By the hypotheses of the theorem, $x_T = By_T$, where $\|y_T\| \le 1$. After some transformations we obtain

$$z_\alpha = -\alpha B[\alpha E + (AB)^* AB]^{-1} y_T.$$

From the equality

$$\|[\alpha E + (AB)^* AB]^{-1}\| = 1/\alpha$$

it follows that

$$z_\alpha \in Y. \tag{2.23}$$

Next we estimate $\|Az_\alpha\|$:

$$\|Az_\alpha\|^2 = \alpha^2\{(AB)^* AB[\alpha E + (AB)^* AB]^{-2} y_T, y_T\} \le \alpha/4. \tag{2.24}$$

From (2.21), (2.23), and (2.24) we have

$$\|z_\alpha\| \le \tilde{\omega}(\sqrt{\alpha}/2) \le \omega(\sqrt{\alpha}/2), \qquad \tilde{\omega}(\alpha) = \sup\|x\|, \quad x \in Y,\ \|Ax\| \le \alpha.$$

It is easy to see that $\|R_\alpha\| \le a/\alpha$, and hence

$$\|R_\alpha f_\varepsilon - x_T\| \le \frac{a}{\alpha}\varepsilon + \omega\left(\frac{\sqrt{\alpha}}{2}\right). \tag{2.25}$$

Setting $\alpha = \tilde{\alpha}(\varepsilon)$ in (2.25), we obtain inequality (2.22).

CHAPTER III

Analytic Continuation

§1. Formulations of problems and classical results

1. A large class of problems of analytic continuation for functions of a single variable in general form can be formulated as follows.

It is known that there exists a function of a complex variable $f(z)$, analytic in a domain D of the complex plane. Some additional information may be given regarding $f(z)$, for example, that $f(z)$ is continuous on the set $\overline{D}$, the closure of D, and is bounded on $\overline{D}$ by a given constant: $|f(z)| \leq c, z \in \overline{D}$. Suppose further that A and B are subsets of $\overline{D}$ such that $A \subset B \subset \overline{D}$, and the values of $f(z)$ are known on A. It is required to determine the values of $f(z)$ on B.

2. Problems of analytic continuation are classical problems of the theory of analytic functions and functions of a complex variable; some uniqueness theorems for these problems were obtained in the last century and are presented in the majority of textbooks.

If we consider functions analytic in D without additional conditions, then problems of analytic continuation are linear, and thus uniqueness of the solution of the problem of analytic continuation is equivalent to the validity of the following assertion.

Let $f(z)$ be a function analytic in a domain D; suppose that $f(z) = 0$ for $z \in A$. Then $f(z) = 0$ for $z \in B$.

We note that in familiar uniqueness theorems for problems of analytic continuation the set B contains the domain of analyticity D: $B \supset D$.

3. We shall present classical results on uniqueness of the solution of problems of analytic continuation (cf. [140] and [168]).

THEOREM 1. *Let $f(z)$ be a function analytic in a domain D, and let D_1 be a subdomain of D. If $f(z) = 0$ for $z \in D_1$, then $f(z) = 0$ for $z \in D$.*

THEOREM 2. *Let $f(z)$ be a function analytic in a domain D, and let A be a set having a limit point belonging to D. If $f(z) = 0$ for $z \in A$, then $f(z) = 0$ for $z \in D$.*

THEOREM 3. *Let $f(z)$ be a function analytic in a domain D, let Γ be the boundary of D, and let $\Gamma_1 \subset \Gamma$ be a rectifiable curve. If $f(z) = 0$ for $z \in \Gamma_1$, then $f(z) = 0$ for $z \in D$.*

4. The concept of a Riemannian manifold is connected with analytic continuation. We shall present some definitions related to this (see [168] and [284]).

DEFINITION 1. An *analytic element* is a pair $\{D, f\}$ consisting of a domain D of the complex plane and a function $f(z)$ analytic in D.

DEFINITION 2. Two analytic elements $\{D_1, f_1\}$ and $\{D_2, f_2\}$ are called a *direct analytic continuation* of one another through a domain $\tilde{D} \subset D_1 \cap D_2$ if $f_1(z) = f_2(z)$ for $z \in \tilde{D}$.

DEFINITION 3. Analytic elements $\{D_0, f_0\}$ and $\{D_n, f_n\}$ are called an *analytic continuation* of one another if there exist analytic elements $\{D_k, f_k\}$ and domains $\tilde{D}_k$, $k = 1, \dots, n-1$, such that the analytic elements $\{D_k, f_k\}$ and $\{D_{k+1}, f_{k+1}\}$ are a direct analytic continuation of one another through $\tilde{D}_k$.

DEFINITION 4. An *analytic function* is a collection of analytic elements which are analytic continuations of a single analytic element. The functions $f_a(z)$ that belong to the same analytic function are called *branches* of this function.

We consider the collection of pairs $\{a, f_a(z)\}$, where a is a point of the complex plane; $f_a(z)$ is a function analytic in some disk U_a with center a:

$$f_a(z) = \sum_{k=0}^{\infty} c_k (z-a)^k.$$

On the set of pairs $\{a, f_a(z)\}$ we introduce a topology: an ε-neighborhood of the element $\{a, f_a(z)\}$ is a collection of elements $\{b, f_b(z)\}$ such that

1) $|a-b| < \varepsilon$, and

2) the analytic element $\{U_b, f_b\}$ is a direct analytic continuation of the analytic element $\{U_a, f_a\}$.

It can be shown that the topological space of pairs $\{a, f_a(z)\}$ is a Hausdorff space, i.e., its neighborhoods satisfy the axiom of separability.

DEFINITION 5. The Hausdorff space of pairs $\{a, f_a(z)\}$, where a is a point of the complex plane, $f_a(z)$ is a function analytic in a disk U_a, and the topology is introduced in the manner indicated above, is called a *Riemannian manifold.*

The mapping (projection) of the Riemannian manifold onto the complex plane

$$\{a, f_a(z)\} \to a$$

is a local homeomorphism, i.e., in some neighborhoods of the pair $\{a, f_a(z)\}$ and the point a this mapping is one-to-one and continuous.

There is a one-to-one correspondence between analytic functions and domains of a Riemannian manifold.

DEFINITION 6. A domain of a Riemannian manifold for which all functions belonging to its points are branches of some analytic function is called the *Riemann surface* of this function.

5. The following formulations of problems of analytic continuation arise in connection with the definitions above.

Suppose that A is a subset of the Riemann surface of an analytic function $f(z)$ and the values of $f(z)$ are known on the set A. It is required to determine

1) the values of $f(z)$ on some set $B \supset A$,

2) some characteristics (topological, for example) of the Riemann surface of the function $f(z)$, and

3) singular points of the function $f(z)$ and the behavior of the function in a neighborhood of these points, in particular, the types of the singular points, and the behavior in a neighborhood of singular points of the Riemann surface.

There is the following result [284].

THEOREM 4. *Suppose a set A contains a limit point within the Riemann surface of an analytic function $f(z)$. Then the Riemann surface of the function $f(z)$ is uniquely determined by the values of $f(z)$ on the set A.*

6. We formulate the problem of analytic continuation for functions of several variables in analogy to that for one variable.

It is known that there exists a function of n complex variables $f(z_1, \dots, z_n)$ analytic in a domain D of n-dimensional complex space; A and B are subsets of the closure of D: $A \subset B \subset \overline{D}$. The values of $f(z_1, \dots, z_n)$ are known on A. It is required to determine the values of $f(z_1, \dots, z_n)$ on B.

Just as for functions of a single variable, questions of analytic continuation for functions of several variables began to be worked out almost from the start of the development of the theory—for functions of several variables that would be the beginning of our century. Analytic continuation occupies considerable space in monographs on the theory of analytic functions of several variables, for example, [74] and [278]. We shall present some concepts and results related to the analytic continuation of functions of several variables (for more details see [278] and [284]).

In the present section we denote the space of n complex variables $(z_1, \dots, z_n)$ by the symbol C^n.

DEFINITION 7. A domain G strictly containing a domain $D \subset C^n$ is called a *holomorphic extension* of D if any function $f(z_1, \dots, z_n)$ analytic in D admits analytic continuation to G.

As is known, domains in the complex plane do not admit holomorphic extension. In spaces of two and more complex variables there are domains admitting holomorphic extension. The existence of a class of such domains is established by the following theorem.

THEOREM 5 (HARTOGS). *Let D be a domain of C^n, let D_1 be a domain of the complex plane, let Γ_1 be the boundary of D_1, and let M be the set*

$$M = [\overline{D} \times \Gamma_1] \cup [(z_1^0, \dots, z_n^0) \times \overline{D}_1],$$

where $(z_1^0, \dots, z_n^0) \in D$. Then any function $f(z_0, z_1, \dots, z_n)$ analytic in a neighborhood of the set M (in the space C^{n+1}) admits analytic continuation to the domain $\tilde{D} = D \times D_1$.

7. Solutions of some problems of analytic continuation depend continuously on the data. For a single variable this is the case when the set A coincides with the boundary Γ of the domain D. For several variables this is the case, for example, for analytic continuation to a holomorphic extension discussed in Theorem 5.

For functions of a single variable, problems of analytic continuation from sets within the domain of regularity and from part of the boundary of the domain are ill-posed in the classical sense; the character of instability in these problems is the same as that mentioned in ill-posed Cauchy problems for the Laplace and heat equations in Chapter I.

8. As is known, solutions of some differential equations inside domains of regularity are analytic functions. In particular, solutions of elliptic equations with analytic coefficients possess this property. In connection with this there arise problems of the following type, which we also classify as problems of analytic continuation.

Let L be a differential operator in the space R^n possessing the property that all solutions of the equation

$$Lu = 0 \tag{3.1}$$

within the domain of regularity are analytic functions of the variables $(x_1, \dots, x_n)$.

In the domain D there exists a regular solution $u(x_1, \dots, x_n)$ of (3.1). Additional information may be given regarding the function $u(x_1, \dots, x_n)$, for example,

$$|u| \leq c, \quad (x_1, \ldots, x_n) \in \overline{D}.$$

Let A and B be subsets of $\overline{D}$, $A \subset B \subset \overline{D}$.

The values of a solution of (3.1) are known on the set A; it is required to determine the values of the solution on B.

§2. Analytic continuation from continua

1. We shall consider from the point of view of the theory of ill-posed problems the classical problem of analytic continuation from an interior subdomain of a domain of regularity.

Thus, let $D \supset D_1 \supset D_2$ be domains of the complex plane, and suppose that their boundaries Γ, Γ_1, and Γ_2 are piecewise smooth curves: $\Gamma_1 \subset D$ and $\Gamma_2 \subset D_1$.

Suppose, further, that it is known that there exists a function $f(z)$ analytic in D and the values of $f(z)$ are known everywhere in D_2. It is required to determine the values of $f(z)$ in D_1.

This problem reduces to an integral equation of the first kind. Indeed, by Cauchy's formula $\{a, f_a(z)\} \to a$, and to solve the problem it suffices to determine the value of $f(z)$ on Γ_1. Thus, solution of the given problem of analytic continuation reduces to solution of the integral equation

$$\frac{1}{2\pi i} \int_{\Gamma_1} \frac{\varphi(\varsigma)}{\varsigma - z}\, d\varsigma = f(z), \quad z \in D_2. \tag{3.2}$$

One of the possible compact sets of well-posedness for this problem is provided by the a priori assumption that $f(z)$ is bounded in the entire domain D by a fixed constant.

2. Estimates of conditional stability of the problem of analytic continuation are based on the following theorem.

THEOREM 1 (HADAMARD). *Let $f(z)$ be a function analytic in the annulus $r < |z| < R$ and continuous in the closed annulus $r \leq |z| \leq R$. Then*

$$|f(z)| \leq \exp\left(\ln m \frac{\ln R - \ln|z|}{\ln R - \ln r} + \ln M \frac{\ln|z| - \ln r}{\ln R - \ln r}\right), \tag{3.3}$$

where

$$M = \max|f(z)|, \quad |z| = R,$$
$$m = \max|f(z)|, \quad |z| = r.$$

PROOF. We consider the function

$$F(z) = f(z) \exp\left(\ln M \frac{\ln R - \ln z}{\ln R - \ln r} + \ln m \frac{\ln z - \ln r}{\ln R - \ln r}\right).$$

Obviously,

$$\max_{|z|=r} |F(z)| = \max_{|z|=R} |F(z)| = Mm.$$

Hence, by the maximum principle,

$$|F(z)| \leq Mm, \quad r < |z| < R,$$

whence we obtain (3.3).

Suppose now that the domain D_2 is a disk (or contains a disk) of radius r_0 with center at the point z_0:

$$D_2 = \{z \colon |z - z_0| < r_0\},$$

and let $\tilde{z}$ be some point of $D_1 : \tilde{z} \in D_1$. We denote by $\{U_k^j\}$ a system of circles with centers at the points z_k and radii r_k^j, $k = 0, \ldots, n$, $j = 0, 1, 2$, satisfying the following relations:

1) $U_0^0 = D_2$, 2) $U_k^0 \subset U_k^1 \subset U_k^2 \subset D$, 3) $U_{k+1}^0 \subset U_k^1$, 4) $\tilde{z} \in U_n^1$.

It is obvious that for any point $\tilde{z} \in D_1$ such systems of circles exist.

Let $f(z)$ be a function analytic in the domain D and satisfying the inequalities

$$|f(z)| \leq M, \quad z \in D,$$
$$|f(z)| \leq \varepsilon, \quad z \in D_2.$$

Applying Theorem 1 n times, we obtain the following estimate of conditional stability of the problem of analytic continuation from an interior subdomain:

$$|f(\tilde{z})| \leq \inf \exp\left(\frac{\ln r_n^2 - \ln r_n^1}{\ln r_n^2 - \ln r_n^0} \ln \varepsilon_{n-1} + \frac{\ln r_n^1 - \ln r_n^0}{\ln r_n^2 - \ln r_n^0} \ln M \right),$$

$$\varepsilon_k = \exp\left(\frac{\ln r_{k-1}^2 - \ln r_{k-1}^1}{\ln r_{k-1}^2 - \ln r_{k-1}^0} \ln \varepsilon_{k-1} + \frac{\ln r_{k-1}^1 - \ln r_{k-1}^0}{\ln r_{k-1}^2 - \ln r_{k-1}^0} \ln M \right),$$
$$\varepsilon_0 = \varepsilon; \ k = 1, \ldots, n-1 \quad (3.4)$$

(the infimum is taken over the set of all chains of disks satisfying conditions 1)–3)).

3. We shall present an example of regularization (construction of a regularizing family) for the problem of analytic continuation from an interior subdomain, which is based on Taylor's formula. We consider the case where D, D_1, and D_2 are concentric circles:

$$D = \{z \colon |z| < R\}, \qquad D_1 = \{z \colon |z| < \rho\}, \qquad D_2 = \{z \colon |z| < r\}.$$

We will use the values of $f(z)$ only on the boundary Γ_2 of D_2. Thus, let

$$f(z)|_{|z|=r} = \varphi(z).$$

We denote by R_n the operator which assigns to the function $\varphi(z)$ defined on Γ_2 the function $f_n(z)$:

$$f_n(z) = \sum_{k=0}^{n} a_k z^k,$$

$$a_k = \frac{1}{2\pi i}\int_{\Gamma_2} \frac{\varphi(\varsigma)\,d\varsigma}{\varsigma^{k+1}} = \frac{1}{2\pi r^k}\int_0^{2\pi} e^{-ik\alpha}\varphi(re^{i\alpha})\,d\alpha.$$

It is obvious that the operators R_n are continuous. The numbers a_n are the coefficients of the expansion of $f(z)$ in a power series, whence it follows that

$$\lim_{n\to\infty} R_n\varphi = \lim_{n\to\infty} f_n(z) = f(z), \quad z \in D,$$

and hence $\{R_n\}$ is a regularizing family for the problem in question.

Suppose now that $|f(z)| \leq M$, $z \in D$. Then, as is known from the theory of analytic functions,

$$|f(z) - f_n(z)| \leq \frac{|z|^{n+1}M}{R^n(R-|z|)}. \tag{3.5}$$

We consider R_n as an operator assigning to the function $\varphi(z)$ on Γ_2 the function $f_n(z)$ on $\overline{D}_1 = \Gamma_1 \cup D_1$, and we assume that $\varphi(z)$ and $f_n(z)$ are elements of function spaces with uniform norm:

$$\|\varphi(z)\| = \max_{z\in\Gamma_2} |\varphi(z)|, \qquad \|f_n(z)\| = \max_{z\in\overline{D}_1} |f_n(z)|.$$

It is not hard to show that the norm of R_n here satisfies the inequality

$$\|R_n\| \leq \frac{\rho^{n+1}}{r^n}\frac{1-(r/\rho)^{n+1}}{\rho - r}. \tag{3.6}$$

From (3.5) and (3.6) we obtain the following estimate of the efficiency of applying the regularizing family $\{R_n\}$ to the problem of finding an approximate solution on the basis of data with error $\varphi_\varepsilon(z)$:

$$\|\varphi_T(z) - \varphi_\varepsilon(z)\| \leq \varepsilon,$$

$$\|f_T - R_n\varphi_\varepsilon\| \leq \frac{\rho^{n+1}}{r^n}\frac{1-(r/\rho)^{n+1}}{\rho - r}\varepsilon + \frac{\rho^n}{R^n(R-\rho)}M.$$

In the case of arbitrary domains D, D_1, and D_2, the regularizing family can be constructed by means of systems of circles $\{U_k^j\}$.

4. We shall formulate the problem of analytic continuation from part of the boundary of a domain.

Suppose that the boundary of the domain D is a piecewise smooth curve (or several curves) $\Gamma = \Gamma_1 \cup \Gamma_2, \Gamma_1 \cap \Gamma_2 = \varnothing$, and the Γ_j are also piecewise smooth curves. It is known that there exists a function $f(z)$ analytic in D

and continuous in $\overline{D}$, and the values of $f(z)$ are known on Γ_1. It is required to determine the values of $f(z)$ in the domain $D_1 \subset D$.

We shall present a theorem on the conditional stability of this problem.

DEFINITION 1. A function $\omega(z) = \omega(z, \Gamma_1, D)$ which is a solution of the Laplace equation in the domain D and which on the boundary Γ assumes the values $\omega(z) = 1$ for $z \in \Gamma_1$ and $\omega(z) = 0$ for $z \in \Gamma_2$ (interior points of Γ_j are intended) is called a *harmonic measure* of the curve Γ_1 in the domain D relative to the point z. It is known that such a function $\omega(z)$ exists if, for example, D is a simply connected domain [168].

THEOREM 2. *Suppose the function $f(z)$ satisfies the inequalities*

$$|f(z)| \leq M, \quad z \in \Gamma_2,$$
$$|f(z)| \leq m, \quad z \in \Gamma_1.$$

Then

$$|f(z)| \leq M^{1-\omega(z)} m^{\omega(z)}. \tag{3.7}$$

The proof is similar to the proof of Theorem 1. Suppose the analytic function $\varphi(z)$ is such that $\operatorname{Re}\varphi(z) = \omega(z)$. We consider the function

$$F(z) = f(z) \exp\{\varphi(z) \ln M + [1 - \varphi(z)] \ln m\}.$$

By the maximum principle $|F(z)| < Mm$, whence we obtain (3.7).

5. One way of regularizing the problem of analytic continuation from part of the boundary of a domain is to use a Carleman function.

DEFINITION 2. A function of two complex arguments and a scalar argument $G(z, \varsigma, \alpha)$ is called a *Carleman function* of a curve Γ_1 in a domain D if

$$1) \quad G(z, \varsigma, \alpha) = \frac{1}{\varsigma - z} + \tilde{G}(z, \varsigma, \alpha),$$

where $\tilde{G}(z, \varsigma, \alpha)$ is a function of the variable ς which is analytic in D and bounded and piecewise continuous in $\overline{D}$, and

$$2) \quad \frac{1}{2\pi} \int_{\Gamma_2} |G(z, \varsigma, \alpha)|\, |d\varsigma| \leq \alpha.$$

We denote by $\tilde{R}_\alpha$ the operator assigning to a function $\varphi(z)$ defined on Γ_1 a function $f_\alpha(z)$ in D according to the formula

$$f_\alpha(z) = \tilde{R}_\alpha \varphi(z) = \frac{1}{2\pi i} \int_{\Gamma_1} G(z, \varsigma, \alpha) \varphi(\varsigma)\, d\varsigma.$$

We shall show that $\{\tilde{R}_\alpha\}$ is a regularizing family for this problem of analytic continuation. Indeed, by Cauchy's formula

$$f(z) = \frac{1}{2\pi i} \int_{\Gamma} G(z, \varsigma, \alpha) f(\varsigma)\, d\varsigma,$$

and hence

$$|f(z) - f_\alpha(z)| = \frac{1}{2\pi}\left|\int_{\Gamma_2} G(z,\varsigma,\alpha) f(\varsigma)\,d\varsigma\right| \le \alpha M,$$

where $M = \max_{z\in\Gamma} |f(z)|$. Thus,

$$\lim_{\alpha\to 0} |f(z) - \tilde{R}_\alpha\varphi(z)| = 0, \qquad \varphi(z) = f(z), \quad z \in \Gamma_1.$$

The following function is an example of a Carleman function:

$$G = \frac{1}{\varsigma - z}\exp\{\lambda[\varphi(\varsigma) - \varphi(z)]\}, \tag{3.8}$$

where $\operatorname{Re}\varphi(z) = \omega(z)$ is a harmonic measure, and

$$\lambda = \frac{1}{\omega(z)}\left[\ln\alpha - \ln\left(\int_{\Gamma_2}\frac{|d\varsigma|}{|\varsigma - z|}\right)\right].$$

The accuracy of an approximate solution obtained by means of the Carleman function (3.8) can be estimated by the inequality

$$|f_\alpha(z) - f(z)| \le M\alpha + \int_{\Gamma_1}\frac{|d\varsigma|}{|\varsigma - z|}\exp\left[\frac{\omega(z)}{1-\omega(z)}\ln\int\frac{|d\varsigma|}{|\varsigma - z|} - \frac{1-\omega(z)}{\omega(z)}\ln\alpha\right].$$

6. The results presented in this section admit generalizations to analytic functions of several variables. The problem of analytic continuation of a function of several variables from an interior subdomain reduces to the problem for a function of one variable. Indeed, suppose $f(z_1,\dots,z_n)$ is a function analytic in a domain D of n-dimensional complex space. The values of $f(z_1,\dots,z_n)$ are known in some domain $D_1 \subset D$. It is required to find the value of $f(\tilde{z}_1,\dots,\tilde{z}_n)$ for $(\tilde{z}_1,\dots,\tilde{z}_n) \in D\backslash D_1$.

Let $(z_1^0,\dots,z_n^0) \in D_1$. Then there exist functions $\varphi_k(\varsigma)$, $k = 1,\dots,n$, analytic in some neighborhood of the segment $[0,1]$ of the real axis, such that $\varphi_k(0) = z_k^0$, $\varphi_k(1) = \tilde{z}_k$ and the curve in n-dimensional complex space defined by the equations

$$z_k = \varphi_k(\varsigma), \quad k = 1,\dots,n;\ \varsigma \in [0,1],$$

lies entirely inside the domain D.

As is known from the theory of analytic functions, the function

$$F(\varsigma) = f[\varphi_1(\varsigma),\dots,\varphi_n(\varsigma)],$$

is analytic in some neighborhood of $[0,1]$. For the functions φ_k it is possible to take, for example, polynomials of sufficiently high degree.

We present a generalization of Theorem 2 to the case of functions of several variables.

THEOREM 3. *Let $f(z_1, \dots, z_n)$ be a function analytic in the polycylindrical domain $D = D_1 \times \dots \times D_n$ and continuous in the closure $\overline{D}$ of D; D_k is a domain of the complex z_k-plane with piecewise smooth boundary Γ_k. Let Γ'_k be a part of the curve Γ_k. If the function $f(z_1, \dots, z_n)$ satisfies the inequalities*

$$|f(z_1, \dots, z_n)| \le M, \quad (z_1, \dots, z_n) \in \overline{D},$$
$$|f(z_1, \dots, z_n)| \le m, \quad (z_1, \dots, z_n) \in \Gamma'_1 \times \dots \times \Gamma'_n,$$

then

$$|f(z_1, \dots, z_n)| \le \exp\{[1 - \omega_1(z_1) \cdots \omega_n(z_n)] \ln M + \omega_1(z_1) \cdots \omega_n(z_n) \ln m\}, \quad (3.9)$$

where $\omega_k(z_k)$ is a harmonic measure of the curve Γ'_k in the domain D_k relative to the point z_k.

PROOF. We suppose that the assertion of the theorem is true in complex spaces of dimension less than n. Then

$$|f(z_1, \dots, z_{n-1}, \varsigma)| \le \exp\{[1 - \omega_1(z_1) \cdots \omega_n(z_n)] \ln M + \omega_1(z_1) \cdots \omega_n(z_n) \ln m\}, \quad \varphi_n \in \Gamma'_n. \quad (3.10)$$

Inequality (3.9) follows from (3.10) and Theorem 2.

§3. Analytic continuation from classes of sets including discrete sets

1. We shall consider the problem of analytic continuation of functions of one complex variable from a class of sets inside the domain of regularity including discrete sets. We restrict ourselves to the case where the domain in which the desired function is analytic is the unit disk. The problem for an arbitrary simply connected domain can be reduced to this problem by conformal mapping.(8)

Thus, let $\{a_k\}$, $k = 1, 2, \dots$, be a sequence of complex numbers, $|a_k| < 1$. The solution of the problem of analytic continuation from the set $A = \{z : z = a_k\}$ is unique if the sequence $\{a_k\}$ has a limit point a_0 with $|a_0| < 1$. More general sufficient conditions for uniqueness of the solution of the problem of analytic continuation connected with the concept of a Blaschke function [168] are known.

We shall present an estimate of conditional stability of the problem in the class of bounded functions in the case where the sequence has a limit point inside the unit disk. We first introduce a certain characteristic of sets of points of the real axis.

(8)The results of this section were published in [141] and [145].

Let A be a set of points of the real axis, let $\{\Delta_k\}$ be a system of n intervals that contains the set A, $A \subset \bigcup_1^n \Delta_k$, and let l_k be the length of the interval Δ_k.

We denote by $\mu_n(A)$ the infimum of the sum of the lengths l_k over all systems of n intervals containing A,

$$\mu_n(A) = \inf\left(\sum_{k=1}^{n} l_k\right).$$

We observe some obvious properties of $\mu_n(A)$—a function of the set A and the integral argument n:

1) $\mu_n(A) = \mu_n(\hat{A})$, where $\hat{A}$ is the closure of A;

2) $\mu_n(A)$ is a nonincreasing function of the argument n, and $\lim_{n\to\infty} \mu_n(A) \geq \mu_l(A)$, where $\mu_l(A)$ is the outer measure of the set A;

3) If the sets A_1 and A_2 are closed and $A_1 \cap A_2 = \varnothing$, then $\mu_n(A_1 \cup A_2) \geq \mu_n(A_1) + \mu_n(A_2)$;

4) If $n = n_1 + n_2$, then $\mu_n(A_1 \cup A_2) \leq \mu_{n_1}(A_1) + \mu_{n_2}(A_2)$.

We set

$$D = \{z\colon |z| < 1\}, \qquad D_1 = \left\{z\colon |z| \leq \frac{1}{2}\right\}, \qquad D_2 = \left\{z\colon |z| \leq \frac{1}{4}\right\}.$$

THEOREM 1. *Suppose the set $A \subset D_2$, $\overline{A}$ is the projection of A onto the real axis, and $f(z)$ is a function analytic in the disk D, where*

$$\begin{aligned} |f(z)| &\leq 1, \quad z \in D, \\ |f(z)| &\leq \varepsilon, \quad z \in A, \end{aligned} \qquad 0 < \varepsilon \leq \frac{\mu_n(A)}{4}. \tag{3.11}$$

Then one of the two inequalities

$$|f(z)| \leq \varepsilon^{4/25}, \qquad |f(z)| \leq (6/7)^n, \quad z \in D_2, \tag{3.12}$$

holds, where n is the number determined by the relation

$$\left(\frac{\mu_{n+1}(\overline{A})}{2}\right)^{n+1} < 2\varepsilon \leq \left(\frac{\mu_n(\overline{A})}{2}\right)^n. \tag{3.13}$$

PROOF. We consider the function $\varphi(z) = \ln|f(z)|$. As we know, this function is subharmonic in the variables (x, y), $z = x + iy$, at those points D at which $f(z) \neq 0$.

Let $f(a) \neq 0$ but $|f(a)| \leq \varepsilon$. We denote by Γ_a the level line of the function $\varphi(z)$ passing through the point a:

$$|f(z)| = |f(a)| \leq \varepsilon, \quad z \in \Gamma_a.$$

The curve Γ_a can be either closed or nonclosed in the disk D_1. If Γ_a is not closed in D_1, then it intersects the annulus $1/4 \leq |z| \leq 1/2$. It can be shown that in this case the harmonic measure of the curve Γ_a in the domain $D\backslash\Gamma_a$ satisfies the inequality

$$\omega(z, \Gamma_a) \geq \frac{1}{4}\frac{1-|z|^2}{1+|z|^2}.$$

The validity of the first of inequalities (3.12) follows from this by Theorem 1 of §2.

Thus, the assertion of the theorem to be proved is valid if at least one of the curves Γ_a is not closed in D_1.

We now consider the case where all curves Γ_a are closed in D_1. If the domain bounded by Γ_a contains no zeros of $f(z)$, the function $\varphi(z)$ is regular in this domain and is hence identically constant. This implies that in the entire domain D we have $|f(z)| = \text{const} \leq \varepsilon$.

Suppose now that all curves Γ_a are closed in D_1 and each domain bounded by Γ_a contains zeros of $f(z)$. Suppose that, in D_1, $f(z)$ has m zeros at the points $z_1, \dots, z_m$ (each zero is counted as many times as its multiplicity). We denote by $F(z)$ the function

$$F(z) = \prod_1^m \frac{1-\bar{z}_k z}{z-z_k} f(z).$$

In the disk D_1 the function $F(z)$ has no zeros. Since

$$\left|\frac{1-\bar{z}_k z}{z-z_k}\right| = 1 \quad \text{for } |z| = 1,$$

the function $F(z)$ satisfies the inequality

$$|F(z)| \leq 1, \quad z \in D. \tag{3.14}$$

We shall show that in the disk D_2 there exists a point z^* at which the modulus of $F(z)$ satisfies the inequality

$$|F(z^*)| \leq 2\varepsilon \left(\frac{2}{\mu_m(\overline{A})}\right)^m. \tag{3.15}$$

We denote by B the set $\{z\colon |f(z)| \leq \varepsilon, z \in \overline{D}_2\}$, and by $\overline{B}$ the projection of B onto the real axis. Obviously,

$$\sup_{z\in B}\left|\prod_1^m (z-z_k)\right| \geq \sup_{x\in \overline{B}} \prod_1^m |x-x_k|, \qquad x_k = \operatorname{Re} z_k.$$

Because of the closeness of the curves Γ_a in D_1 and the relations $B \supset A$ and $\overline{B} \supset \overline{A}$, the set $\overline{B}$ consists of at most m segments $\{\Delta_k\}$, the sum of whose lengths satisfies the inequality

$$l = \sum_{k=1}^r l_k \geq \mu_m(\overline{A}), \qquad r \leq m.$$

We shall show that the supremum of the product $\prod_1^m |x - x_k| = P(x)$ on the set $\overline{B}$ is not less than the maximum of the Tchebycheff polynomial of degree m which deviates least from zero on a segment of length l. We denote the end points of the segments Δ_k by ξ_k' and ξ_k'',

$$\Delta_k = \{x\colon \xi_k' \le x \le \xi_k''\},$$

and we suppose that

$$\xi_k'' < \xi_{k+1}', \qquad x_p \le \xi_r', \qquad x_{p+1} > \xi_r', \qquad x_k \le x_{k+1}.$$

We consider the product

$$\prod_1^p |x - x_k| \prod_{p+1}^m |x - x_k + \xi_{m_1}'' - \xi_{m_1}'| = P_1(x)$$

on the set

$$\overline{B}_1 = \left(\bigcup_{k=1}^{r-1} \Delta_k\right) \cup \Delta_r', \qquad \Delta_r' = \{x\colon \xi_{r-1}'' \le x \le \xi_{r-1}'' + l_r\}.$$

It is obvious that at any point $x \in \Delta_k$, $k = 1, \dots, r-1$, all factors in $P_1(x)$ are no greater than the factors in the product $P(x)$. At any point $x \in \Delta_r'$ the factors in $P_1(x)$ are also no greater than the factors in $P(x)$:

$$x_1 = x - \xi_r' + \xi_{r-1}''.$$

Varying successively the sets $\overline{B}_j$ and the products $P_j(x)$ r times, we obtain the validity of our assertion.

By what has been said, on the set B there exists a point z^* such that

$$\left|\prod_1^m (z^* - z_k)\right| \ge 2\left(\frac{\mu_m(\overline{A})}{4}\right)^m,$$

whence we obtain (3.15).

We now consider the function $\psi(z) = \ln|F(z)|$. Since the function $F(z)$ has no zeros in D_1, $\psi(z)$ is a regular harmonic function in D_1, and hence

$$\psi(z) = \frac{2}{\pi}\int_{|\varsigma|=1/2} \frac{1/4 - |z|^2}{|z-\varsigma|}\psi(\varsigma)\,|d\varsigma|. \tag{3.16}$$

By (3.14) and (3.15),

$$\begin{aligned} &\psi(z) \le 0, \quad z \in D_1, \\ &\psi(z^*) \le \ln 2\varepsilon + m \ln \frac{2}{\mu_m(\overline{A})}. \end{aligned} \tag{3.17}$$

It follows from (3.16) and (3.17) that in the disk D_2 the function $\psi(z)$ satisfies the inequality

$$\psi(z) \leq \frac{1}{12}\left(\ln 2\varepsilon + m \ln \frac{2}{\mu_m(\overline{A})}\right). \tag{3.18}$$

Exponentiating (3.18), we obtain

$$|F(z)| \leq (2\varepsilon)^{1/12}\left(\frac{2}{\mu_m(\overline{A})}\right)^{m/12},$$

$$|f(z)| \leq \left(\frac{6}{7}\right)^m (2\varepsilon)^{1/12}\left(\frac{2}{\mu_m(\overline{A})}\right)^{m/12}. \tag{3.19}$$

On the other hand, from (3.14) and the obvious inequality

$$\left|\frac{z - z_k}{1 - \bar{z}_k z}\right| \leq \frac{6}{7}, \quad z \in D_2,$$

we have

$$|f(z)| \leq (6/7)^m, \quad z \in D_2. \tag{3.20}$$

The validity of (3.12) follows from (3.19) and (3.20).

2. We present two examples of the application of Theorem 1 for concrete sets lying on the real axis.

1. Suppose the set A is the segment

$$A = \{x \colon |x| \leq l \leq 1/4\}.$$

Then $\mu_n(A) = 2l$, and from (3.12) and (3.13) it follows that

$$|f(z)| \leq (2\varepsilon)^{(\ln 6 - \ln 7)/\ln l}.$$

2. Let $A = \{x \colon x = 1/4^k,\ k = 1, 2, \ldots\}$. Then

$$\mu_n(A) = 1/4^n, \qquad |f(z)| \leq (6/7)^{\sqrt{\ln|\varepsilon|}}. \tag{3.21}$$

3. We formulate a theorem on the stability of the problem of analytic continuation which is more general than Theorem 1.

THEOREM 2. *Let D be a domain of the complex plane with piecewise smooth boundary Γ, and let D_1 be a convex domain, $\overline{D}_1 \subset D$, where $\overline{D}_1$ is the closure of D_1; let A be a set with $A \subset D_1$, and let $\overline{A}$ be the projection of A onto the real axis. Then there exist a continuous function $\omega(z)$ with $\omega(z) > 0$ for $z \in D$ and $\omega(z) = 0$ for $z \in \Gamma$, and a constant $r > 0$ such that any function $f(z)$ analytic in D and satisfying the inequalities $|f(z)| \leq 1$ for $z \in D$ and $|f(z)| \leq \varepsilon$ for $z \in A$ also satisfies the inequality*

$$|f(z)| \leq \exp\left[-\frac{\ln \varepsilon}{\ln \mu_n(\overline{A})}\omega(z)\right], \tag{3.22}$$

where n is the number determined by the relation

$$\left(\frac{\mu_{n+1}(\overline{A})}{r}\right)^{n+1} < \varepsilon \leq \left(\frac{\mu_n(\overline{A})}{r}\right)^n.$$

The validity of inequality (3.22) in D_1 is proved in a manner altogether analogous to (3.12) in Theorem 1. The validity of (3.22) in all of D follows from (3.22) in D_1 and Theorem 2 of §2.

4. We shall present estimates of conditional stability for the problem of analytic continuation of functions of two real variables.

Let A be a bounded set of points of the plane (x, ξ). We denote by $\{H_k\}$ a system of disjoint strips

$$H_k = \left\{(x, \xi) \colon |\xi - \xi_k| \leq \frac{h}{2}\right\},$$

for which $\mu_n(\overline{A}_k) \geq \delta$, where $A_k = H_k \cap A; \overline{A}_k$ is the projection of A_k onto the line $\xi = \xi_k$.

We denote by B the projection of the union of the strips H_k onto the ξ-axis:

$$B = \{\xi \colon (x, \xi) \subset \cup H_k\}.$$

We denote by γ the supremum of the quantity $\mu_m(B)$ over all systems $\{H_k\}$. The quantity γ depends on the set A, the two continuous parameters h and δ, and the two integral parameters n and m:

$$\gamma = \gamma(A, h, \delta, n, m).$$

THEOREM 3. *Suppose that D is the following domain in the space of the two complex variables (z, ς):*

$$D = \{(z, \varsigma) \colon x^2 + \xi^2 < 1, |\xi| < \xi_0, |\eta| < \eta_0\},$$

and let A be the set of points of the real plane lying in the disk $\tilde{D}_\rho$,

$$\tilde{D}_\rho = \{(x, \xi) \colon x^2 + \xi^2 \leq \rho^2 < 1, x = \operatorname{Re} z, \xi = \operatorname{Re} \varsigma\}.$$

Then there exist a function

$$\begin{aligned} \omega(z, \varsigma) &> 0, \qquad (z, \varsigma) \in D, \\ \omega(z, \varsigma) &= 0, \qquad |z| = 1 \quad \cup \quad |\varsigma| = 1 \end{aligned}$$

and constants r_j such that any function $f(z, \varsigma)$ analytic in D and satisfying the inequalities

$$\begin{aligned} |f(z, \varsigma)| &\leq 1, \qquad (z, \varsigma) \in D, \\ |f(x, \xi)| &\leq \varepsilon, \qquad (x, \xi) \in A, \end{aligned} \tag{3.23}$$

also satisfies the inequality

$$|f(z,\varsigma)| \le \inf_{h,\delta} \exp\left[-\frac{\ln(\varepsilon + r_0 h)\omega(z,\varsigma)}{\ln\delta \cdot \ln\gamma(A,h,\delta,n,m)}\right], \tag{3.24}$$

where the parameters n and m are determined by the relations

$$(\delta/r_1)^{n+1} < \varepsilon + r_2 h \le (\delta/r_1)^n,$$

$$\left[\frac{\gamma(A,h,\delta,n,m)}{r_3}\right]^{m+1} < \exp\left[-\frac{\ln(\varepsilon + r_2 h)\omega}{\ln\delta \cdot \ln\gamma}\right] \le \left[\frac{\gamma(A,h,\delta,n,m)}{r_3}\right]^m. \tag{3.25}$$

PROOF. From (3.23) and known properties of analytic functions it follows that

$$|f'_\xi(x,\xi)| \le r_2, \quad (x,\xi) \in D_\rho, \tag{3.26}$$

where r_2 is a constant depending on ξ_0, η_0, and ρ. Let $(\tilde{x},\tilde{\xi}) \in A_k$. Then by (3.23) and (3.26) for $|\xi - \xi_k| \le h/2$ we have

$$|f(\tilde{x},\xi)| \le \varepsilon + r_2 h. \tag{3.27}$$

We consider the function $f(z,\xi)$ for fixed $\xi, |\xi - \xi_k| \le h/2$. It follows from (3.27) and Theorem 2 that if the parameters δ and n occurring in the definition of the system of strips $\{H_k\}$ satisfy (3.25), then

$$|f(z,\xi)| \le \exp\left[-\frac{\ln(\varepsilon + r_2 h)}{\ln\delta}\omega_1(z,\xi)\right]. \tag{3.28}$$

We now consider the function $f(z,\varsigma)$ for fixed z. Applying Theorem 2 again with inequality (3.28), we obtain

$$|f(z,\varsigma)| \le \exp\left[-\frac{\ln(\varepsilon + r_0 h)\omega(z,\varsigma)}{\ln\delta \ln\gamma(A,h,\delta,n,m)}\right], \tag{3.29}$$

where the number m is determined by (3.25). The desired inequality (3.24) obviously follows from (3.29).

§4. Recovery of solutions of elliptic and parabolic equations from their values on sets lying inside the domain of regularity

We consider two problems.

1. Suppose the function $u(x,y)$ is harmonic in the rectangle $P = \{0 < x < \pi,\ 0 < y < a\}$, is continuous in $\overline{P}$, and for $y = 0$, a satisfies a Hölder condition in x of order $\alpha > 1/2$ (we denote the constant contained in the Hölder condition by M). Suppose, moreover, that

$$u(0,y) = u(\pi,y) = 0, \qquad u(\xi,y) = f(y), \tag{3.30}$$

where $0 \le y \le a$; ξ is a given number in $[0,\pi]$; $f(y)$ is a given function.

Our problem consists in recovering the function $u(x, y)$ on the entire rectangle P from the data (3.30).

It is easy to see that the question of uniqueness of the solution of problem (3.30) can be resolved in a positive or negative sense depending on whether the number ξ/π is irrational or rational. It turns out that if ξ/π is not only irrational but is also "poorly approximated" by rational numbers, then the solution of problem (3.30) is not only unique but also admits an estimate of stability of Hölder type with respect to variations of the function $f(y)$.

THEOREM 1. *Suppose that* $|f(y)| \leq \varepsilon$ *for* $0 \leq y \leq a$, *and* ξ *is such that for all natural numbers* n

$$|\sin n\xi| \geq \theta n^{-\sigma}$$

with some $\theta > 0$ *and* $\sigma \geq 1$. *Set* $\Delta = \min(y, a - y)$. *Then for any point* $(x, y) \in P$,

$$|u(x, y)| \leq C(\alpha, \sigma, \theta)\varepsilon^{\omega(\Delta)} M^{1-\omega(\Delta)}, \tag{3.31}$$

where

$$\omega(\Delta) = \frac{\alpha - 1/2}{\alpha + \sigma + 1/2} \exp\left(\frac{\pi}{\arctan \frac{\Delta}{\pi}} \ln 2\right)$$

and C *depends only on the indicated parameters.*

PROOF. We show first of all that under the conditions imposed on $u(x, y)$ for $(x, y) \in P$ we have the equality

$$u(x, y) = \sum_{k=1}^{\infty} [u_k' \sinh k(a - y) + u_k'' \sinh ky] \sinh^{-1} ka \sin kx, \tag{3.32}$$

where

$$u_k' = \frac{2}{\pi} \int_0^{\pi} u(x, 0) \sin kx \, dx, \qquad u_k'' = \frac{2}{\pi} \int_0^{\pi} u(x, a) \sin kx \, dx.$$

The series in (3.32) and the series obtained from it by twofold termwise differentiation converge uniformly with respect to x and y for $0 \leq x \leq \pi$ and $a' \leq y \leq a''$ $(0 < a' < a'' < a)$. From this it follows that this series converges to a function $v(x, y)$ continuous in P and harmonic for $0 \leq x \leq \pi$ and $0 < y < a$. We shall show that $v(x, y)$ is continuous in $\overline{P}$ and

$$\lim_{y \to 0} v(x, y) = u(x, 0), \qquad \lim_{y \to a} v(x, y) = u(x, a). \tag{3.33}$$

For this it suffices to show that the series in (3.32) converges uniformly in $\overline{P}$. Since $u(x, 0)$ and $u(x, a)$ satisfy a Hölder condition of order $\alpha > 1/2$, by a theorem of S. N. Bernstein the series of Fourier coefficients of these functions converge absolutely, and since

$$|[u_k' \sinh k(a - y) + u_k'' \sinh ky] \sinh^{-1} ka \sin kx| \leq |u_k'| + |u_k''|,$$

the series $\sum_1^\infty(|u'_k|+|u''_k|)$ majorizes the series (3.32) on $\overline{P}$. This implies the continuity of $v(x,y)$ on $\overline{P}$, and termwise passage to the limit in (3.32) gives (3.33). Thus, the functions $u(x,y)$ and $v(x,y)$ are harmonic in P, continuous in $\overline{P}$, and assume the same boundary values. Applying the maximum principle to the difference $u-v$, we obtain (3.32).

We proceed to the proof of (3.31). It follows from (3.32) that

$$f(y)=\sum_1^\infty\left(\frac{-u'_ke^{-ka}+u''_k}{e^{ka}-e^{-ka}}e^{ky}+\frac{u'_ke^{ka}-u''_k}{e^{ka}-e^{-ka}}e^{-ky}\right)\sin k\xi.$$

The function $f(y)$, initially defined only for real y, $0\le y\le a$, admits analytic continuation to the rectangle $0<\operatorname{Re} z<a$, $0\le \operatorname{Im} z\le 2\pi$ by means of the equality

$$f(z)=\sum_1^\infty\left(\frac{-u'_ke^{-ka}+u''_k}{e^{ka}-e^{-ka}}e^{kz}+\frac{u'_ke^{ka}-u''_k}{e^{ka}-e^{-ka}}e^{-kz}\right)\sin k\xi, \tag{3.34}$$

and in this rectangle

$$|f(z)|\le\sum_1^\infty(|u'_k|+|u''_k|)\le\pi^\alpha[2^{2-\alpha}+2^{1/2}(1-2^{-\alpha+1/2})^{-1}]M \tag{3.35}$$

(we have used the estimate for the Fourier coefficients of functions of Hölder class from [103], p. 241). For real $z, 0\le z\le a$, by the hypothesis of the theorem,

$$|f(z)|\le\varepsilon, \tag{3.36}$$

and, since $f(z)$ is periodic in $\operatorname{Im} z$ with period 2π, the same inequality holds also on the segment $\operatorname{Im} z=2\pi$, $0\le\operatorname{Re} z\le a$. Using the estimate of the Carleman function analytic in a corner, from (3.35) and (3.36) it is not hard to obtain an estimate of $f(z)$ valid on the segments $\operatorname{Re} z=\Delta$ and $\operatorname{Re} z=a-\Delta$ $(0\le\operatorname{Im} z\le 2\pi,\ 0<\Delta\le a/2)$:

$$|f(z)|\le\{\pi^\alpha[2^{2-\alpha}+2^{1/2}(1-2^{-\alpha+1/2})^{-1}]M\}^{1-\rho(\Delta)}\varepsilon^{\rho(\Delta)},$$

where

$$\rho(\Delta)=\exp\left(-\frac{\pi}{\arctan\frac{\Delta}{\pi}}\ln 2\right).$$

For brevity we denote the right side of the last inequality by $\varepsilon(\Delta)$. From (3.34) we obtain

$$\frac{-u'_ke^{-ka}+u''_k}{e^{ka}-e^{-ka}}\sin k\xi=\frac{1}{2\pi i}\int f(z)e^{-kz}\,dz,$$

$$\frac{u'_ke^{ka}-u''_k}{e^{ka}-e^{-ka}}\sin k\xi=\frac{1}{2\pi i}\int f(z)e^{kz}\,dz,$$

where the first integral is taken over the segment $\operatorname{Re} z = a-\Delta$, $0 \le \operatorname{Im} z \le 2\pi$ and the second over the segment $\operatorname{Re} z = \Delta$, $0 \le \operatorname{Im} z \le 2\pi$ (to justify the last formulas we set $t = e^z$; then the rectangle $0 \le \operatorname{Re} z \le a$, $0 \le \operatorname{Im} z \le 2\pi$ goes over into the annulus $1 \le |t| \le e^a$, and the series (3.34) goes over into a Laurent series in t). Using the estimate of $|f(z)|$, we obtain

$$\left|\frac{-u_k' e^{-ka} + u_k''}{e^{ka} - e^{-ka}} \sin k\xi\right| \le e^{-k(a-\Delta)}\varepsilon(\Delta),$$

$$\left|\frac{u_k' e^{ka} - u_k''}{e^{ka} - e^{-ka}} \sin k\xi\right| \le e^{k\Delta}\varepsilon(\Delta).$$

From this, using the condition on ξ, we find that

$$\left|\frac{-u_k' e^{-ka} + u_k''}{e^{ka} - e^{-ka}}\right| \le \theta^{-1} k^\sigma e^{-k(a-\Delta)}\varepsilon(\Delta), \qquad \left|\frac{u_k' e^{ka} - u_k''}{e^{ka} - e^{-ka}}\right| \le \theta^{-1} k^\sigma e^{k\Delta}\varepsilon(\Delta)$$

and for $0 \le x \le \pi$, $\Delta \le y \le a - \Delta$,

$$\begin{aligned} |u(x,y)| &= \left|\sum_1^\infty \left(\frac{-u_k' e^{-ka} + u_k''}{e^{ka} - e^{-ka}} e^{ky} + \frac{u_k' e^{ka} - u_k''}{e^{ka} - e^{-ka}} e^{-ky}\right) \sin kx\right| \\ &\le 2\theta^{-1}\varepsilon(\Delta) \sum_1^{N-1} k^\sigma + \sum_N^\infty (|u_k'| + |u_k''|) \\ &\le 2\theta^{-1}(1+\sigma)^{-1}\varepsilon(\Delta) N^{\sigma+1} + \pi^\alpha 2^{-1/2}(1 - 2^{-\alpha+1/2})^{-1} M N^{-\alpha+1/2} \end{aligned}$$

(we have again used the estimate for the Fourier coefficients of functions of Hölder class of [103]). Minimizing the right side of this inequality with respect to N, we obtain

$$|u(x,y)| \le C(\alpha,\sigma,\theta) \exp\left[\frac{\alpha - 1/2}{\alpha+\sigma+1/2} \ln\varepsilon(\Delta) + \left(1 - \frac{\alpha-1/2}{\alpha+\sigma+1/2}\right) \ln M\right].$$

The variable y and the parameter Δ in this inequality are related by the single condition $\Delta \le y \le a - \Delta$. It is therefore possible to set $\Delta = \min(y, a-y)$. Recalling, finally, the expression for $\varepsilon(\Delta)$, we arrive at (3.31) with some new constant $C(\alpha,\sigma,\theta)$. The theorem is proved.

We can thus guarantee Hölder stability of the original problem if ξ belongs to a set $X(\sigma)$ of numbers satisfying the condition

$$|\sin n\xi| \ge \theta(\xi) n^{-\sigma}$$

for all $n = 1, 2, \ldots$ with some $\theta(\xi) > 0$ and $\sigma \ge 1$. Using some facts from the theory of Diophantine approximations (available, for example, in [281]), it can be shown that ($|\cdot|$ is the cardinality, μ is Lebesgue measure) $|X(1)| = c$, $\mu X(1) = 0$, and $X(1)$ is dense on $[0,\pi]$; if, however, $\sigma > 1$, then $\mu X(\sigma) = \pi$.

We consider the question of the magnitude of the coefficient $\theta(\xi)$. Observe that for any $\sigma \geq 1$,

$$\inf \theta(\xi) = 0, \quad \xi \in X(\sigma),$$

but as $\theta \to 0$, for the coefficient $C(\alpha, \sigma, \theta)$ in (3.31) we have $C(\alpha, \sigma, \theta) \to \infty$ (as is not hard to see).

There naturally arises the question of the measure of the set of $\xi \in X(\sigma)$ for which $\theta(\xi)$ is bounded below by a positive constant.

THEOREM 2. *Let $\sigma > 1$ and $0 < \delta < 1$. Let $X(\sigma, \delta)$ be the set of those $\xi \in [0, \pi]$ such that $|\sin \xi| \geq 2\delta n^{-\sigma}/\pi$ for each $n = 1, 2, \dots$. There exists a measurable set $\tilde{X}(\sigma, \delta)$ such that*

$$X(\sigma, \delta) \supset \tilde{X}(\sigma, \delta), \qquad \mu\tilde{X}(\sigma, \delta) \geq \pi - 2\delta \sum_1^\infty n^{-\sigma}.$$

PROOF. It suffices to show that there exists a measurable set $\tilde{Y}(\sigma, \delta)$ such that

$$[0, \pi] \backslash X(\sigma, \delta) \subset \tilde{Y}(\sigma, \delta), \qquad \mu\tilde{Y}(\sigma, \delta) \leq 2\delta \sum_1^\infty n^{-\sigma}. \tag{3.37}$$

Let

$$Y_n = [0, \delta n^{-\sigma-1}) \cup \left(\frac{\pi}{n} - \delta n^{-\sigma-1}, \frac{\pi}{n} + \delta n^{-\sigma-1}\right) \cup \cdots \cup (\pi - \delta n^{-\sigma-1}, \pi].$$

We set $\tilde{Y}(\sigma, \delta) = \bigcup_1^\infty Y_n$. Obviously $\tilde{Y}(\sigma, \delta)$ is measurable, and

$$\mu\tilde{Y}(\sigma, \delta) \leq \sum_1^\infty \mu Y_n = 2\delta \sum_1^\infty n^{-\sigma}.$$

Let $\xi \in [0, \pi] \backslash X(\sigma, \delta)$. Then for some natural number n,

$$|\sin n\xi| < 2\delta n^{-\sigma}/\pi,$$

which is possible only if for some k, $0 \leq k \leq n$,

$$\left|\xi - \pi\frac{k}{n}\right| < \delta n^{-\sigma-1},$$

i.e., $\xi \in Y_n \subset \tilde{Y}(\sigma, \delta)$. The inclusion (3.37) is proved.

Recalling that for irrational ξ/π the solution of problem (3.30) is unique, it can be said that problem (3.30) possesses the uniqueness property "with probability 1" and admits a uniform estimate of stability of Hölder type "with probability arbitrarily close to 1".

2. Suppose that in the rectangle $P = \{0 \leq s \leq \pi,\ 0 \leq x \leq X\}$ there is defined a solution $u(s, x) \in C(P)$ of the problem

$$u_x - u_{ss} = 0, \qquad u(0, x) = u(\pi, x) = 0, \quad 0 \leq x \leq X. \tag{3.38}$$

We consider a curve $s = \alpha(x)$ lying in P and assume that on this curve there are given the values of the solution $u(s, x)$ of (3.38):

$$u(\alpha(x), x) = f(x). \tag{3.39}$$

We pose the question: does the function $f(x)$ determine the solution $u(s, x)$ of problem (3.38), (3.39) in the entire rectangle P? If $\alpha(x) = \text{const}$, then the problem in question has properties completely analogous to the properties of the preceding problem and, in particular, does not have the uniqueness property, generally speaking. Indeed if $\alpha(x) = (n/m)\pi, 0 < n/m < 1$, then the function

$$u(s, x) = \sum_{k=1}^{n} u_{mk} e^{-m^2k^2x} \sin mks$$

with arbitrary u_{mk} satisfies (3.38) and for it $f(x) = u(n\pi/m, x) = 0$. It turns out, however, that this case of nonuniqueness ($\alpha(x) = \text{const}$) is in some sense exceptional. Namely, we have the following result.

THEOREM 3. *Suppose a real-valued function $\alpha(x)$ on $[0, \infty)$ admits analytic continuation to the domain $|\arg z| < \pi\theta, 0 < \theta \leq 1$, $z = x + iy$, is continuous up to the boundary of this domain, and satisfies the inequality*

$$|\alpha(z)| < Ce^{k|z|}, \qquad k < 3 \sin \pi(\theta - 1/2).$$

Suppose that on the segment $[0, X]$

$$f(x) = u(\alpha(x), x) \equiv 0.$$

If $\alpha(x) \neq \text{const}$, then $u(s, x) \equiv 0$ on P.

To prove the theorem we need two auxiliary lemmas, the first of which is well known.

LEMMA 1. *Suppose that a function $\alpha(z)$ is analytic in the domain $|\arg z| < \pi\theta$, $0 < \theta < 1$, is continuous up to the boundary, and satisfies the inequalities $|\alpha(z)| < Ce^{k|z|}$ and $|\alpha(x)| < Ce^{-3x}$. If $k < 3 \sin \pi(\theta - 1/2)$, then $\alpha(z) \equiv 0$.*

LEMMA 2. *Suppose a function $\alpha(x)$ real-valued and continuous on $[0, \infty)$ possesses the property that either $\lim_{x\to\infty} \alpha(x)$ does not exist or*

$$|\alpha(x) - \lim_{x\to\infty} \alpha(x)|e^{3x} \to \infty.$$

Then for any $k = 1, 2, \dots$ there exists a sequence $\{x_n^{(k)}\}$ such that as $n \to \infty$

$$x_n^{(k)} \to +\infty, \qquad \exp[-(2k+1)x_n^{(k)}]|\sin k\alpha(x_n^{(k)})|^{-1} \to 0. \tag{3.40}$$

PROOF. We show first of all that for any $k = 1, 2, \dots$ there exists a sequence $\{x_n^{(k)}\}$ such that as $n \to \infty$,

$$x_n^{(k)} \to +\infty, \qquad \min_{m\in M} \left|\frac{1}{\pi}\alpha(x_n^{(k)}) - \frac{m}{k}\right| \exp[(2k+1)x_n^{(k)}] \to \infty \tag{3.41}$$

(M is the set of all integers). Suppose first that

$$\varliminf_{x\to\infty} \alpha(x) < \varlimsup_{x\to\infty} \alpha(x).$$

Then the function $\alpha(x)/\pi$ as $x \to \infty$ assumes any value of the interval

$$\left(\frac{1}{\pi}\varliminf \alpha(x),\ \frac{1}{\pi}\varlimsup \alpha(x)\right)$$

infinitely often. Let ξ be an irrational number in this interval, and let $\{x_n^{(k)}\}$ be such that $x_n^{(k)} \to \infty$ $(n \to \infty)$ and $\alpha(x_n^{(k)}) = \pi\xi$. Then

$$\min_{m\in M}\left|\frac{1}{\pi}\alpha(x_n^{(k)}) - \frac{m}{k}\right| = C(k) > 0,$$

and hence (3.41) is satisfied. If $\alpha(x) \to \pm\infty$ $(x \to +\infty)$, then there exist an irrational ξ and a sequence $x_n^{(k)} \to +\infty$ $(n \to \infty)$ such that $\alpha(x_n^{(k)}) = \pi(\xi \pm n)$. It is easy to see that for such $\{x_n^{(k)}\}$ the previous estimate holds, and hence (3.41) is satisfied.

Suppose now that

$$\lim_{x\to\infty} \alpha(x) = a\pi, \qquad \varepsilon(x) = \alpha(x) - a\pi.$$

Suppose first that a is irrational. Then for $m \in M$

$$|a - (m/k)| \geq C(k) > 0$$

and for sufficiently large x

$$\left|\frac{1}{\pi}\alpha(x) - \frac{m}{k}\right| = \left|a - \frac{m}{k} + \frac{1}{\pi}\varepsilon(x)\right| \geq \left|\left|a - \frac{m}{k}\right| - \frac{1}{\pi}|\varepsilon(x)|\right|$$
$$\geq C(k) - \frac{1}{\pi}|\varepsilon(x)| > \frac{1}{2}C(k), \quad (3.42)$$

thus, (3.41) holds for any sequence $x_n^{(k)} \to +\infty$ $(n \to \infty)$. Suppose a is rational. Then there exists a set $M_a \subset M$ (possibly $M_a = \varnothing$) such that

$$|a - (m/k)| = 0 \text{ for } m \in M_a, \qquad |a - (m/k)| \geq C(k) > 0 \text{ for } m \in M\backslash M_a.$$

Then for $m \in M_a$

$$\left|\frac{1}{\pi}\alpha(x) - \frac{m}{k}\right| = \left|a - \frac{m}{k} + \frac{1}{\pi}\varepsilon(x)\right| = \frac{1}{\pi}|\varepsilon(x)|,$$

and for $m \in M\backslash M_a$ and sufficiently large x the estimate (3.42) holds. Thus, if x is sufficiently large, then

$$\min_{m\in M}\left|\frac{1}{\pi}\alpha(x) - \frac{m}{k}\right| \geq \min\left(\frac{1}{2}C(k), \frac{1}{\pi}|\varepsilon(x)|\right) = \frac{1}{\pi}|\varepsilon(x)|,$$

and, since $k \geq 1$, it follows that

$$\begin{aligned}\min_{m\in M}\left|\frac{1}{\pi}\alpha(x)-\frac{m}{k}\right|e^{(2k+1)x} &\geq \frac{1}{\pi}|\varepsilon(x)|e^{(2k+1)x}\\ &= \frac{1}{\pi}|\alpha(x)-\lim\alpha(x)|e^{3x}e^{2(k-1)x}\\ &\geq \frac{1}{\pi}|\alpha(x)-\lim\alpha(x)|e^{3x}\to\infty, \quad x\to\infty.\end{aligned}$$

Thus, in this case as well (3.41) holds for any sequence $x_n^{(k)} \to +\infty$ ($n \to \infty$).

The proof of the lemma can now be completed easily. Indeed, if $x_n^{(k)} \to \infty$ as $n \to \infty$ and (3.41) holds, then (m_0 is the integer nearest to the number $\pi^{-1}k\alpha(x_n^{(k)})$)

$$\begin{aligned}|\sin k\alpha(x_n^{(k)})| &= |\sin[k\alpha(x_n^{(k)})-m_0\pi]| = \sin|k\alpha(x_n^{(k)})-m_0\pi|\\ &\geq \frac{2}{\pi}|k\alpha(x_n^{(k)})-m_0\pi| = 2k\left|\frac{1}{\pi}\alpha(x_n^{(k)})-\frac{m_0}{k}\right|\\ &= 2k\min_{m\in M}\left|\frac{1}{\pi}\alpha(x_n^{(k)})-\frac{m}{k}\right|\end{aligned}$$

and from (3.50) we obtain

$$\begin{aligned}&|\sin k\alpha(x_n^{(k)})|\exp[(2k+1)x_n^{(k)}]\\ &\qquad\geq 2k\min_{m\in M}\left|\frac{1}{\pi}\alpha(x_n^{(k)})-\frac{m}{k}\right|\exp[(2k+1)x_n^{(k)}]\to\infty.\end{aligned}$$

The lemma is proved.

PROOF OF THEOREM 3. Since $u(s,x)$ is continuous on P, for $x > 0$ we have the equality

$$u(s,x) = \sum_{k=1}^{\infty} u_k e^{-k^2 x}\sin ks$$

and the sequence $|u_k|$ is bounded. By the hypothesis of the theorem, for $0 < x \leq X$

$$\sum_{k=1}^{\infty} u_k e^{-k^2 x}\sin k\alpha(x) = 0. \tag{3.43}$$

Since $\alpha(z)$ is analytic for $x > 0$ and $\alpha(x)$ is real, there exists a domain Ω of the half-plane $x > 0$ containing the real semiaxis and such that $|\operatorname{Im}\alpha(z)| < \varepsilon$ for $z \in \Omega$. In the domain $\Omega \cap \{z\colon x > \delta > 0\}$ the series (3.43) has the convergent majorant

$$2\sum_{k=1}^{\infty}|u_k|\exp(k\varepsilon - k^2\delta),$$

which implies analyticity of the sum of the series (3.43) in Ω. Hence, equality (3.43), originally satisfied only for $x \in (0, X]$, holds for all $x \in (0, \infty)$.

We shall show that the function $\alpha(x)$ satisfies the conditions of Lemma 2. Indeed, if

$$|\alpha(x) - \lim \alpha(x)|e^{3x} < C,$$

then the function $\bar{\alpha}(x) = \alpha(x) - \lim \alpha(x)$ satisfies the conditions of Lemma 1, and hence $\bar{\alpha}(x) \equiv 0$, i.e., $\alpha(x) = \lim \alpha(x) = \text{const}$, which contradicts the hypothesis of the theorem.

We shall now show that if (3.43) is satisfied for all $x \in (0, \infty)$ then $u_k = 0$ for all $k = 1, 2, \ldots$. Indeed, suppose that $u_1 = \cdots = u_{k-1} = 0$. Then from (3.43) we get

$$u_k \sin k\alpha(x) = -\sum_{i=k+1}^{\infty} u_i \exp[-(i^2 - k^2)x] \sin i\alpha(x),$$

whence for $x > \delta$

$$\begin{aligned} |u_k|\,|\sin k\alpha(x)| &\leq \exp[-(2k+1)x] \sum_{i=k+1}^{\infty} |u_i| \exp[-(i^2 - 2k - 1)x] \\ &\leq e^{-(2k+1)x} \sum_{i=k+1}^{\infty} |u_i| \exp[-(i^2 - 2k - 1)\delta] \\ &= \exp[-(2k+1)x] C(k, \delta). \end{aligned} \tag{3.44}$$

Suppose now that $\{x_n^{(k)}\}$ is the sequence constructed in Lemma 2. Then by (3.44)

$$|u_k| \leq C(k, \delta) \exp[-(2k+1)x_n^{(k)}]|\sin k\alpha(x_n^{(k)})|^{-1}.$$

Passing to the limit as $n \to \infty$ in this inequality, we obtain $u_k = 0$. The theorem is proved.

CHAPTER IV

Boundary Value Problems for Differential Equations

In this chapter we consider questions related to uniqueness and stability of boundary value problems for equations of mathematical physics that are ill-posed in the Hadamard sense. We shall mainly be concerned with local Cauchy problems, i.e, Cauchy problems with initial data having compact support. Exceptions are §2, devoted to a mixed problem for a parabolic equation with decreasing time, and the part of §3 in which we consider the problem of extending a solution of a pseudoparabolic equation from a "timelike" half-line. Estimates of integral norms of solutions with a weight increasing in a neighborhood of the support of the data form the means of obtaining existence and stability theorems in this chapter, and these estimates themselves we obtain by means of certain combinations of partial integrations. This makes our considerations altogether elementary, although very laborious. We use the Laplace transform to obtain an estimate only in one case.

Uniqueness theorems for the Cauchy problem for elliptic equations and the noncharacteristic Cauchy problem for parabolic equations of second order are proved in [298], [300], [317], and [321]. The Hölder property for the Cauchy problem for elliptic equations is established in [141], [142], and [161]. The Hölder property for the noncharacteristic Cauchy problem for a general parabolic equation of second order was apparently first proved in [8]. Earlier this result had been obtained only for an equation with a single spatial variable [160]. Other questions connected with the topic of §1 are discussed in [145], [163], [289], [310], and [311]. Theorem 1 of §2 is well known. The reader can find a detailed consideration of related questions (especially within the framework of operator differential equations) in [135], [136], and [329]. The introductory article to [329] contains a detailed bibliography. Questions of the approximate solution of ill-posed Cauchy problems for evolution equations are studied in [111], [138], [164], [285], [291], [311], and [324]. Theorem 1 of §3 is proved in [287]; a "timelike" Cauchy problem for a pseudoparabolic equation is considered in [290] and [326].

Cauchy problems with data on a timelike manifold for equations containing hyperbolic and ultrahyperbolic operators are studied in [76], [77], [123], [124], [286], [289], [315], [316], [318], and [328]. Our exposition is based on the method of [286]. The example of F. John demonstrating logarithmic stability of these problems was taken from [315].

§1. The noncharacteristic Cauchy problem for a parabolic equation. The Cauchy problem for an elliptic equation

Suppose that for $\eta > 0, 0 < X < 1$ and $0 < T < 1$

$$\Omega_\eta = \left\{ (t,x) \colon 0 < x_1 < \frac{1}{2}\left(2\eta - X^{-2}\sum_{i=2}^{n} x_i^2 - T^{-2}t^2\right)\right\}.$$

In the domain $\Omega_{1/2}$ we consider the differential inequality

$$\left| u_t - \sum_{i,j=1}^{n} a^{ij}(t,x)u_{ij}\right| < a(|\nabla u| + |u|), \tag{4.1}$$

whose coefficients $a^{ij}(t,x)$ satisfy the conditions

$$a^{ij}(t,x) \in C^2(\overline{\Omega}_{1/2}), \qquad \sum_{i,j=1}^{n} a^{ij}(t,x)\xi_i\xi_j \geq \mu \sum_{i=1}^{n} \xi_i^2, \tag{4.2}$$

where μ is a positive constant. Regarding a solution of (4.1), we shall assume that

$$u(t,x) \in C^2(\Omega_{1/2}) \cap C^1(\overline{\Omega}_{1/2}). \tag{4.3}$$

Suppose that Cauchy data for the function $u(t,x)$ are given on the intersection of $\overline{\Omega}_{1/2}$ with the plane $x_1 = 0$:

$$u(t,0,x_2,\dots,x_n), \qquad u_1(t,0,x_2,\dots,x_n).$$

We shall show that these Cauchy data determine a solution of (4.1) in unique fashion, and under a certain additional condition on $u(t,x)$ we shall obtain an a priori estimate of the solution of Hölder type. To this end we prove an inequality for the L^2-norm of the solution with weight

$$\varphi(t,x) = \exp[\lambda\psi^{-\nu}(t,x)],$$

where λ and ν are positive parameters on which additional conditions will be imposed in the process of obtaining the estimates, while the function $\psi(t,x)$ has the form

$$\psi(t,x) = x_1 + \frac{1}{2X^2}\sum_{i=2}^{n} x_i^2 + \frac{1}{2T^2}t^2 + \alpha, \quad 0 < \alpha < \frac{1}{2}.$$

If $(t,x) \in \Omega_\eta$ and $\eta \le \frac{1}{2}$, then obviously $\alpha < \psi(t,x) < \eta + \alpha < 1$, and if we set $\partial\Omega_\eta = \omega_\eta^{(1)} \cup \omega_\eta^{(2)}$, where

$$\omega_\eta^{(1)} = \left\{ (t,x)\colon x_1 = 0, \frac{1}{2X^2}\sum_{i=2}^{n} x_i^2 + \frac{1}{2T^2}t^2 \le \eta \right\},$$

$$\omega_\eta^{(2)} = \left\{ (t,x)\colon x_1 \ge 0, x_1 = \eta - \frac{1}{2X^2}\sum_{i=2}^{n} x_i^2 - \frac{1}{2T^2}t^2 \right\},$$

then

$$\psi(t,x) \begin{cases} \ge \alpha \text{ for } (t,x) \in \omega_\eta^{(1)}, \\ = \alpha + \eta \text{ for } (t,x) \in \omega_\eta^{(2)}. \end{cases}$$

Everywhere in what follows we use the symbol C to denote constants (distinct, generally speaking) depending only on the dimension n and the C^2-norm of the coefficients $a^{ij}(t,x)$.

LEMMA 1. *If λ and ν exceed some positive absolute constant, then*

$$\mu\varphi^2|\nabla u|^2 - \left(\lambda\nu\frac{1}{T}\psi^{-\nu-1} + C\lambda^2\frac{1}{X^2}\psi^{-2\nu-2}\right)\varphi^2 u^2$$

$$\le \varphi^2\left(u_t - \sum_{i,j=1}^{n} a^{ij}u_{ij}\right)u - \frac{1}{2}(\varphi^2u^2)_t + \operatorname{div} U,$$

where the modulus of the vector U satisfies the inequality

$$|U| \le C\varphi^2\left(|\nabla u|^2 + \lambda\nu\frac{1}{X}\psi^{-\nu-1}u^2\right).$$

PROOF. We have

$$\varphi^2\left(u_t - \sum_{i,j=1}^{n} a^{ij}u_{ij}\right)u = -\varphi\varphi_t u^2 + \varphi^2\sum_{i,j=1}^{n} a^{ij}u_iu_j$$

$$+ \sum_{i,j=1}^{n}(\varphi^2a^{ij})_j uu_i + \frac{1}{2}(\varphi^2u^2)_t - \sum_{i,j=1}^{n}(\varphi^2a^{ij}u_iu)_j$$

$$= -\varphi\varphi_t u^2 + \varphi^2\sum_{i,j=1}^{n} a^{ij}u_iu_j - \frac{1}{2}\sum_{i,j=1}^{n}(\varphi^2a^{ij})_{ij}u^2$$

$$+ \frac{1}{2}(\varphi^2u^2)_t = \sum_{i,j=1}^{n}(\varphi^2a^{ij}u_iu)_j + \frac{1}{2}\sum_{i,j=1}^{n}[(\varphi^2a^{ij})_ju^2]_i.$$

We shall estimate the terms in the last expression. Because of condition (4.2),

$$\varphi^2\sum_{i,j=1}^{n} a^{ij}u_iu_j \ge \mu\varphi^2|\nabla u|^2. \tag{4.4}$$

Further,

$$\begin{aligned} -\varphi\varphi_t u^2 &= \lambda\nu\psi^{-\nu-1}\psi_t\varphi^2 u^2 = \lambda\nu\frac{t}{T^2}\psi^{-\nu-1}\varphi^2 u^2 \\ &\geq -\lambda\nu\frac{1}{T}\psi^{-\nu-1}\varphi^2 u^2, \end{aligned} \tag{4.5}$$

$$\begin{aligned} -\frac{1}{2}&\sum_{i,j=1}^{n}(\varphi^2 a^{ij})_{ij}u^2 \\ &= -\frac{1}{2}\sum_{i,j=1}^{n}(2\varphi\varphi_{ij}a^{ij} + 2\varphi_i\varphi_j a^{ij} + 4\varphi\varphi_i a^{ij}_j + 2\varphi\varphi_j a^{ij}_i + \varphi^2 a^{ij}_{ij})u^2 \\ &\geq -C\sum_{i,j=1}^{n}(\varphi|\varphi_{ij}| + |\varphi_i|\,|\varphi_j| + \varphi|\varphi_i| + \varphi^2)u^2. \end{aligned} \tag{4.6}$$

For the derivatives of the function φ we have $(i, j > 1, i \neq j)$

$$\begin{aligned} \varphi_1 &= -\lambda\nu\psi^{-\nu-1}\varphi, \qquad \varphi_{11} = \lambda^2\nu^2\psi^{-2\nu-2}\varphi + \lambda\nu(\nu+1)\psi^{-\nu-2}\varphi, \\ \varphi_i &= -\lambda\nu\frac{x_i}{X^2}\psi^{-\nu-1}\varphi, \\ \varphi_{ii} &= \lambda^2\nu^2\frac{x_i^2}{X^4}\psi^{-2\nu-2}\varphi + \lambda\nu(\nu+1)\frac{x_i^2}{X^4}\psi^{-\nu-2}\varphi - \lambda\nu\frac{1}{X^2}\psi^{-\nu-1}\varphi, \\ \varphi_{1i} &= \lambda^2\nu^2\frac{x_i}{X^2}\psi^{-2\nu-2}\varphi + \lambda\nu(\nu+1)\frac{x_i}{X^2}\psi^{-\nu-2}\varphi, \\ \varphi_{ij} &= \lambda^2\nu^2\frac{x_i x_j}{X^4}\psi^{-2\nu-2}\varphi + \lambda\nu(\nu+1)\frac{x_i x_j}{X^4}\psi^{-\nu-2}\varphi. \end{aligned}$$

Since $|x_i| < X$ and $0 < \psi < 1$ in the domain $\Omega_{1/2}$, assuming that the parameters λ and ν exceed some positive absolute constant, for these derivatives we obtain the estimates

$$\begin{aligned} |\varphi_1| &\leq \lambda\nu\psi^{-\nu-1}\varphi, & |\varphi_{11}| &\leq 2\lambda^2\nu^2\psi^{-2\nu-2}\varphi, \\ |\varphi_i| &\leq \lambda\nu\tfrac{1}{X}\psi^{-\nu-1}\varphi, & |\varphi_{ii}| &\leq 2\lambda^2\nu^2\tfrac{1}{X^2}\psi^{-2\nu-2}\varphi, \\ |\varphi_{1i}| &\leq 2\lambda^2\nu^2\tfrac{1}{X}\psi^{-2\nu-2}\varphi, & |\varphi_{ij}| &\leq 2\lambda^2\nu^2\tfrac{1}{X^2}\psi^{-2\nu-2}\varphi. \end{aligned}$$

Using these estimates, from (4.6) we easily find that

$$-\frac{1}{2}\sum_{i,j=1}^{n}(\varphi^2 a^{ij})_{ij}u^2 \geq -C\lambda^2\nu^2\frac{1}{X^2}\psi^{-2\nu-2}\varphi^2 u^2. \tag{4.7}$$

We proceed to estimate the divergence terms. We have

$$\begin{aligned} -&\sum_{i,j=1}^{n}(\varphi^2 a^{ij}u_i u)_j + \frac{1}{2}\sum_{i,j=1}^{n}[(\varphi^2 a^{ij})_j u^2]_i \\ &= \sum_{i=1}^{n}\left[-\varphi^2\sum_{j=1}^{n}a^{ij}u_j u + \frac{1}{2}\sum_{j=1}^{n}(\varphi^2 a^{ij})_j u^2\right]_i = -\operatorname{div} U. \end{aligned}$$

For the modulus of the vector U we obtain the estimate

$$\begin{aligned} |U| &\le \sum_{i=1}^{n} \left| -\varphi^2 \sum_{j=1}^{n} a^{ij} u_j u + \frac{1}{2} \sum_{j=1}^{n} (\varphi^2 a^{ij})_j u^2 \right| \\ &= \sum_{i=1}^{n} \left| -\varphi^2 \sum_{j=1}^{n} a^{ij} u_j u + \frac{1}{2} \sum_{j=1}^{n} (2\varphi\varphi_j a^{ij} + \varphi^2 a^{ij}_j) u^2 \right| \\ &\le C\varphi^2 \left(|\nabla u|^2 + \lambda\nu \frac{1}{X} \psi^{-\nu-1} u^2 \right). \end{aligned} \tag{4.8}$$

Collecting the estimates (4.4), (4.5) and (4.7), we arrive at the inequality

$$\begin{aligned} \varphi^2 \left(u_t - \sum_{i,j=1}^{n} a^{ij} u_{ij} \right) u &\ge \mu\varphi^2 |\nabla u|^2 \\ &\quad - \left(\lambda\nu \frac{1}{T} \psi^{-\nu-1} + C\lambda^2\nu^2 \frac{1}{X^2} \psi^{-2\nu-2} \right) \varphi^2 u^2 + \frac{1}{2}(\varphi^2 u^2)_t - \operatorname{div} U, \end{aligned}$$

where the modulus of the vector U satisfies (4.8). The lemma is proved.

LEMMA 2. *If λ and ν exceed some absolute constant and, moreover, for some $\sigma = \sigma(\alpha)$*

$$\nu \ge \sigma(\ln(1/X) + \ln(1/T)),$$

then

$$\begin{aligned} \psi^{\nu+2}\varphi^2 \left(u_t - \sum_{i,j=1}^{n} a^{ij} u_{ij} \right)^2 &\ge -C\lambda\nu \frac{1}{X^2} \varphi^2 |\nabla u|^2 \\ &\quad + 4\lambda^3\nu^4\psi^{-2\nu-2} \left(\mu^2 |\nabla\psi|^4 - \frac{C}{\nu} |\nabla\psi|^3 - \frac{C}{\nu}\frac{1}{X^2} |\nabla\psi|^2 - \frac{C}{\lambda\nu} \right) \varphi^2 u^2 \\ &\quad + \tilde{U}_t + \operatorname{div} U, \end{aligned}$$

where the function $\tilde{U}$ and the vector U admit the estimates

$$|\tilde{U}| \le C\varphi^2 \left(|\nabla u|^2 + \lambda^2\nu^2 \frac{1}{X^2} \psi^{-\nu} u^2 \right),$$

$$|U| \le C\varphi^2 \left(u_t^2 + \lambda\nu \frac{1}{X} |\nabla u|^2 + \lambda^3\nu^3 \frac{1}{X^3} \psi^{-2\nu-2} u^2 \right).$$

PROOF. We introduce the auxiliary function $v(t,x) = \varphi(t,x)u(t,x)$. Since

$$\begin{aligned} u_t &= \varphi^{-1}(v_t + \lambda\nu\psi^{-\nu-1}\psi_t v), \\ u_{ij} &= \varphi^{-1}\{v_{ij} + \lambda\nu\psi^{-\nu-1}\psi_i v_j + \lambda\nu\psi^{-\nu-1}\psi_j v_i \\ &\quad + [\lambda^2\nu^2\psi^{-2\nu-2}\psi_i\psi_j - \lambda\nu(\nu+1)\psi^{-\nu-2}\psi_i\psi_j + \lambda\nu\psi^{-\nu-1}\psi_{ij}]v\}, \end{aligned}$$

it follows that

$$\psi^{\nu+2}\varphi^2\left(u_t-\sum_{i,j=1}^n a^{ij}u_{ij}\right)^2$$
$$=\psi^{\nu+2}\left\{v_t-\sum_{i,j=1}^n a^{ij}v_{ij}-2\lambda\nu\psi^{-\nu-1}\sum_{i,j=1}^n a^{ij}\psi_i v_j\right.$$
$$-\sum_{i,j=1}^n[\lambda^2\nu^2\psi^{-2\nu-2}\psi_i\psi_j-\lambda\nu(\nu+1)\psi^{-\nu-2}\psi_i\psi_j$$
$$\left.+\lambda\nu\psi^{-\nu-1}\psi_{ij}]a^{ij}v+\lambda\nu\psi^{-\nu-1}\psi_t v\right\}^2.$$

Denoting the terms in braces by $z_1,\ldots,z_5$, we have

$$\psi^{\nu+2}\varphi^2\left(u_t-\sum_{i,j=1}^n a^{ij}u_{ij}\right)^2=\psi^{\nu+2}[z_1^2+(z_2+z_4+z_5)^2+z_3^2$$
$$+2z_1z_2+2z_1z_3+2z_1(z_4+z_5)+2z_2z_3+2z_3(z_4+z_5)].$$

The proof of the lemma consists in estimating each term in the last expression. This procedure is elementary but very tedious. To make it sequential, we break the proof of the lemma into several parts, in each of which we deal only with one term (or with a small group of terms).

1) $2\psi^{\nu+2}z_1z_2$. We have

$$2\psi^{\nu+2}z_1z_2=-2\sum_{i,j=1}^n\psi^{\nu+2}a^{ij}v_{ij}v_t=\sum_{i,j=1}^n[-(\psi^{\nu+2}a^{ij})_t v_i v_j$$
$$+2(\psi^{\nu+2}a^{ij})_i v_t v_j]+\sum_{i,j=1}^n[(\psi^{\nu+2}a^{ij}v_iv_j)_t=2(\psi^{\nu+2}a^{ij}v_tv_j)_i].$$

We estimate the terms in the first sum. Since

$$\frac{\partial}{\partial s}(\psi^{\nu+2}a^{ij})=(\nu+2)\psi^{\nu+1}\psi_s a^{ij}+\psi^{\nu+2}a_s^{ij},$$

$\psi_1=1$, $\psi_i=x_i/X^2$, $\psi_t=t/T^2$, it follows that

$$\sum_{i,j=1}^n[-(\psi^{\nu+2}a^{ij})_t v_iv_j+2(\psi^{\nu+2}a^{ij})_i v_tv_j]$$
$$=-\sum_{i,j=1}^n[(\nu+2)\frac{t}{T^2}\psi^{\nu+1}a^{ij}+\psi^{\nu+2}a_t^{ij}]v_iv_j$$

(*Continued*)

(*Continued*)

$$+ 2v_t \sum_{i,j=1}^{n} [(\nu+2)\psi_i \psi^{\nu+1} a^{ij} + \psi^{\nu+2} a_i^{ij}] v_j$$

$$\geq - C\nu \frac{1}{T} \psi^{\nu+1} |\nabla v|^2 + 2(\nu+2)\psi^{\nu+1} v_t \sum_{i,j=1}^{n} a^{ij} \psi_i v_j + 2\psi^{\nu+2} v_t \sum_{i,j=1}^{n} a_i^{ij} v_j.$$

We shall estimate the divergence terms. Since

$$v_i = \varphi(u_i - \lambda\nu\psi^{-\nu-1}\psi_i u), \qquad v_t = \varphi(u_t - \lambda\nu\psi^{-\nu-1}\psi_t u),$$

it follows that

$$\left| \sum_{i,j=1}^{n} \psi^{\nu+2} a^{ij} v_i v_j \right| \leq C\psi^{\nu+2}\varphi^2 \left(|\nabla u|^2 + \lambda^2 \nu^2 \frac{1}{X^2} \psi^{-2\nu-2} u^2 \right).$$

If we set

$$2\sum_{i=1}^{n} \left(- \sum_{j=1}^{n} \psi^{\nu+2} a^{ij} v_i v_j \right)_i = \operatorname{div} U,$$

then for the modulus of the vector U we easily obtain

$$\begin{aligned} |U| &\leq 2 \sum_{i=1}^{n} \left| \sum_{j=1}^{n} \psi^{\nu+2} a^{ij} v_t v_j \right| \\ &\leq C\psi^{\nu+2}\varphi^2 \left[u_t^2 + |\nabla u|^2 + \lambda^2\nu^2 \left(\frac{1}{X^2} + \frac{1}{T^2} \right) \psi^{-2\nu-2} u^2 \right]. \end{aligned}$$

Thus,

$$\begin{aligned} 2\psi^{\nu+2} z_1 z_2 \geq{}& 2(\nu+2)\psi^{\nu+1} v_t \sum_{i,j=1}^{n} a^{ij} \psi_j v_i \\ &+ 2\psi^{\nu+2} v_t \sum_{i,j=1}^{n} a_i^{ij} v_j - C\nu \frac{1}{T} \psi^{\nu+1} |\nabla u|^2 + \tilde{U}_t + \operatorname{div} U, \end{aligned}$$

where the function $\tilde{U}$ and the vector U satisfy the inequalities

$$\begin{aligned} |\tilde{U}| &\leq C\psi^{\nu+2}\phi^2 \left(|\nabla u|^2 + \lambda^2\nu^2 \frac{1}{X^2} \psi^{-2\nu-2} u^2 \right), \\ |U| &\leq C\psi^{\nu+2}\varphi^2 \left[u_t^2 + |\nabla u|^2 + \lambda^2\nu^2 \left(\frac{1}{X^2} + \frac{1}{T^2} \right) \psi^{-2\nu-2} u^2 \right]. \end{aligned}$$

2) $\psi^{\nu+2}(z_1^2+z_3^2+2z_1z_2+2z_1z_3)$. Using the results of the preceding part, we obtain

$$\begin{aligned}
&\psi^{\nu+2}(z_1^2+z_3^2+2z_1z_2+2z_1z_3)\\
&\quad\geq \psi^{\nu+2}(z_1^2+z_3^2+2z_1z_3)+2(\nu+2)\psi^{\nu+1}v_t\sum_{i,j=1}^{n}a^{ij}\psi_jv_i\\
&\qquad+2\psi^{\nu+2}v_t\sum_{i,j=1}^{n}a_i^{ij}v_j-C\nu\frac{1}{T}\psi^{\nu+1}|\nabla v|^2+\tilde{U}_t+\operatorname{div}U\\
&\quad=\psi^{\nu+2}\left\{z_1^2+z_3^2+2z_1\left[z_3+(\nu+2)\psi^{-1}\sum_{i,j=1}^{n}a^{ij}\psi_jv_i+\sum_{i,j=1}^{n}a_i^{ij}v_j\right]\right\}\\
&\qquad\qquad-C\nu\frac{1}{T}\psi^{\nu+1}|\nabla v|^2+\tilde{U}_t+\operatorname{div}U.
\end{aligned}$$

Applying the Schwarz-Bunyakovskiĭ inequality to the third term in braces, we obtain

$$\begin{aligned}
&\psi^{\nu+2}(z_1^2+z_3^2+2z_1z_3+2z_1z_2)\\
&\quad\geq\psi^{\nu+2}\Bigg[-2z_3(\nu+2)\psi^{-1}\sum_{i,j=1}^{n}a^{ij}\psi_jv_i\\
&\qquad\qquad-(\nu+2)^2\psi^{-2}\left(\sum_{i,j=1}^{n}a^{ij}\psi_jv_i\right)^2-2z_3\sum_{i,j=1}^{n}a_i^{ij}v_j\\
&\qquad\qquad-2(\nu+2)\psi^{-1}\sum_{i,j=1}^{n}a^{ij}\psi_jv_i\sum_{i,j=1}^{n}a_j^{ij}v_i-\left(\sum_{i,j=1}^{n}a_i^{ij}v_j\right)^2\Bigg]\\
&\qquad\qquad-C\nu\frac{1}{T}\psi^{\nu+1}|\nabla v|^2+\tilde{U}_t+\operatorname{div}U.
\end{aligned}$$

We recall that $z_3=-2\lambda\nu\psi^{-\nu-1}\sum_{i,j=1}^{n}a^{ij}\psi_jv_i$. Hence,

$$\begin{aligned}
&\psi^{\nu+2}(z_1^2+z_3^2+2z_1z_3+2z_1z_2)\\
&\quad\geq\psi^{\nu+2}\Bigg\{[4\lambda\nu(\nu+2)\psi^{-\nu-2}-(\nu+2)^2\psi^{-2}]\left(\sum_{i,j=1}^{n}a^{ij}\psi_jv_i\right)^2\\
&\qquad\qquad+[4\lambda\nu\psi^{-\nu-1}-2(\nu+2)\psi^{-1}]\left(\sum_{i,j=1}^{n}a^{ij}\psi_jv_i\right)\left(\sum_{i,j=1}^{n}a_i^{ij}v_j\right)\Bigg\}
\end{aligned}$$

(Continued)

(*Continued*)

$$
\begin{aligned}
&-\psi^{\nu+2}\left(\sum_{i,j=1}^{n} a_i^{ij} v_j\right)^2 - C\nu\frac{1}{T}\psi^{\nu+1}|\nabla v|^2 + \tilde{U}_t + \operatorname{div} U \\
&\geq \psi^{\nu+2}\Bigg\{[4\lambda\nu(\nu+2)\psi^{-\nu-2} - (\nu+2)^2\psi^{-2} - 2\varepsilon\lambda\nu\psi^{-\nu-1} \\
&\qquad + \varepsilon(\nu+2)\psi^{-1}]\left(\sum_{i,j=1}^{n} a^{ij}\psi_j v_i\right)^2 \\
&\qquad - \frac{1}{\varepsilon}[2\lambda\nu\psi^{-\nu-1} - (\nu+2)\psi^{-1}]\left(\sum_{i,j=1}^{n} a_i^{ij} v_j\right)^2\Bigg\} \\
&\qquad\qquad - C\nu\frac{1}{T}\psi^{\nu+1}|\nabla v|^2 + \tilde{U}_t + \operatorname{div} U
\end{aligned}
$$

(we have used the inequality $2ab \geq -\varepsilon a^2 - b^2/\varepsilon$). Choosing $\varepsilon = (\nu+2)\psi^{-1}$, for the expression in question we obtain the following lower bound:

$$
\begin{aligned}
&2\lambda\nu(\nu+2)\left(\sum_{i,j=1}^{n} a^{ij}\psi_j v_i\right)^2 - (2\lambda\psi^2 - \psi^{\nu+2})\left(\sum_{i,j=1}^{n} a_i^{ij} v_j\right)^2 \\
&\qquad -C\nu\frac{1}{T}\psi^{\nu+1}|\nabla v|^2 + \tilde{U}_t + \operatorname{div} U \\
&\qquad\qquad \geq -C\left(\lambda\psi^2 + \nu\frac{1}{T}\psi^{\nu+1}\right)|\nabla v|^2 + \tilde{U}_t + \operatorname{div} U.
\end{aligned}
$$

To complete this part it remains for us to return to the function $u(t,x)$ in the first term of the last expression. We have

$$
\begin{aligned}
&-C\left(\lambda\psi^2 + \nu\frac{1}{T}\psi^{\nu+1}\right)|\nabla v|^2 \\
&\quad = -C\left(\lambda\psi^2 + \nu\frac{1}{T}\psi^{\nu+1}\right)\left[\varphi^2|\nabla u|^2 + \sum_{i=1}^{n}(2\varphi\varphi_i u u_i + \varphi_i^2 u^2)\right] \\
&\quad = -C\left(\lambda\psi^2 + \nu\frac{1}{T}\psi^{\nu+1}\right)\left(\varphi^2|\nabla u|^2 + \sum_{i=1}^{n}\varphi_i^2 u^2\right) \\
&\qquad + C\sum_{i=1}^{n}\left[\left(\lambda\psi^2 + \nu\frac{1}{T}\psi^{\nu+1}\right)\varphi\varphi_i\right]_i u^2 \\
&\qquad - C\sum_{i=1}^{n}\left[\left(\lambda\psi^2 + \nu\frac{1}{T}\psi^{\nu+1}\right)\varphi\varphi_i u^2\right]_i
\end{aligned}
$$

$$\begin{aligned}
= & -C\left(\lambda\psi^2+\nu\frac{1}{T}\psi^{\nu+1}\right)\varphi^2|\nabla u|^2 \\
& +C\sum_{i=1}^{n}\Bigg\{\left(\lambda\psi^2+\nu\frac{1}{T}\psi^{\nu+1}\right) \\
& \qquad\times\left[\lambda^2\nu^2\psi^{-2\nu-2}\psi_i^2+\lambda\nu(\nu+1)\psi^{-\nu-2}\psi_i^2-\lambda\nu\psi^{-\nu-1}\psi_{ii}\right] \\
& \qquad-\left[2\lambda\psi+\nu(\nu+1)\frac{1}{T}\psi^{\nu}\right]\lambda\nu\psi^{-\nu-1}\psi_i^2\Bigg\}\varphi^2u^2+\operatorname{div}U \\
\geq & -C\left(\lambda\psi^2+\nu\frac{1}{T}\psi^{\nu+1}\right)\varphi^2|\nabla u|^2 \\
& -C\sum_{i=1}^{n}\Bigg\{\left(\lambda\psi^2+\nu\frac{1}{T}\psi^{\nu+1}\right)\lambda\nu\psi^{-\nu-1}\psi_{ii} \\
& \qquad+\left[2\lambda\psi+\nu(\nu+1)\frac{1}{T}\psi^{\nu}\right]\lambda\nu\psi^{-\nu-1}\psi_i^2\Bigg\}\varphi^2u^2+\operatorname{div}U.
\end{aligned}$$

Since $\psi_{11}=0$ and $\psi_{ii}=1/X^2$ for $i\geq 2$, assuming that $\nu\geq 1$, we obtain

$$\begin{aligned}
-C\left(\lambda\psi^2+\nu\frac{1}{T}\psi^{\nu+1}\right)|\nabla v|^2 \geq & -C\left(\lambda\psi^2+\nu\frac{1}{T}\psi^{\nu+1}\right)\varphi^2|\nabla u|^2 \\
& -C\Bigg[\lambda\nu\left(\lambda\psi^2+\nu\frac{1}{T}\psi^{\nu+1}\right)\psi^{-\nu-1}\frac{1}{X^2} \\
& \qquad+\lambda\nu\left(\lambda\psi+\nu^2\frac{1}{T}\psi^{\nu}\right)\psi^{-\nu-1}|\nabla\psi|^2\Bigg]\varphi^2u^2+\operatorname{div}U,
\end{aligned}$$

and, as is easily seen,

$$|U|\leq C(\lambda\psi^{-\nu}+\nu/T)\lambda\nu\phi^2u^2/X.$$

Thus,

$$\begin{aligned}
\psi^{\nu+2}(z_1^2+z_3^2+2z_1z_3+2z_1z_2)\geq & -C\left(\lambda\psi^2+\nu\frac{1}{T}\psi^{\nu+1}\right)\varphi^2|\nabla u|^2 \\
& -C\Bigg[\lambda\nu\left(\lambda\psi^2+\nu\frac{1}{T}\psi^{\nu+1}\right)\psi^{-\nu-1}\frac{1}{X^2} \\
& \qquad+\lambda\nu\left(\lambda\psi+\nu^2\frac{1}{T}\psi^{\nu}\right)\psi^{-\nu-1}|\nabla\psi|^2\Bigg]\varphi^2u^2+\tilde{U}_t+\operatorname{div}U,
\end{aligned}$$

where the function $\tilde{U}$ and the vector U satisfy the inequalities

$$|\tilde{U}|\leq C\psi^{\nu+2}\varphi^2\left(|\nabla u|^2+\lambda^2\nu^2\frac{1}{X^2}\psi^{-2\nu-2}u^2\right),$$

$$\begin{aligned}|U| &\le C\psi^{\nu+2}\varphi^2\left[u_t^2+|\nabla u|^2+\lambda^2\nu^2\left(\frac{1}{X^2}+\frac{1}{T^2}\right)\psi^{-2\nu-2}u^2\right]\\ &\quad +C\left(\lambda\psi^{-\nu}+\nu\frac{1}{T}\right)\lambda\nu\frac{1}{X}\varphi^2u^2\\ &\le C\psi^{\nu+2}\varphi^2\left[u_t^2+|\nabla u|^2+\lambda^2\nu^2\left(\frac{1}{X^2}+\frac{1}{T^2}\right)\psi^{-2\nu-2}u^2\right]\end{aligned}$$

with some new constant C.

3) $2\psi^{\nu+2}z_1(z_4+z_5)$. We have ($\delta_i^j$ is the Kronecker symbol)

$$\begin{aligned}2\psi^{\nu+2}z_1(z_4+z_5) = -2\Bigg\{&[\lambda^2\nu^2\psi^{-\nu}-\lambda\nu(\nu+1)]a^{11}\\ &+2\sum_{i=2}^{n}\left[\lambda^2\nu^2\frac{x_i}{X^2}\psi^{-\nu}-\lambda\nu(\nu+1)\frac{x_i}{X^2}\right]a^{1i}\\ &+\sum_{i,j=2}^{n}\left[\lambda^2\nu^2\frac{x_ix_j}{X^4}\psi^{-\nu}-\lambda\nu(\nu+1)\frac{x_ix_j}{X^4}+\delta_i^j\lambda\nu\frac{1}{X^2}\psi\right]a^{ij}\\ &-\lambda\nu\frac{t}{T^2}\psi\Bigg\}vv_t\end{aligned}$$

$$\begin{aligned}=\Bigg\{&-\lambda^2\nu^3\frac{t}{T^2}\psi^{-\nu-1}a^{11}-2\sum_{i=2}^{n}\lambda^2\nu^3\frac{x_i}{X^2}\frac{t}{T^2}\psi^{-\nu-1}a^{1i}\\ &+\sum_{i,j=2}^{n}\left(-\lambda^2\nu^3\frac{x_ix_j}{X^4}\frac{t}{T^2}\psi^{-\nu-1}+\delta_i^j\lambda\nu\frac{1}{X^2}\frac{t}{T^2}\right)a^{ij}\\ &-\lambda\nu\frac{t^2}{T^4}-\lambda\nu\frac{1}{T^2}\psi+[\lambda^2\nu^2\psi^{-\nu}-\lambda\nu(\nu+1)]a_t^{11}\\ &+2\sum_{i,j=2}^{n}\left[\lambda^2\nu^2\frac{x_i}{X^2}\psi^{-\nu}-\lambda\nu(\nu+1)\frac{x_i}{X^2}\right]a_t^{1i}\\ &+\sum_{i,j=2}^{n}\left[\lambda^2\nu^2\frac{x_ix_j}{X^4}\psi^{-\nu}-\lambda\nu(\nu+1)\frac{x_ix_j}{X^4}+\delta_i^j\lambda\nu\frac{1}{X^2}\psi\right]a_t^{ij}\Bigg\}v^2+\tilde{U}_t\\ \ge -C&\left(\lambda^2\nu^3\frac{1}{X^2T}\psi^{-\nu-1}+\lambda\nu\frac{1}{T^2}\right)\varphi^2u^2+\tilde{U}_t,\end{aligned}$$

where

$$|\tilde{U}|\le C\left(\lambda^2\nu^2\frac{1}{X^2}\psi^{-\nu}+\lambda\nu\frac{1}{T}\right)\varphi^2u^2.$$

4) $2\psi^{\nu+2}z_2z_3$. We have

$$\begin{aligned}
2\psi^{\nu+2}z_2z_3 &= 4\lambda\nu \sum_{i,j=1}^{n} \psi a^{ij}v_{ij} \sum_{k,l=1}^{n} a^{kl}\psi_k v_l \\
&= 2\lambda\nu \sum_{i,j=1}^{n}\sum_{l=1}^{n}\Bigg[\left(\psi a^{ij}\sum_{k=1}^{n} a^{kl}\psi_k\right)_l v_iv_j - 2\left(\psi a^{ij}\sum_{k=1}^{n} a^{kl}\psi_k\right)_i v_lv_j \\
&\qquad - \left(\psi a^{ij}\sum_{k=1}^{n} a^{kl}\psi_k v_iv_j\right)_l + 2\left(\psi a^{ij}\sum_{k=1}^{n} a^{kl}\psi_k v_lv_i\right)_j\Bigg] \\
&= 2\lambda\nu \sum_{i,j=1}^{n}\sum_{l=1}^{n}\Bigg[\Bigg(a^{ij}\sum_{k=1}^{n} a^{kl}\psi_k\psi_l + \psi a^{ij}\sum_{k=1}^{n} a^{kl}\psi_{kl} \\
&\qquad + \psi a^{ij}_l\sum_{k=1}^{n} a^{kl}\psi_k + \psi a^{ij}\sum_{k=1}^{n} a^{kl}_l\psi_k\Bigg) v_iv_j \\
&\qquad - 2\Bigg(\psi_i a^{ij}\sum_{k=1}^{n} a^{kl}\psi_k + \psi a^{ij}\sum_{k=1}^{n} a^{kl}\psi_{ki} \\
&\qquad + \psi a^{ij}_i\sum_{k=1}^{n} a^{kl}\psi_k + \psi a^{ij}\sum_{k=1}^{n} a^{kl}_i\psi_k\Bigg) v_lv_j\Bigg] + \operatorname{div} U.
\end{aligned}$$

Since

$$\sum_{i,j=1}^{n}\sum_{l=1}^{n} a^{ij}\sum_{k=1}^{n} a^{kl}\psi_k\psi_l v_iv_j = \sum_{k,l=1}^{n} a^{kl}\psi_k\psi_l \sum_{i,j=1}^{n} a^{ij}v_iv_j \geq 0,$$

$$\sum_{i,j=1}^{n}\sum_{l=1}^{n} \psi a^{ij}\sum_{k=1}^{n} a^{kl}\psi_{kl} v_iv_j = \psi\sum_{l=2}^{n} a^{ll}\psi_{ll} \sum_{i,j=1}^{n} a^{ij}v_iv_j \geq 0,$$

$$\left|\sum_{i,j=1}^{n}\sum_{l=1}^{n} \psi_i a^{ij}\sum_{k=1}^{n} a^{kl}\psi_k v_lv_j\right| \leq C\frac{1}{X^2}|\nabla v|^2,$$

$$\left|\sum_{i,j=1}^{n}\sum_{l=1}^{n} \psi a^{ij}\sum_{k=1}^{n} a^{kl}\psi_{ki} v_lv_j\right| = \left|\psi\sum_{\substack{i=2\\ j=1}}^{n} a^{ij}\sum_{l=1}^{n} a^{il}\psi_{ii} v_lv_j\right| \leq C\frac{1}{X^2}|\nabla v|^2,$$

$$\begin{aligned}
&\Bigg|\sum_{i,j=1}^{n}\sum_{l=1}^{n}\Bigg[\left(\psi a^{ij}_l\sum_{k=1}^{n} a^{kl}\psi_k + \psi a^{ij}\sum_{k=1}^{n} a^{kl}_l\psi_k\right) v_iv_j \\
&\qquad -2\left(\psi a^{ij}_i\sum_{k=1}^{n} a^{kl}\psi_k + \psi a^{ij}\sum_{k=1}^{n} a^{kl}_i\psi_k\right) v_lv_j\Bigg]\Bigg| \leq C\frac{1}{X}|\nabla v|^2,
\end{aligned}$$

it follows that

$$2\psi^{\nu+2}z_2z_3 \geq -C\lambda\nu\frac{1}{X^2}|\nabla v|^2 + \operatorname{div} U.$$

Just as in part 3, we return to the function u. We have

$$\begin{aligned}-C\lambda\nu\frac{1}{X^2}|\nabla v|^2 &= -C\lambda\nu\frac{1}{X^2}\left\{\varphi^2|\nabla u|^2 - \sum_{i=1}^{n}[\varphi\varphi_{ii}u^2 - (\varphi\varphi_i u^2)_i]\right\}\\ &\geq -C\lambda\nu\frac{1}{X^2}\varphi^2|\nabla u|^2 - C\lambda^2\nu^2\frac{1}{X^4}\psi^{-\nu-1}\varphi^2u^2 + \operatorname{div} U.\end{aligned}$$

Thus,

$$2\psi^{\nu+2}z_2z_3 \geq -C\lambda\nu\frac{1}{X^2}\varphi^2|\nabla u|^2 - C\lambda^2\nu^2\frac{1}{X^4}\psi^{-\nu-1}\varphi^2u^2 + \operatorname{div} U,$$

where the vector U admits the following estimate:

$$\begin{aligned}|U| &\leq C\lambda\nu\psi|\nabla\psi|\,|\nabla v|^2 + C\lambda\nu\frac{1}{X^2}\varphi|\nabla\varphi|u^2\\ &\leq C\lambda\nu\frac{1}{X}\sum_{i=1}^{n}(\varphi u_i + \varphi_i u)^2 + C\lambda^2\nu^2\frac{1}{X^3}\psi^{-\nu-1}\varphi^2u^2\\ &\leq C\lambda\nu\frac{1}{X}\varphi^2|\nabla u|^2 + C\lambda^2\nu^2\frac{1}{X^2}\psi^{-\nu-1}\varphi^2|\nabla u|\,|u|\\ &\quad + C\lambda^3\nu^3\frac{1}{X^3}\psi^{-2\nu-2}\varphi^2u^2 \leq C\lambda\nu\frac{1}{X}\varphi^2|\nabla u|^2 + C\lambda^3\nu^3\frac{1}{X^3}\psi^{-2\nu-2}\varphi^2u^2.\end{aligned}$$

5) $2\psi^{\nu+2}z_3(z_4+z_5)$. We have

$$\begin{aligned}&2\psi^{\nu+2}z_3(z_4+z_5)\\ &= 4\lambda^2\nu^2\sum_{i,j=1}^{n}a^{ij}\psi_i v_j\left\{[\lambda\nu\psi^{-2\nu-1} - (\nu+1)\psi^{-\nu-1}]\sum_{i,j=1}^{n}a^{ij}\psi_i\psi_j v\right.\\ &\qquad\left. + \frac{1}{X^2}\psi^{-\nu}\sum_{i=2}^{n}a^{ii}v - \frac{t}{T^2}\psi^{-\nu}v\right\}.\end{aligned}$$

We consider the expression

$$\begin{aligned}&4\lambda^2\nu^2\sum_{i,j=1}^{n}\sum_{k,l=1}^{n}[\lambda\nu\psi^{-2\nu-1} - (\nu+1)\psi^{-\nu-1}]a^{ij}a^{kl}\psi_i\psi_k\psi_l v_j v\\ &= -2\lambda^2\nu^2\sum_{i,j=1}^{n}\sum_{k,l=1}^{n}\{[-\lambda\nu(2\nu+1)\psi^{-2\nu-2} + (\nu+1)^2\psi^{-\nu-2}]a^{ij}a^{kl}\psi_i\psi_j\psi_k\psi_l\\ &\qquad + [\lambda\nu\psi^{-2\nu-1} - (\nu+1)\psi^{-\nu-1}][(a^{ij}_j a^{kl} + a^{ij}a^{kl}_j)\psi_i\psi_k\psi_l\\ &\qquad\qquad + a^{ij}a^{kl}(\psi_{ij}\psi_k\psi_l + \psi_i\psi_{kj}\psi_l + \psi_i\psi_k\psi_{lj})]\}v^2\end{aligned}$$

$$+ 2\lambda^2\nu^2 \sum_{i,j=1}^{n} \sum_{k,l=1}^{n} \{[\lambda\nu\psi^{-2\nu-1} - (\nu+1)\psi^{-\nu-1}]a^{ij}a^{kl}\psi_i\psi_k\psi_l v^2\}_j$$

$$= 2\lambda^2\nu^2[\lambda\nu(2\nu+1)\psi^{-2\nu-2} - (\nu+1)^2\psi^{-\nu-2}] \left(\sum_{i,j=1}^{n} a^{ij}\psi_i\psi_j\right)^2 v^2$$

$$- 2\lambda^2\nu^2[\lambda\nu\psi^{-2\nu-1} - (\nu+1)\psi^{-\nu-1}]$$

$$\times \left(\sum_{k,l=1}^{n} a^{kl}\psi_k\psi_l \sum_{i,j=1}^{n} a^{ij}_j\psi_i + \sum_{i,j=1}^{n} a^{ij}\psi_i \sum_{k,l=1}^{n} a^{kl}_j\psi_k\psi_l\right) v^2$$

$$- 2\lambda^2\nu^2[\lambda\nu\psi^{-2\nu-1} - (\nu+1)\psi^{-\nu-1}]\frac{1}{X^2}\sum_{i=2}^{n} a^{ii} \sum_{k,l=1}^{n} a^{kl}\psi_k\psi_l v^2$$

$$- 4\lambda^2\nu^2[\lambda\nu\psi^{-2\nu-1} - (\nu+1)\psi^{-\nu-1}]\frac{1}{X^2}\sum_{j=2}^{n}\sum_{i,l=1}^{n} a^{ij}a^{jl}\psi_i\psi_l v^2 + \operatorname{div} U$$

$$\geq 2\lambda^3\nu^3\psi^{-2\nu-2}\left[2\nu\mu^2|\nabla\psi|^4 - C|\nabla\psi|^3 - C\frac{1}{X^2}|\nabla\psi|^2\right] v^2 + \operatorname{div} U$$

$$= 2\lambda^3\nu^3\psi^{-2\nu-2}\left[2\nu\mu^2|\nabla\psi|^4 - C|\nabla\psi|^3 - C\frac{1}{X^2}|\nabla\psi|^2\right] \varphi^2 u^2 + \operatorname{div} U,$$

and, as is easily seen, $|U| \leq C\lambda^3\nu^3(1/X^3)\psi^{-2\nu-1}\varphi^2u^2$. Further,

$$4\lambda^2\nu^2\frac{1}{X^2}\sum_{i,j=1}^{n}\sum_{k=2}^{n}\psi^{-\nu}\psi_i a^{ij}a^{kk}v_j v - 4\lambda^2\nu^2\frac{t}{T^2}\sum_{i,j=1}^{n}\psi^{-\nu}\psi_i a^{ij}v_j v$$

$$= -2\lambda^2\nu^2\frac{1}{X^2}\sum_{i,j=1}^{n}\sum_{k=2}^{n}(-\nu\psi^{-\nu-1}\psi_i\psi_j a^{ij}a^{kk}$$

$$+ \psi^{-\nu}\psi_{ij}a^{ij}a^{kk} + \psi^{-\nu}\psi_i a^{ij}_j a^{kk} + \psi^{-\nu}\psi_i a^{ij}a^{kk}_j)v^2$$

$$+ 2\lambda^2\nu^2\frac{t}{T^2}\sum_{i,j=1}^{n}(-\nu\psi^{-\nu-1}\psi_i\psi_j a^{ij} + \psi^{-\nu}\psi_{ij}a^{ij} + \psi^{-\nu}\psi_i a^{ij}_j)v^2$$

$$+ 2\lambda^2\nu^2\frac{1}{X^2}\sum_{i,j=1}^{n}\sum_{k=2}^{n}(\psi^{-\nu}\psi_i a^{ij}a^{kk}v^2)_j$$

$$- 2\lambda^2\nu^2\frac{1}{T^2}\sum_{i,j=1}^{n}(t\psi^{-\nu}\psi_i a^{ij}v^2)_j$$

(Continued)

(Continued)

$$= 2\lambda^2\nu^2\frac{1}{X^2}\psi^{-\nu-1}\left[\nu\sum_{k=2}^n a^{kk}\sum_{i,j=1}^n a^{ij}\psi_i\psi_j - \psi\frac{1}{X^2}\left(\sum_{i=2}^n a^{ii}\right)^2 - \psi\sum_{i,j=1}^n a_j^{ij}\psi_i\sum_{k=2}^n a^{kk} - \psi\sum_{i,j=1}^n a^{ij}\psi_i\sum_{k=2}^n a_j^{kk}\right]v^2$$

$$+ 2\lambda^2\nu^2\frac{t}{T^2}\psi^{-\nu-1}\left(-\nu\sum_{i,j=1}^n a^{ij}\psi_i\psi_j + \frac{1}{X^2}\psi\sum_{i=2}^n a^{ii} + \psi\sum_{i,j=1}^n a_j^{ij}\psi_i\right)v^2 + \operatorname{div} U$$

$$\geq \left\{2\lambda^2\nu^2\frac{1}{X^2}\psi^{-\nu-1}\left[\nu(n-1)\mu^2|\nabla\psi|^2 - C\frac{1}{X^2}\right] - C\lambda^2\nu^2\frac{1}{T}\psi^{-\nu-1}\left(\nu|\nabla\psi|^2 + \frac{1}{X^2}\right)\right\}\varphi^2u^2 + \operatorname{div} U,$$

and it is easy to see that

$$|U| \leq C\lambda^2\nu^2\psi^{-\nu}\left(\frac{1}{X^3} + \frac{1}{TX}\right)\varphi^2u^2.$$

Collecting the estimates obtained in this part, we find

$$2\psi^{\nu+2}z_3(z_4+z_5)$$
$$\geq 2\lambda^3\nu^3\psi^{-2\nu-2}\left\{2\nu\mu^2|\nabla\psi|^4 - C|\nabla\psi|^3 - C\frac{1}{X^2}|\nabla\psi|^2 + \frac{1}{\lambda\nu X^2}\psi^{\nu+1}\left[\nu(n-1)\mu^2|\nabla\psi|^2 - C\frac{1}{X^2}\right] - C\frac{1}{\lambda\nu T}\psi^{\nu+1}\left(\nu|\nabla\psi|^2 + \frac{1}{X^2}\right)\right\}\varphi^2u^2 + \operatorname{div} U$$

$$\geq 2\lambda^3\nu^3\psi^{-2\nu-2}\left[2\nu\mu^2|\nabla\psi|^4 - C|\nabla\psi|^3 - C\frac{1}{X^2}|\nabla\psi|^2 - C\frac{1}{\lambda}\left(\frac{1}{X^4} + \frac{1}{T^2}\right)\psi^{\nu+1}\right]\varphi^2u^2 + \operatorname{div} U$$

(we have used the inequality $-2/(TX^2) \geq -1/X^4 - 1/T^2$), where

$$|U| \leq C\lambda^3\nu^3\psi^{-2\nu-1}\left(\frac{1}{X^3} + \frac{1}{\lambda\nu T^2}\psi^{\nu+1}\right)\varphi^2u^2.$$

Thus, we have considered all terms in the representation

$$\psi^{\nu+2}\varphi^2\left(u_t - \sum_{i,j=1}^n a^{ij}u_{ij}\right)^2.$$

Collecting the inequalities obtained for them in parts 2–5, we find that

$$
\begin{aligned}
&\psi^{\nu+2}\varphi^2\left(u_t-\sum_{i,j=1}^n a^{ij}u_{ij}\right)^2\\
&\geq -C\lambda\nu\frac{1}{X^2}\left(1+\frac{1}{\nu}X^2\psi^2+\frac{1}{\lambda T}X^2\psi^{\nu+1}\right)\varphi^2|\nabla u|^2\\
&\quad+4\lambda^3\nu^4\psi^{-2\nu-2}\left[\mu^2|\nabla\psi|^4-C\frac{1}{\nu}|\nabla\psi|^3-C\frac{1}{\nu X^2}|\nabla\psi|^2\right.\\
&\quad-C\frac{1}{\lambda\nu}\left(\frac{1}{X^4}+\frac{1}{T^2}\right)\psi^{\nu+1}-C\frac{1}{\lambda^2\nu^3}\left(\lambda\psi^2+\nu\frac{1}{T}\psi^{\nu+1}\right)\psi^{\nu+1}\frac{1}{X^2}\\
&\quad-C\frac{1}{\lambda^2\nu^3}\left(\lambda\psi+\nu^2\frac{1}{T}\psi^\nu\right)\psi^{\nu+1}|\nabla\psi|^2\\
&\quad-C\frac{1}{\lambda^3\nu^4}\left(\lambda^2\nu^3\frac{1}{X^2T}\psi^{-\nu-1}+\lambda\nu\frac{1}{T^2}\right)\psi^{2\nu+2}\\
&\qquad\qquad\left.-C\frac{1}{\lambda\nu^2X^4}\psi^\nu\right]\varphi^2u^2+\tilde{U}_t+\operatorname{div}U,
\end{aligned}
$$

where

$$
\begin{aligned}
|\tilde{U}|&\leq C\varphi^2\left[\psi^{\nu+2}|\nabla u|^2+\left(\lambda^2\nu^2\frac{1}{X^2}\psi^{-\nu}+\lambda\nu\frac{1}{T}\right)u^2\right],\\
|U|&\leq C\varphi^2\left[\psi^{\nu+2}u_t^2+\lambda\nu\frac{1}{X}|\nabla u|^2\right.\\
&\qquad\left.+\lambda^3\nu^3\psi^{-2\nu-2}\left(\frac{1}{X^3}+\frac{1}{\lambda\nu T^2}\psi^{\nu+1}\right)u^2\right].
\end{aligned}
$$

All these inequalities hold for all λ and ν exceeding some absolute constant. We simplify these inequalities by imposing additional conditions on ν. Since $X<1$ and $\lambda,\nu>1$ and $\psi<\frac{1}{2}+\alpha$, it follows that

$$
1+\frac{1}{\nu}X^2\psi^2+\frac{1}{\lambda T}X^2\psi^{\nu+1}\leq 1+1+\frac{1}{T}\left(\frac{1}{2}+\alpha\right)^\nu,
$$

and, since $0<\alpha<\frac{1}{2}$, there exists $\sigma_1=\sigma_1(\alpha)$ such that the condition $\nu\geq\sigma_1\ln(1/T)$ implies

$$
1+\frac{1}{\nu}X^2\psi^2+\frac{1}{\lambda T}X^2\psi^{\nu+1}\leq 3.
$$

Similarly,

$$\begin{aligned}
&C\frac{1}{\lambda\nu}\left(\frac{1}{X^4}+\frac{1}{T^2}\right)\psi^{\nu+1}+C\frac{1}{\lambda^2\nu^3}\left(\lambda\psi^2+\nu\frac{1}{T}\psi^{\nu+1}\right)\psi^{\nu+1}\frac{1}{X^2}\\
&\quad+C\frac{1}{\lambda^2\nu^3}\left(\lambda\psi+\nu^2\frac{1}{T}\psi^{\nu}\right)\psi^{\nu+1}|\nabla\psi|\\
&\quad+C\frac{1}{\lambda^3\nu^4}\left(\lambda^2\nu^3\frac{1}{X^2T}\psi^{-\nu-1}+\lambda\nu\frac{1}{T^2}\right)\psi^{2\nu+2}+C\frac{1}{\lambda\nu^2X^4}\psi^{\nu}\\
&\le C\frac{1}{\lambda\nu}\left(\frac{1}{X^4}+\frac{1}{T^2}\right)\left(\frac{1}{2}+\alpha\right)^{\nu}\le C\frac{1}{\lambda\nu}
\end{aligned}$$

under the condition

$$\nu\ge\sigma_2(\alpha)(\ln(1/X)+\ln(1/T)).$$

Increasing σ_1 and σ_2, if necessary, we reduce the estimates for $\tilde{U}$ and U to the form

$$\begin{aligned}
|\tilde{U}|&\le C\varphi^2\left(|\nabla u|^2+\lambda^2\nu^2\frac{1}{X^2}\psi^{-\nu}u^2\right),\\
|U|&\le C\varphi^2\left(u_t^2+\lambda\nu\frac{1}{X}|\nabla u|^2+\lambda^3\nu^3\frac{1}{X^3}\psi^{-2\nu-2}u^2\right).
\end{aligned}\tag{4.9}$$

Thus, if with some constant $\sigma(\alpha)$

$$\nu\ge\sigma(\alpha)\left(\ln(1/X)+\ln(1/T)\right),$$

then

$$\begin{aligned}
\psi^{\nu+2}\varphi^2\left(u_t-\sum_{i,j=1}^{n}a^{ij}u_{ij}\right)^2&\ge -C\lambda\nu\frac{1}{X^2}\varphi^2|\nabla u|^2\\
&\quad+4\lambda^3\nu^4\psi^{-2\nu-2}\left(\mu^2|\nabla\psi|^2-C\frac{1}{\nu}|\nabla\psi|^3\right.\\
&\qquad\left.-C\frac{1}{\nu X^2}|\nabla\psi|^2-C\frac{1}{\lambda\nu}\right)\varphi^2u^2+\tilde{U}_t+\operatorname{div}U,
\end{aligned}$$

where $\tilde{U}$ and U satisfy (4.9). The lemma is proved.

LEMMA 3. *There exist an absolute constant σ_0, a constant σ_1 (depending only on α), and a constant σ_2 depending on the coefficients $a^{ij}(t,x)$ such that for $\lambda\ge\sigma_0$ and $\nu\ge\max(\sigma_1\ln T^{-1},\sigma_2/X^4)$ the inequality*

$$\begin{aligned}
\lambda\nu\frac{1}{X^2}\varphi^2|\nabla u|^2+\lambda^3\nu^4\psi^{-2\nu-2}\varphi^2u^2&\le C\lambda\nu\frac{1}{X^2}\varphi^2\left(u_t-\sum_{i,j=1}^{n}a^{ij}u_{ij}\right)u\\
&\quad+C\psi^{\nu+2}\varphi^2\left(u_t-\sum_{i,j=1}^{n}a^{ij}u_{ij}\right)^2+\tilde{U}_t+\operatorname{div}U,
\end{aligned}$$

holds, where

$$|\tilde{U}| \le C\varphi^2\left(|\nabla u|^2 + \lambda^2\nu^2\frac{1}{X^2}\psi^{-\nu}u^2\right),$$
$$|U| \le C\varphi^2\left(u_t^2 + \lambda\nu\frac{1}{X}|\nabla u|^2 + \lambda^3\nu^3\frac{1}{X^3}\psi^{-2\nu-2}u^2\right).$$

PROOF. We note that under the condition $\nu \ge \sigma(\ln(1/X) + \ln(1/T))$ with sufficiently large $\sigma = \sigma(\alpha)$ the inequality of Lemma 1 implies

$$\varphi^2\left(u_t - \sum_{i,j=1}^n a^{ij}u_{ij}\right)u \ge \mu\varphi^2|\nabla u^2| - C\lambda^2\nu^2\frac{1}{X^2}\psi^{-2\nu-2}\varphi^2u^2 + \tilde{U}_t + \operatorname{div} U,$$

where

$$|\tilde{U}| \le \varphi^2u^2, \qquad |U| \le C\varphi^2\left(|\nabla u|^2 + \lambda\nu\frac{1}{X}\psi^{-\nu-1}u^2\right).$$

Multiplying this inequality by $\lambda\nu\theta X^{-2}$, where $\theta > 0$ will be chosen later, and adding it to the inequality of Lemma 2, we obtain

$$\begin{aligned}\lambda\nu\frac{1}{X^2}(\theta\mu - C)\varphi^2|\nabla u|^2 + 4\lambda^3\nu^4\psi^{-2\nu-2}\Bigg(\mu^2|\nabla\psi|^4 \\ - C\frac{\theta}{\nu}\frac{1}{X^4} - C\frac{1}{\nu}|\nabla\psi|^3 - C\frac{1}{\nu}\frac{1}{X^2}|\nabla\psi|^2 - C\frac{1}{\lambda\nu}\Bigg)\varphi^2u^2 \\ \le \lambda\nu\frac{1}{X^2}\theta\varphi^2\left(u_t - \sum_{i,j=1}^n a^{ij}u_{ij}\right)u \\ + \psi^{\nu+2}\varphi^2\left(u_t - \sum_{i,j=1}^n a^{ij}u_{ij}\right)^2 + \tilde{U}_t + \operatorname{div} U,\end{aligned}$$

where

$$|\tilde{U}| \le C\varphi^2\left[|\nabla u|^2 + \left(\lambda\nu\frac{\theta}{X^2} + \lambda^2\nu^2\frac{1}{X^2}\psi^{-\nu}\right)u^2\right]$$

and

$$\begin{aligned}|U| \le C\varphi^2\Bigg[u_t^2 + \lambda\nu\left(\frac{1}{X} + \frac{\theta}{X^2}\right)|\nabla u|^2 \\ + \left(\lambda^2\nu^2\frac{\theta}{X^2}\psi^{-\nu-1} + \lambda^3\nu^3\frac{1}{X^3}\psi^{-2\nu-2}\right)u^2\Bigg].\end{aligned}$$

Since

$$|\nabla\psi| = \left(1 + \frac{1}{X^4}\sum_{i=2}^n x_i^2\right)^{1/2},$$

it follows that $1 \le |\nabla\psi| \le C/X$. Hence,

$$\begin{aligned}
&\lambda\nu\frac{1}{X^2}(\theta\mu - C)\varphi^2|\nabla u|^2 + 4\lambda^3\nu^4\psi^{-2\nu-2}\Big(\mu^2|\nabla\psi|^4 \\
&\qquad - C\frac{\theta}{\nu}\frac{1}{X^4} - C\frac{1}{\nu}|\nabla\psi|^3 - C\frac{1}{\nu X^2}|\nabla\psi|^2 - C\frac{1}{\lambda\nu}\Big)\varphi^2u^2 \\
&\quad \ge \lambda\nu\frac{1}{X^2}(\theta\mu - C)\varphi^2|\nabla u|^2 \\
&\qquad + 4\lambda^3\nu^4\psi^{-2\nu-2}\left(\mu^2 - C\frac{\theta}{\nu X^4} - C\frac{1}{\nu X^3} - C\frac{1}{\nu X^4} - C\frac{1}{\lambda\nu}\right)\varphi^2u^2 \\
&\quad \ge \lambda\nu\frac{1}{X^2}(\theta\mu - C)\varphi^2|\lambda u|^2 + 4\lambda^3\nu^4\psi^{-2\nu-2}\left[\mu^2 - \frac{C(\theta+1)}{\nu X^4}\right]\varphi^2u^2.
\end{aligned}$$

We choose θ from the condition $\theta\mu - C = \mu^2 - \frac{C(\theta+1)}{\nu X^4}$, i.e., we set

$$\theta = \left[\mu^2 + C\left(1 - \frac{1}{\nu X^4}\right)\right]\left(\mu + \frac{C}{\nu X^4}\right)^{-1}.$$

Then

$$\theta\mu - C = \left[\mu^3 - (\mu + C)\frac{C}{\nu X^4}\right]\left(\mu + \frac{C}{\nu X^4}\right)^{-1}.$$

It is easy to see that there exists a constant σ_2 such that the condition $\nu \ge \sigma_2/X^4$ (and it may be assumed that $\sigma_2/X^4 \ge \sigma\ln(1/X)$) implies the inequalities

$$\theta > 0, \qquad \theta\mu - C = \mu^2 - \frac{C(\theta+1)}{\nu X^4} \ge \frac{\mu^2}{2}.$$

Noting that $\theta \le (\mu^2 + C)/\mu$, we obtain

$$\begin{aligned}
&\lambda\nu\frac{1}{X^2}\frac{\mu^2}{2}\varphi^2|\nabla u|^2 + \lambda^3\nu^4\frac{4\mu^2}{2}\psi^{-2\nu-2}\varphi^2u^2 \\
&\quad \le \lambda\nu\frac{1}{X^2}\frac{\mu^2+C}{\mu}\varphi^2\left(u_t - \sum_{i,j=1}^n a^{ij}u_{ij}\right)u + \psi^{\nu+2}\varphi^2\left(u_t - \sum_{i,j=1}^n a^{ij}u_{ij}\right)^2 \\
&\qquad + \tilde{U}_t + \operatorname{div} U,
\end{aligned}$$

whence the lemma follows.

THEOREM 1. *Suppose the function $u(t,x)$ and the operator*

$$\frac{\partial}{\partial t} - \sum_{i,j=1}^n a^{ij}(t,x)\frac{\partial^2}{\partial x_i \partial x_j}$$

satisfy conditions (4.1)–(4.3). *Suppose, moreover,*

$$|u|, |u_t|, |\nabla u| \le \begin{cases} \varepsilon & \text{for } (t,x) \in \omega^{(1)}_{1/2}, \\ 1 & \text{for } (t,x) \in \omega^{(2)}_{1/2}. \end{cases} \tag{4.10}$$

Then for any $0 < \eta < \frac{1}{2}$

$$\frac{1}{TX^n} \int_{\Omega_\eta} u^2\, dt\, dx \le \gamma_1 \varepsilon^{2\gamma_2},$$

where $\gamma_1 > 0$ *and* $0 < \gamma_2 < 1$ (γ_1 *and* γ_2 *depend on the coefficients* $a^{ij}(t,x)$ *and the quantities* T, X, *and* η).

PROOF. Applying Lemma 3 and using (4.1), we obtain

$$\begin{aligned} &\lambda\nu \frac{1}{X^2} \varphi^2 |\nabla u|^2 + \lambda^3 \nu^4 \psi^{-2\nu-2} \varphi^2 u^2 \\ &\le aC\lambda\nu \frac{1}{X^2} \varphi^2 (|\nabla u| + |u|)|u| + a^2 C \psi^{\nu+2} \varphi^2 (|\nabla u| + |u|)^2 \\ &\quad + \tilde{U}_t + \operatorname{div} U, \end{aligned} \tag{4.11}$$

where

$$\lambda \ge \sigma_0, \qquad \nu \ge \max\left(\sigma_1 \ln \frac{1}{T}, \frac{\sigma_2}{X^4}\right),$$

$$|\tilde{U}| \le C\varphi^2 \left[|\nabla u|^2 + \lambda^2 \nu^2 \frac{1}{X^2} \psi^{-\nu} u^2 \right],$$

$$|U| \le C\varphi^2 \left(u_t^2 + \lambda\nu \frac{1}{X^2} |\nabla u|^2 + \lambda^3 \nu^3 \frac{1}{X^3} \psi^{-2\nu-2} u^2 \right).$$

From (4.11) we have

$$\begin{aligned} \lambda\nu \frac{1}{X^2} \varphi^2 |\nabla u|^2 + \lambda^3 \nu^4 \psi^{-2\nu-2} \varphi^2 u^2 &\le aC\lambda\nu \frac{1}{X^2} \varphi^2 |\nabla u|\, |u| + aC\lambda\nu \frac{1}{X^2} \varphi^2 u^2 \\ &\quad + 2a^2 C \psi^{\nu+2} \varphi^2 |\nabla u|^2 + 2a^2 C \psi^{\nu+2} \varphi^2 u^2 + \tilde{U}_t + \operatorname{div} U \\ \le \left(\frac{1}{2} aC\lambda\nu^{1/2} \frac{1}{X^2} + 2a^2 C \psi^{\nu+2} \right) \varphi^2 |\nabla u|^2 & \\ + \left(\frac{1}{2} aC\lambda\nu^{3/2} \frac{1}{X^2} + aC\lambda\nu \frac{1}{X^2} + 2a^2 C \psi^{\nu+2} \right) \varphi^2 u^2 + \tilde{U}_t + \operatorname{div} U & \\ \le (a + a^2) C\lambda\nu^{1/2} \frac{1}{X^2} \varphi^2 |\nabla u|^2 & \\ + (a + a^2) C\lambda\nu^{3/2} \frac{1}{X^2} \varphi^2 u^2 + \tilde{U}_t + \operatorname{div} U & \end{aligned}$$

with some new constant C. From the last inequality we get

$$\begin{aligned} &\lambda\nu \frac{1}{X^2} \left(1 - \frac{a^2 + a}{\sqrt{\nu}} C \right) \varphi^2 |\nabla u|^2 \\ &\quad + \lambda^3 \nu^4 \psi^{-2\nu-2} \left(1 - \frac{a^2 + a}{\lambda^2 \sqrt{\nu^5}} C \frac{1}{X^2} \psi^{2\nu+2} \right) \varphi^2 u^2 \\ &\le \tilde{U}_t + \operatorname{div} U, \end{aligned}$$

whence

$$\lambda\nu\frac{1}{X^2}\left(1-\frac{a^2+a}{\sqrt{\nu}}C\right)\varphi^2|\nabla u|^2+\lambda^3\nu^4\psi^{-2\nu-2}\left(1-\frac{a^2+a}{\sqrt{\nu^5}X^2}C\right)\varphi^2u^2 \le \tilde{U}_t+\operatorname{div}U.$$

If σ_2 is sufficiently large, then

$$\frac{a^2+a}{\sqrt{\nu}}C\le\frac{1}{2},\qquad \frac{a^2+a}{\sqrt{\nu^5}X^2}\le\frac{1}{2}.$$

Assuming that σ_2 has been chosen in this manner, we obtain

$$\lambda\nu\frac{1}{X^2}\varphi^2|\nabla u|^2+\lambda^3\nu^4\psi^{-2\nu-2}\varphi^2u^2\le\tilde{U}_t+\operatorname{div}U.$$

Disregarding the first term on the left and integrating the resulting inequality over $\Omega_{1/2}$, we have

$$\begin{aligned}\lambda^3\nu^4\int_{\Omega_{1/2}}\psi^{-2\nu-2}\varphi^2u^2\,dt\,dx&\le\int_{\omega_{1/2}^{(1)}}|U|\,ds+\int_{\omega_{1/2}^{(2)}}(|\tilde{U}|+|U|)\,ds\\&\le C\int_{\omega_{1/2}^{(1)}\cup\omega_{1/2}^{(2)}}\left(u_t^2+\lambda\nu\frac{1}{X^2}|\nabla u|^2+\lambda^3\nu^3\frac{1}{X^3}\psi^{-2\nu-2}u^2\right)\varphi^2\,ds.\end{aligned}$$

Using (4.10), we obtain

$$\begin{aligned}\lambda^3\nu^4\int_{\Omega_{1/2}}\psi^{-2\nu-2}\varphi^2u^2\,dt\,dx&\le C\lambda^3\nu^3\alpha^{-2\nu-2}\exp(2\lambda\alpha^{-\nu})TX^{n-4}\varepsilon^2\\&\quad+C\lambda^3\nu^3(\alpha+1/2)^{-2\nu-2}\exp\left[2\lambda(\alpha+1/2)^{-\nu}\right]TX^{n-4}.\end{aligned}$$

Suppose now that $\eta<1/2$. Then $\Omega_\eta\subset\Omega_{1/2}$, and by the last inequality

$$\begin{aligned}\lambda^3\nu^4\int_{\Omega_\eta}\psi^{-2\nu-2}\varphi^2u^2\,dt\,dx&\le C\lambda^3\nu^3\alpha^{-2\nu-2}\exp(2\lambda\alpha^{-\nu})TX^{n-4}\varepsilon^2\\&\quad+C\lambda^3\nu^3(\alpha+1/2)^{-2\nu-2}\exp[2\lambda(\alpha+1/2)^{-\nu}]TX^{n-4}.\end{aligned}$$

Since $\inf_{\Omega_\eta}\psi(t,x)=\alpha+\eta$, it follows that

$$\begin{aligned}&\lambda^3\nu^4(\alpha+\eta)^{-2\nu-2}\exp[2\lambda(\alpha+\eta)^{-\nu}]\int_{\Omega_\eta}u^2\,dt\,dx\\&\quad\le C\lambda^3\nu^3\alpha^{-2\nu-2}\exp(2\lambda\alpha^{-\nu})TX^{n-4}\varepsilon^2\\&\qquad+C\lambda^3\nu^3(\alpha+1/2)^{-2\nu-2}\exp[2\lambda(\alpha+1/2)^{-\nu}]TX^{n-4},\end{aligned}$$

whence

$$\begin{aligned}\int_{\Omega_\eta}u^2\,dt\,dx&\le C\left(\frac{\alpha+\eta}{\alpha}\right)^{2\nu+2}\exp\{2\lambda[\alpha^{-\nu}-(\alpha+\eta)^{-\nu}]\}\frac{TX^n}{\nu X^4}\varepsilon^2\\&\quad+C\left(\frac{\alpha+\eta}{\alpha+1/2}\right)^{2\nu+2}\exp\{-2\lambda\left[(\alpha+\eta)^{-\nu}-(\alpha+1/2)^{-\nu}\right]\}\frac{TX^n}{\nu X^4}.\end{aligned}$$

Hence, with some new constant C,

$$\frac{1}{TX^n}\int_{\Omega_\eta} u^2\,dt\,dx \le C\left(\frac{\alpha+\eta}{\alpha}\right)^{2\nu}\exp\{2\lambda[\alpha^{-\nu}-(\alpha+\eta)^{-\nu}]\}\varepsilon^2$$
$$+C\left(\frac{\alpha+\eta}{\alpha+1/2}\right)^{2\nu}\exp\{-2\lambda\left[(\alpha+\eta)^{-\nu}-(\alpha+1/2)^{-\nu}\right]\}.$$

Assuming ε sufficiently small, we choose λ from the condition

$$\left(\frac{\alpha+\eta}{\alpha}\right)^{2\nu}\exp\{2\lambda[\alpha^{-\nu}-(\alpha+\eta)^{-\nu}]\}\varepsilon^2$$
$$=\left(\frac{\alpha+\eta}{\alpha+1/2}\right)^{2\nu}\exp\{-2\lambda\left[(\alpha+\eta)^{-\nu}-(\alpha+1/2)^{-\nu}\right]\},$$

i.e., we set

$$\lambda=\frac{1}{2}\left[\frac{2\nu}{\alpha^{-\nu}-(\alpha+1/2)^{-\nu}}\ln\frac{\alpha}{\alpha+1/2}-\frac{2}{\alpha^{-\nu}-(\alpha+1/2)^{-\nu}}\ln\varepsilon\right].$$

Then

$$\frac{1}{TX^n}\int_{\Omega_\eta} u^2\,dt\,dx \le 2C\left\{\frac{\alpha+\eta}{\alpha}\exp\left[\frac{\alpha^{-\nu}-(\alpha+\eta)^{-\nu}}{\alpha^{-\nu}-(\alpha+1/2)^{-\nu}}\ln\frac{\alpha}{\alpha+1/2}\right]\right\}$$
$$\times\exp\left\{2\left[1-\frac{\alpha^{-\nu}-(\alpha+\eta)^{-\nu}}{\alpha^{-\nu}-(\alpha+1/2)^{-\nu}}\right]\ln\varepsilon\right\},$$

whence

$$\frac{1}{TX^n}\int_{\Omega_\eta} u^2\,dt\,dx \le C\left(\frac{\alpha+\eta}{\alpha}\right)^{2\nu}\exp\left\{2\left[1-\frac{\alpha^{-\nu}-(\alpha+\eta)^{-\nu}}{\alpha^{-\nu}-(\alpha+1/2)^{-\nu}}\right]\ln\varepsilon\right\}.$$

The theorem is proved.

THEOREM 2. *Suppose the functions $u^{(1)}(t,x)$ and $u^{(2)}(t,x)$ satisfy condition* (4.3), *and*

$$u_t^{(l)}-\sum_{i,j=1}^{n}a^{ij}(t,x)u_{ij}^{(l)}=\bar{a}(t,x)\nabla u^{(l)}+b(t,x)u^{(l)},\quad l=1,2,$$

where $\bar{a}(t,x)$ is a bounded vector on $\overline{\Omega}_{1/2}$, $b(t,x)$ is a bounded function on $\overline{\Omega}_{1/2}$, and the $a^{ij}(t,x)$ satisfy condition (4.2). *If, moreover,*

$$|u^{(1)}-u^{(2)}|,\ |u_1^{(1)}-u_1^{(2)}|=0\quad \textit{on } \omega_{1/2}^{(1)},$$

then $u_1\equiv u_2$ on $\overline{\Omega}_{1/2}$.

For the proof it suffices to apply Theorem 1 with $\varepsilon=0$ to the difference u_1-u_2.

Assuming that $u(t,x)$ does not depend on t, we easily obtain analogous theorems for an elliptic inequality (or equation).

THEOREM 3. *Let*

$$\Omega_\eta = \left\{x\colon 0 < x_1 < \frac{1}{2}\left(2\eta - X^{-2}\sum_{i=2}^{n} x_i^2\right)\right\},$$

$$\omega_\eta^{(1)} = \left\{x\colon x_1 = 0,\ \frac{1}{2}X^{-2}\sum_{i=2}^{n} x_i^2 \le \eta\right\},$$

$$\omega_\eta^{(2)} = \left\{x\colon x_1 \ge 0,\ x_1 = \eta - \frac{1}{2}X^{-2}\sum_{i=2}^{n} x_i^2\right\},$$

$$a^{ij}(x) \in C^2(\overline{\Omega}_{1/2}), \qquad \sum_{i,j=1}^{n} a^{ij}(x)\xi_i\xi_j \ge \mu\sum_{i=1}^{n}\xi_i^2, \quad \mu > 0,$$
$$u(x) \in C^2(\Omega_{1/2}) \cap C^1(\overline{\Omega}_{1/2}).$$

If in the domain $\Omega_{1/2}$

$$\left|\sum_{i,j=1}^{n} a^{ij}(x)u_{ij}\right| \le a(|\nabla u| + |u|),$$

$$|u|, |\nabla u| \le \begin{cases} \varepsilon & \text{for } x \in \omega_{1/2}^{(1)}, \\ 1 & \text{for } x \in \omega_{1/2}^{(2)}, \end{cases}$$

then for any $0 < \eta < 1/2$,

$$\frac{1}{X^n}\int_{\Omega_\eta} u^2\,dx \le \gamma_1\varepsilon^{2\gamma_2},$$

where $\gamma_1 > 0$ *and* $0 < \gamma_2 < 1$ (γ_1 *and* γ_2 *depend on the coefficients* $a^{ij}(x)$ *and the quantities* X *and* η).

THEOREM 4. *Suppose the functions* $u^{(1)}(x)$ *and* $u^{(2)}(x)$ *satisfy the smoothness conditions of the preceding theorem and*

$$\sum_{i,j=1}^{n} a^{ij}(x)u_{ij}^{(l)} = \bar{a}(x)\nabla u^{(l)} + b(x)u^{(l)}, \quad l = 1, 2,$$

where $\bar{a}(x)$ *is a bounded vector on* $\overline{\Omega}_{1/2}$, $b(x)$ *is a bounded function on* $\overline{\Omega}_{1/2}$, *and the* $a^{ij}(x)$ *satisfy the conditions of Theorem 3. If, moreover,*

$$|u^{(1)} - u^{(2)}|,\ |u_1^{(1)} - u_1^{(2)}| = 0 \quad \text{on } \omega_{1/2}^{(1)},$$

then $u_1 \equiv u_2$ *on* $\overline{\Omega}_{1/2}$.

§2. A mixed problem for a parabolic equation with decreasing time

Suppose that Ω is a domain in the space R^n with boundary of class C^1 and $\operatorname{Cyl}(\tau) = \{(t,x)\colon -\tau < t < 0, x \in \Omega\}$. In the domain $\operatorname{Cyl}(T)$ we consider the inequality

$$\left| u_t - \sum_{i,j=1}^{n} a^{ij}(t,x) u_{ij} \right| \le a(|\nabla u| + |u|), \tag{4.12}$$

the coefficients $a^{ij}(t,x)$ of which satisfy the conditions

$$a^{ij}(t,x) \in C^1(\overline{\operatorname{Cyl}}(T)), \qquad \sum_{i,j=1}^{n} a^{ij}(t,x)\xi_i\xi_j \ge \mu \sum_{i=1}^{n} \xi_i^2, \tag{4.13}$$

where μ is a positive constant. Suppose a solution $u(t,x)$ of this inequality satisfies the conditions

$$u(t,x) \in C^2(\operatorname{Cyl}(T)) \cap C^1(\overline{\operatorname{Cyl}}(T)), \tag{4.14}$$

$$u(t,x)|_{-T<t<0, x\in\partial\Omega} = 0 \text{ or } \sum_{i,j=1}^{n} a^{ij} u_i \cos(k, x_j)|_{-T<t<0,x\in\partial\Omega} = 0 \tag{4.15}$$

(k is the outer normal to $\partial\Omega$).

Suppose that on the intersection of $\overline{\operatorname{Cyl}}(T)$ with the plane $t = 0$ the function $u(0, x_1, \dots, x_n)$ is given.

We shall show that a solution of inequality (4.12) under conditions (4.15) is determined by this in unique fashion, and we shall obtain for $u(t,x)$ an a priori estimate of Hölder type. The method used to this end is altogether analogous to the method of the preceding section.

LEMMA 1. *Let $\lambda \ge 1, \alpha > 0$ and $(t,x) \in \operatorname{Cyl}(T)$. Then*

$$\begin{aligned}
&(\alpha - t)^{-2\lambda}\left(u_t - \sum_{i,j=1}^{n} a^{ij} u_{ij} \right) u \\
&\ge \frac{\mu}{2}(\alpha - t)^{-2\lambda}|\nabla u|^2 - [\lambda + C(\alpha - t)](\alpha - t)^{-2\lambda - 1}u^2 \\
&\quad + \left[\frac{1}{2}(\alpha - t)^{-2\lambda}u^2\right] - (\alpha - t)^{-2\lambda}\sum_{i,j=1}^{n}(a^{ij}u_i u)_j,
\end{aligned}$$

where C is a constant depending on the coefficients $a^{ij}(t,x)$. (Just as in the preceding section, we use a single symbol C to denote various constants depending on the coefficients of the basic operator.)

PROOF. We have

$$
\begin{aligned}
&(\alpha-t)^{-2\lambda}\left(u_t-\sum_{i,j=1}^{n}a^{ij}u_{ij}\right)u \\
&= -\lambda(\alpha-t)^{-2\lambda-1}u^2+(\alpha-t)^{-2\lambda}\sum_{i,j=1}^{n}a^{ij}u_iu_j \\
&\quad +(\alpha-t)^{-2\lambda}\sum_{i,j=1}^{n}a_j^{ij}u_iu+\left[\frac{1}{2}(\alpha-t)^{-2\lambda}u^2\right]-(\alpha-t)^{-2\lambda}\sum_{i,j=1}^{n}(a^{ij}u_iu)_j \\
&\geq \mu(\alpha-t)^{-2\lambda}|\nabla u|^2-\lambda(\alpha-t)^{-2\lambda-1}u^2-C(\alpha-t)^{-2\lambda}|\nabla u|\,|u| \\
&\quad +\left[\frac{1}{2}(\alpha-t)^{-2\lambda}u^2\right]_t-(\alpha-t)^{-2\lambda}\sum_{i,j=1}^{n}(a^{ij}u_iu)_j.
\end{aligned}
$$

Applying the inequality

$$
-C|\nabla u|\,|u|\geq -\frac{\mu}{2}|\nabla u|^2-\frac{C^2}{2\mu}|u|^2
$$

to the product $-C|\nabla u|\,|u|$, we obtain

$$
\begin{aligned}
&(\alpha-t)^{-2\lambda}\left(u_i-\sum_{i,j=1}^{n}a^{ij}u_{ij}\right)u \\
&\geq \frac{\mu}{2}(\alpha-t)^{-2\lambda}|\nabla u|^2-[\lambda+C(\alpha-t)](\alpha-t)^{-2\lambda-1}u^2 \\
&\quad +\left[\frac{1}{2}(\alpha-t)^{-2\lambda}u^2\right]_t-(\alpha-t)^{-2\lambda}\sum_{i,j=1}^{n}(a^{ij}u_iu)_j
\end{aligned}
$$

with some new constant C. The lemma is proved.

LEMMA 2.

$$
\begin{aligned}
(\alpha-t)^{-2\lambda}\left(u_t-\sum_{i,j=1}^{n}a^{ij}u_{ij}\right)^2 &\geq -C(\alpha-t)^{-2\lambda}|\nabla u|^2+\lambda(\alpha-t)^{-2\lambda-2}u^2 \\
&\quad +\left\{(\alpha-t)^{-2\lambda}\left[-\lambda(\alpha-t)^{-1}u^2+\sum_{i,j=1}^{n}a^{ij}u_iu_j\right]\right\}_t \\
&\quad -2(\alpha-t)^{-2\lambda}\sum_{i,j=1}^{n}\{a^{ij}u_i[u_t+\lambda(\alpha-t)^{-1}u]\}_j.
\end{aligned}
$$

PROOF. We introduce the function

$$
v(t,x)=(\alpha-t)^{-\lambda}u(t,x).
$$

Then

$$u_t = (\alpha - t)^\lambda v_t = \lambda(\alpha - t)^{\lambda-1} v.$$

Hence

$$(\alpha - t)^{-2\lambda}\left(u_t - \sum_{i,j=1}^n a^{ij}u_{ij}\right)^2 = \left[v_t - \lambda(\alpha-t)^{-1}v - \sum_{i,j=1}^n a^{ij}v_{ij}\right]^2$$

$$\geq v_t^2 - 2\lambda(\alpha - t)^{-1}v_t v - 2v_t\sum_{i,j=1}^n a^{ij}v_{ij} = v_t^2 + \lambda(\alpha-t)^{-2}v^2$$

$$+ \sum_{i,j=1}^n(-a_t^{ij}v_iv_j + 2a_j^{ij}v_iv_t) - [\lambda(\alpha-t)^{-1}v^2]_t$$

$$+ \sum_{i,j=1}^n [(a^{ij}v_iv_j)_t - 2(a^{ij}v_iv_t)_j]$$

$$\geq v_t^2 + \lambda(\alpha-t)^{-2}v^2 - C|\nabla v|^2 - C|\nabla v|\,|v_t| - [\lambda(\alpha-t)^{-1}v^2]_t$$

$$+ \sum_{i,j=1}^n [(a^{ij}v_iv_j)_t - 2(a^{ij}v_iv_t)_j].$$

Applying the inequality $-C|\nabla v|\,|v_t| \geq -v_t^2 - C^2|\nabla v|^2/4$ to the product $-C|\nabla v|\,|v_t|$, we obtain

$$(\alpha - t)^{-2\lambda}\left(u_t - \sum_{i,j=1}^n a^{ij}u_{ij}\right)^2 \geq -C|\nabla v|^2 + \lambda(\alpha-t)^{-2}v^2$$

$$-[\lambda(\alpha-t)^{-1}v^2]_t + \sum_{i,j=1}^n [(a^{ij}u_iv_j)_t - 2(a^{ij}v_iv_t)_j]$$

with some new constant C.

We return to the function $u(t,x)$. We obtain

$$(\alpha - t)^{-2\lambda}\left(u_t - \sum_{i,j=1}^n a^{ij}u_{ij}\right)^2 \geq -C(\alpha-t)^{-2\lambda}|\nabla u|^2 + \lambda(\alpha-t)^{-2\lambda-2}u^2$$

$$+\left\{(\alpha-t)^{-2\lambda}\left[-\lambda(\alpha-t)^{-1}u^2 + \sum_{i,j=1}^n a^{ij}u_iu_j\right]\right\}_t$$

$$-2(\alpha-t)^{-2\lambda}\sum_{i,j=1}^n\{a^{ij}u_i[u_t + \lambda(\alpha-t)^{-1}u]\}_j.$$

LEMMA 3. *If the function $u(t,x)$ satisfies conditions (4.12) and (4.14), then there exists a constant $\sigma > 0$, depending on the coefficients $a^{ij}(t,x)$,* *such that for* $\lambda \geq 1$ *and* $\alpha + T \leq \sigma$,

$$(\alpha - t)^{-2\lambda}|\nabla u|^2 + \frac{\lambda}{2}(\alpha - t)^{-2\lambda-2}u^2$$
$$\leq \left\{(\alpha - t)^{-2\lambda}\left[-\sum_{i,j=1}^{n} a^{ij}u_iu_j + \left[\lambda(\alpha - t)^{-1} - C\right]u^2\right]\right\}_t$$
$$+2(\alpha - t)^{-2\lambda}\sum_{i,j=1}^{n}\{a^{ij}u_i[u_t + [\lambda(\alpha - t)^{-1} - C]u]\}_j.$$

PROOF. We multiply the inequality of Lemma 1 by $\theta > 0$ and add it to the inequality of Lemma 2:

$$(\alpha - t)^{-2\lambda}\left(u_t - \sum_{i,j=1}^{n} a^{ij}u_{ij}\right)^2 + \theta(\alpha - t)^{-2\lambda}\left(u_t - \sum_{i,j=1}^{n} a^{ij}u_{ij}\right)u$$
$$\geq \left(\frac{\theta\mu}{2} - C\right)(\alpha - t)^{-2\lambda}|\nabla u|^2$$
$$+ [\lambda - \theta\lambda(\alpha - t) - \theta C(\alpha - t)^2](\alpha - t)^{-2\lambda-2}u^2$$
$$+ \left\{(\alpha - t)^{-2\lambda}\left[\sum_{i,j=1}^{n} a^{ij}u_iu_j - \left[\lambda(\alpha - t)^{-1} - \frac{\theta}{2}\right]u^2\right]\right\}_t$$
$$- 2(\alpha - t)^{-2\lambda}\sum_{i,j=1}^{n}\left\{a^{ij}u_i\left[u_t + \left[\lambda(\alpha - t)^{-1} + \frac{\theta}{2}\right]u\right]\right\}_j.$$

Since

$$\left(u_t - \sum_{i,j=1}^{n} a^{ij}u_{ij}\right)^2 + \theta\left(u_t - \sum_{i,j=1}^{n} a^{ij}u_{ij}\right)u$$
$$\leq a^2(|\nabla u| + |u|)^2 + \theta a(|\nabla u| + |u|)|u|$$
$$\leq 2a^2|\nabla u|^2 + \theta|\nabla u|\,|u| + (2a^2 + \theta a)|u|^2$$
$$\leq 2a^2|\nabla u|^2 + \theta\left(\frac{\mu}{4}|\nabla u|^2 + \frac{1}{\mu}u^2\right) + (2a^2 + \theta a)u^2$$
$$= \left(\frac{\theta\mu}{4} + 2a^2\right)|\nabla u|^2 + \left(\frac{\theta}{\mu} + \theta a + 2a^2\right)u^2,$$

from the preceding inequality we obtain

$$\left(\frac{\theta\mu}{4} - 2a^2 - C\right)(\alpha - t)^{-2\lambda}|\nabla u|^2 + \left[\lambda - \theta\lambda(\alpha - t) - \theta C(\alpha - t)^2\right.$$
$$\left. - \left(\frac{\theta}{\mu} + \theta a + 2a^2\right)(\alpha - t)^2\right](\alpha - t)^{-2\lambda-2}u^2$$

$$\le \left\{(\alpha-t)^{-2\lambda}\left[-\sum_{i,j=1}^{n} a^{ij}u_iu_j + \left[\lambda(\alpha-t)^{-1}-\frac{\theta}{2}\right]u^2\right]\right\}_t$$
$$+\,2(\alpha-t)^{-2\lambda}\sum_{i,j=1}^{n}\left\{a^{ij}u_i\left[u_t+\left[\lambda(\alpha-t)^{-1}+\frac{\theta}{2}\right]u\right]\right\}_j.$$

We set $\theta\mu/4 = 2a^2 + C + 1$, i.e., $\theta = 4\mu^{-1}(2a^2+C+1)$. Then from the last inequality we have

$$(\alpha-t)^{-2\lambda}|\nabla u|^2 + \lambda[1-C(\alpha-t)-C(\alpha-t)^2](\alpha-t)^{-2\lambda-2}u^2$$
$$\le \left\{(\alpha-t)^{-2\lambda}\left[-\sum_{i,j=1}^{n} a^{ij}u_iu_j + [\lambda(\alpha-t)^{-1}-C]u^2\right]\right\}_t$$
$$+2(\alpha-t)^{-2\lambda}\sum_{i,j=1}^{n}\{a^{ij}u_i[u_t+[\lambda(\alpha-t)^{-1}+C]u]\}_j,$$

where C is some new constant, depending on the coefficients $a^{ij}(t,x)$. Choosing α and T so small that $C[\alpha+T+(\alpha+T)^2]\le 1/2$, we arrive at the assertion of the lemma.

THEOREM 1. *Suppose conditions* (4.12)–(4.15) *are satisfied and*

$$|u(0,x)|\le\varepsilon;\qquad |u(-T,x)|,\ |\nabla u(-T,x)|\le 1\quad (x\in\Omega).$$

If $\alpha+T\le\sigma$ *and* $0<\tau<T$, *then*

$$\frac{1}{\operatorname{meas}\Omega}\int_{\mathrm{Cyl}(\tau)} u^2\,d\tau\,dx \le C\exp\left[2\left(1-\frac{\ln(\alpha+\tau)-\ln\alpha}{\ln(\alpha+T)-\ln\alpha}\right)\ln\varepsilon\right].$$

PROOF. Integrating the inequality of Lemma 3 and using condition (4.15), we obtain

$$\frac{\lambda}{2}\int_{\mathrm{Cyl}(T)}(\alpha-t)^{-2\lambda-2}u^2\,dt\,dx \le \lambda\alpha^{-2\lambda-1}\int_\Omega u^2(0,x)\,dx$$
$$+(\alpha+T)^{-2\lambda}\int_\Omega\sum_{i,j=1}^{n}a^{ij}(-T,x)u_i(-T,x)u_j(-T,x)\,dx$$
$$+(\alpha+T)^{-2\lambda}C\int_\Omega u^2(-T,x)\,dx \le C\operatorname{meas}\Omega[\lambda\alpha^{-2\lambda-1}\varepsilon^2+(\alpha+T)^{-2\lambda}],$$

whence

$$\int_{\mathrm{Cyl}(T)}(\alpha-t)^{-2\lambda-2}u^2\,dt\,dx \le C\operatorname{meas}\Omega[\alpha^{-2\lambda-1}\varepsilon^2+(\alpha+T)^{-2\lambda}].$$

If $0 < \tau < T$, then

$$(\alpha+\tau)^{-2\lambda-2}\int_{\mathrm{Cyl}(\tau)} u^2\,dt\,dx \le \int_{\mathrm{Cyl}(\tau)} (\alpha-t)^{-2\lambda-2}u^2\,dt\,dx$$
$$\le \int_{\mathrm{Cyl}(T)} (\alpha-t)^{-2\lambda-2}u^2\,dt\,dx.$$

Hence,

$$\int_{\mathrm{Cyl}(\tau)} u^2\,dt\,dx \le C\operatorname{meas}\Omega\left[\frac{(\alpha+\tau)^{2\lambda+2}}{\alpha^{2\lambda+1}}\varepsilon^2 + \frac{(\alpha+\tau)^{2\lambda+2}}{(\alpha+T)^{2\lambda}}\right]$$
$$\le C\operatorname{meas}\Omega\left[\left(\frac{\alpha+\tau}{\alpha}\right)^{2\lambda+1}\varepsilon^2 + \left(\frac{\alpha+\tau}{\alpha+T}\right)^{2\lambda+1}\right].$$

Assuming ε sufficiently small and choosing λ from the condition

$$\left(\frac{\alpha+\tau}{\alpha}\right)^{2\lambda+1}\varepsilon^2 = \left(\frac{\alpha+\tau}{\alpha+T}\right)^{2\lambda+1},$$

we obtain the inequality

$$\int_{\mathrm{Cyl}(\tau)} u^2\,dt\,dx \le C\operatorname{meas}\Omega\exp\left[2\left(1-\frac{\ln(\alpha-\tau)-\ln\alpha}{\ln(\alpha+T)-\ln\alpha}\right)\ln\varepsilon\right].$$

The theorem is proved.

If the right side of (4.12) is zero, the coefficients do not depend on t, and the operator is selfadjoint in the spatial variables, then it is convenient to interpret our problem as a Cauchy problem for an operator differential equation in a suitable space. We consider the question of approximately solving this problem in just this form.

Let A be a selfadjoint positive operator acting in a Hilbert space H, and let $u(t)$ be a function with values in H which is defined and continuously differentiable in the norm of H on the interval $[0,T]$. If the operator A is unbounded (and only this case is of interest to us), then the Cauchy problem

$$\frac{du}{dt} = Au, \quad u(0) = u_0,\ 0 \le t \le T, \tag{4.16}$$

is not well-posed in the Hadamard sense. However, from the existence of the a priori estimate [135]

$$\|u(t)\| \le \|u(0)\|^{1-t/T}\|u(T)\|^{t/T} \tag{4.17}$$

it follows that this problem is conditionally well-posed in a class of solutions bounded in norm, say, by a constant M:

$$\|u(t)\| \le M \qquad \text{for } t \in [0,T]. \tag{4.18}$$

(The stability properties of the original mixed problem and its abstract analogue are thus completely analogous.) In the case where only some approximation u_0^δ to the initial value u_0 is known,

$$\|u_0^\delta - u_0\| \le \delta, \tag{4.19}$$

the solution of problem (4.16) consists in constructing a regularizing family of operators $R(t,\delta)$ with the following properties:

$$\begin{cases} \text{each of the operators } R(t,\delta) \text{ is bounded in } H; \\ \|R(t,\delta)u_0^\delta - u(t)\| \to 0, \;\; t \in [0,T), \;\; \text{as } \delta \to 0. \end{cases} \tag{4.20}$$

Suppose that for $t \in [0,T)$

$$\mu(t) = (1 - t/T)^{-2+t/T}\frac{\delta}{M}, \qquad R(t,\delta) = e^{At}[I + \mu(t)e^{AT}]^{-t/T} \tag{4.21}$$

(I is the identity operator). It is easy to see that the operators $R(t,\delta)$ are bounded in H. We show that (4.20) holds. Using the spectral representation for functions of the operator A, we obtain

$$\begin{aligned} \|u(t) - R(t,\delta)u_0^\delta\| &= \|e^{At}u_0 - R(t,\delta)u_0^\delta\| \\ &\le \|[e^{A(t-T)} - e^{-AT}R(t,\delta)]e^{AT}u_0\| + \|R(t,\delta)(u_0 - u_0^\delta)\| \\ &\le \max_{0\le\lambda<\infty} e^{\lambda(t-T)}\{1 - [1 + \mu(t)e^{\lambda T}]^{-t/T}\}\|e^{AT}u_0\| \\ &\quad + \max_{0\le\lambda<\infty} e^{\lambda t}[1 + \mu(t)e^{\lambda T}]^{-t/T}\|u_0 - u_0^\delta\|. \end{aligned}$$

To estimate the first term we use the inequality

$$(1+x)^\tau - 1 \le \tau x(1+x)^\tau(1+\tau x)^{-1},$$

which for $x \ge 0$ and $\tau \in [0,1]$ is easily justified by differential calculus. Noting also that

$$\|e^{AT}u_0\| = \|u(T)\| \le M,$$

we obtain

$$\begin{aligned} &\max_{0\le\lambda<\infty} e^{\lambda(t-T)}\{1 - [1 + \mu(t)e^{\lambda T}]^{-t/T}\}\|e^{AT}u_0\| \\ &\quad\le [\mu(t)]^{1-t/T} \max_{\mu(t)\le x<\infty} x^{t/T-1}[1 - (1+x)^{-t/T}]M \\ &\quad\le [\mu(t)]^{1-t/T} \max_{\mu(t)\le x<\infty} (t/T)x^{t/T}(1 + tx/T)^{-1}M \\ &\quad\le (t/T)(1 - t/T)^{1-t/T}[\mu(t)]^{1-t/T}M. \end{aligned}$$

Finding the maximum of the function in the second term, we have

$$\|R(t,\delta)u_0^\delta - u(t)\| \le (t/T)(1 - t/T)^{1-t/T}[\mu(t)]^{1-t/T}M + [\mu(t)]^{-t/T}\delta,$$

or

$$\|R(t,\delta)u_0^\delta - u(t)\| \le (1-t/T)^{-(1-t/T)^2}\delta^{1-t/T}M^{t/T}. \qquad (4.22)$$

The function $(1-t/T)^{-(1-t/T)^2}$ achieves its maximum value $e^{1/2e} \approx 1.21$ for $t/T = 1 - 1/\sqrt{e} \approx 0.39$. It is obvious that (4.20) is satisfied. Thus, the operators $R(t,\delta)$ defined by (4.21) form a regularizing family. Here the estimate (4.22) of the deviation of the approximate solution of problem (4.16) from the exact solution differs only inessentially from the a priori estimate (4.17). In this sense our method of solving problem (4.16) is optimal.

We consider the question of calculating the elements $R(t,\delta)u_{u_0}^\delta$. Let $Q_n(x;\mu,s)$ be the nth polynomial of best approximation to the function $(x+\mu)^{-s}$ on the segment $0 \le x \le 1$. To calculate

$$R(t,\delta)u_0^\delta = [e^{-AT} + \mu(t)]^{t/T}u_0^\delta,$$

it is expedient to set

$$R(t,\delta)u_0^\delta \approx Q_n[e^{-AT};\mu(t),t/T]u_0^\delta.$$

The operator on the right is a polynomial in e^{-AT}, and to calculate

$$Q_n[e^{-AT};\mu(t),t/T]u_0^\delta,$$

it suffices to be able to compute the elements $e^{-kAT}u_0^\delta$ for $k = 1, 2, \dots$. But $e^{-kAT}u_0^\delta$ is the solution of the well-posed Cauchy problem

$$\frac{dv}{dt} = -Av, \qquad v(0) = u_0^\delta$$

for $t = kT$. We have thus reduced solution of the ill-posed problem (4.16) to solution of a sequence of well-posed problems.

We shall estimate the error incurred hereby:

$$\|u(t) - Q_n[e^{-AT};\mu(t),t/T]u_0^\delta\| \le \|u(t) - R(t,\delta)u_0^\delta\| + \|R(t,\delta)u_0^\delta - Q_n[e^{-AT};\mu(t),t/T]u_0^\delta\|.$$

The second term does not exceed

$$\max_{0\le\lambda<\infty} |e^{\lambda t}[1+\mu(t)e^{\lambda T}]^{-t/T} - Q_n[e^{-\lambda T};\mu(t),t/T]|\,\|u_0^\delta\| = \max_{0\le x\le 1}|[x+\mu(t)]^{-t/T} - Q_n[x;\mu(t),t/T]|\,\|u_0^\delta\|.$$

As shown in Lemma 4 (see below),

$$\max_{0\le x\le 1}|(x+\mu)^{-s} - Q_n(x;\mu,s)| \le \frac{2^{2s-1}}{\Gamma(s)}(n+1)^{s-1}\frac{\alpha^{n+1+s}}{(1-\alpha^2)^{1+s}},$$

where $\alpha = 1 + 2\mu - 2\sqrt{\mu + \mu^2}$. Thus,

$$\|u(t) - Q_n[e^{-AT}; \mu(t), t/T]u_0^\delta\| \leq (1 - t/T)^{-(1+t/T)^2}\delta^{1-t/T}M^{t/T} + \frac{2^{2t/T+1}}{\Gamma(t/T)}(n+1)^{t/T-1}\frac{[\alpha(t)]^{n+1+t/T}}{[1-\alpha^2(t)]^{1+t/T}}\|u_0^\delta\|. \quad (4.23)$$

Using (4.23), it is easy to see that, for example, for $\delta\|u_0^\delta\|^{-1} = 0.1$ and $\|u_0^\delta\|M^{-1} = 0.1$, to obtain the accuracy $2\delta^{1-t/T}M^{t/T}$, it suffices to use the polynomial Q_2; if, however, these two quantities are equal to 0.05 and 0.1 respectively, then to obtain the same accuracy it suffices to use the polynomial Q_4, i.e, in cases of interest in practice the degree of the polynomials Q_n is small.

Since $(x + \mu)^{-s} = 2^s(a - z)^{-s}$, where $z = 1 - 2x$ and $a = 1 + 2\mu$, it follows that

$$Q_n(x; \mu, s) = 2^s P_n(z; a, s).$$

Here $P_n(z; a, s)$ is the nth polynomial of best approximation to the function $(a - z)^{-s}$, where $-1 \leq z \leq 1$, $a > 1$, and $0 \leq s < 1$. S. N. Bernstein ([45], pp. 90–96) established an estimate for the deviation of $P_n(z; a, s)$ from $(a - z)^{-s}$ which is asymptotic with respect to n:

$$\max_{-1\leq z\leq 1}|(a-z)^{-s} - P_n(z; a, s)| \sim \frac{(n+1)^{s-1}(a - \sqrt{a^2-1})^n}{\Gamma(s)(a^2-1)^{(s+1)/2}}. \quad (4.24)$$

Later ([29], Russian pp. 325–326) the following estimate was obtained:

$$\max_{-1\leq z\leq 1}|(a-z)^{-s} - P_n(z; a, s)| \leq \frac{(n+1)^{s-1}}{\Gamma(s)}\frac{(a - \sqrt{a^2-1})^n}{(a^2-1)^{(s+1)/2}}\left[1 + \frac{5(1-s)}{2(n+2)\sqrt{a^2-1}}\right],$$

which is satisfied for $n > 2 + 5(1-s)/(2\sqrt{a^2-1})$. For our purposes the latter estimate is useless, since we are interested in the case of small n and a close to one. It turns out, however, that the following lemma holds.

LEMMA 4. *In S. N. Bernstein's estimate* (4.24) *the symbol* $\sim$ *can be replaced by the sign* $\leq$ *for all* $n = 0, 1, 2\dots$.

PROOF. Below we use a number of formulas which can be found in Akhiezer's book just cited.

The polynomials $P_n(z; a, s)$ can be expressed in terms of the polynomials $L_n(z; a)$ of best approximation to the function $(z - a)^{-1}$ in the following way:

$$P_n(z; a, s) = \frac{\sin \pi s}{\pi}\int_{+\infty}^{a} L_n(z; t)(t-a)^{-s}\,dt. \quad (4.25)$$

As is known,

$$(z-t)^{-1} - L_n(z;t) = M(\tau)\Phi(v,\tau), \tag{4.26}$$

where

$$v = e^{i\theta} = z - \sqrt{z^2-1}, \qquad \theta \geq 0, \qquad \tau = t - \sqrt{t^2-1},$$
$$M(\tau) = \frac{4\tau^{n+2}}{(1-\tau^2)^2}, \qquad \Phi(v,\tau) = \frac{1}{2}\left(v^n \frac{\tau - v}{1-\tau v} + v^{-n}\frac{1-\tau v}{\tau - v}\right).$$

It is not hard to verify that

$$\Phi(v,\tau) = -\frac{\tau^2\cos(n-1)\theta - 2\tau\cos n\theta + \cos(n+1)\theta}{\tau^2 - 2\tau\cos\theta + 1}.$$

Multiplying (4.26) by

$$-\frac{dt}{(t-a)^s} = \frac{2^{s-1}\alpha^s(1-\tau^2)\,dt}{(\alpha-\tau)^s(1-\alpha\tau)^s\tau^{2-s}},$$

where $\alpha = a - \sqrt{a^2-1}$ and integrating with respect to t from $+\infty$ to a, we obtain

$$\int_{+\infty}^a \frac{L_n(z;t)}{(t-a)^s}\,dt - \int_{+\infty}^a \frac{dt}{(z-t)(t-a)^s}$$
$$= 2^{s+1}\alpha^s \int_0^\alpha \frac{1-\tau^2}{(\alpha-\tau)^s(1-\alpha\tau)^s\tau^{2-s}} M(\tau)\Phi(v,\tau)\,d\tau,$$

or

$$\int_{+\infty}^a \frac{L_n(z;t)}{(t-a)^s}\,dt - \frac{\pi}{\sin\pi s}\frac{1}{(a-z)^s}$$
$$= -2^{s+1}\alpha^s \int_0^\alpha \frac{\tau^2\cos(n-1)\theta - 2\tau\cos n\theta + \cos(n+1)\theta}{\tau^2 - 2\tau\cos\theta + 1}$$
$$\times \frac{\tau^{n+s}(\alpha-\tau)^s}{(1-\alpha\tau)^s(1-\tau^2)}\,d\tau.$$

From this and (4.25) we get

$$P_n(z;a,s) - (a-z)^{-s} = -2^{s+1}\alpha\frac{\sin\pi s}{\pi}$$
$$\times \int_0^\alpha \frac{\tau^2\cos(n-1)\theta - 2\tau\cos n\theta + \cos(n+1)\theta}{\tau^2 - 2\tau\cos\theta + 1}\,\frac{\tau^{n+s}(\alpha-\tau)^{-s}}{(1-\alpha\tau)^s(1-\tau^2)}\,d\tau \tag{4.27}$$

and hence

$$\max_{-1\leq z\leq 1}|P_n(z;a,s) - (a-z)^{-s}| \leq 2^{s+1}\alpha\frac{\sin\pi s}{\pi}$$
$$\times \int_0^\alpha \max_{\theta\geq 0}\left|\frac{\tau^2\cos(n-1)\theta - 2\tau\cos n\theta + \cos(n+1)\theta}{\tau^2 - 2\tau\cos\theta + 1}\right|\frac{\tau^{n+s}(\alpha-\tau)^{-s}}{(1-\alpha\tau)^s(1-\tau^2)}\,d\tau.$$

We shall establish the inequalities

$$-1 \leq \frac{\tau^2 \cos(n-1)\theta - 2\tau \cos n\theta + \cos(n+1)\theta}{\tau^2 - 2\tau\cos\theta + 1} \leq 1.$$

Since $\tau^2 - 2\tau\cos + 1 \geq 0$, it suffices to show that

$$\begin{aligned} -\left(\tau^2 - 2\tau\cos\theta + 1\right) &\leq \tau^2\cos(n-1)\theta - 2\tau\cos n\theta + \cos(n+1)\theta \\ &\leq \tau^2 - 2\tau\cos + 1. \end{aligned} \tag{4.28}$$

The left inequality is equivalent to

$$[1+\cos(n-1)\theta]\tau^2 - 2(\cos n\theta + \cos\theta)\tau + 1 + \cos(n+1)\theta \geq 0. \tag{4.29}$$

The discriminant of the trinomial (in τ) on the left side is

$$\begin{aligned} &4(\cos n\theta + \cos\theta)^2 - 4[1+\cos(n-1)\theta][1+\cos(n+1)\theta] \\ &\quad = 4(\cos n\theta + \cos\theta)^2 - 4[1 + \cos n\theta\cos\theta + \sin n\theta \sin\theta] \\ &\qquad \times [1 + \cos n\theta\cos\theta - \sin n\theta\sin\theta] \\ &\quad = 4(\cos n\theta + \cos\theta)^2 - 4[(1+\cos n\theta\cos\theta)^2 - \sin^2 n\theta \sin^2\theta] \\ &\quad = 4(\cos n\theta + \cos\theta)^2 - 4[(1+\cos n\theta\cos\theta)^2 \\ &\qquad\qquad - (1-\cos^2 n\theta)(1-\cos^2\theta)] \\ &\quad = 4(\cos n\theta + \cos\theta)^2 - 4(2\cos n\theta\cos\theta + \cos^2 n\theta + \cos^2\theta) = 0, \end{aligned}$$

and, since the coefficient of τ^2 on the left side of (4.29) is nonnegative, (4.29) holds for all n and θ. The right inequality of (4.28) is proved similarly.

Thus,

$$\max_{-1\leq z\leq 1} |P_n(z;a,s) - (a-z)^{-s}| = 2^{s+1}\alpha^s \frac{\sin\pi s}{\pi} \int_0^\alpha \frac{\tau^{n+s}(\alpha-\tau)^{-s}}{(1-\alpha\tau)^s(1-\tau^2)}\,d\tau. \tag{4.30}$$

(It is possible to verify that the given maximum value is attainable, by setting $\theta = 0$ in (4.27).) Noting that

$$(1-\alpha\tau)^{-s}(1-\tau^2)^{-1} \leq (1-\alpha^2)^{-s-1},$$

from (4.30) we obtain the required estimate

$$\begin{aligned} \max_{-1\leq z\leq 1} |P_n(z;a,s) - (a-z)^{-s}| &\leq \frac{2^{s+1}}{\Gamma(s)}(n+1)^{s-1}\frac{\alpha^{n+s+1}}{(1-\alpha^2)^{s+1}} \\ &= \frac{(n+1)^{s-1}}{\Gamma(s)} \frac{(a+\sqrt{a^2-1})^n}{(a^2-1)^{(s+1)/2}}, \end{aligned}$$

valid for all $n = 0, 1, \ldots$. The lemma is proved.

To construct the polynomials $Q_n(x;\mu,s)$ the following device can be suggested. Proceeding from the formula ([256], §2.11.1)

$$L_n(z;t) = \frac{1}{z-t} - \frac{\cos[n\arccos z + \varphi(z)]}{(t^2-1)(t+\sqrt{t^2-1})^n}, \tag{4.31}$$

where $\varphi(z) = \arccos((tz-1)/(z-t))$, it can be shown that the following recursion formula holds:

$$L_{n+1}(z;t) = \frac{2}{t+\sqrt{t^2-1}} zL_n(z;t) - \frac{1}{t+\sqrt{t^2-1}} L_{n-1}(z;t) - \frac{2}{t\sqrt{t^2-1}}. \tag{4.32}$$

Setting $n = 0, 1, \dots$ in (4.31), we obtain

$$L_0(z;t) = -\frac{t}{t^2-1}, \tag{4.33}$$

$$L_1(z;t) = -\frac{1}{t^2-1} z - \frac{1}{\sqrt{t^2-1}}. \tag{4.34}$$

Formulas (4.32)–(4.34) make it possible to successively compute the polynomials $L_n(z;t)$ and with them also $P_n(z;a,s)$ and $Q_n(x;\mu,s)$. We shall not present the very cumbersome general formula for the coefficients of $Q_n(x;\mu,s)$. We observe only that these coefficients are linear combinations of the integrals

$$J_1(a;s,m) = \frac{\sin \pi s}{\pi} \int_{+\infty}^{a} \frac{dt}{(t+\sqrt{t^2-1})^m (t-a)^s},$$

$$J_2(a;s,m) = \frac{\sin \pi s}{\pi} \int_{+\infty}^{a} \frac{dt}{(t^2-1)(t+\sqrt{t^2-1})^m (t-a)^s},$$

where $a = 1+2\mu$, $s = tT^{-1}$ and $m = 1, \dots, n$.

§3. Cauchy problems with data on a segment of the time axis for degenerate parabolic and pseudoparabolic equations

We consider the question of uniqueness of the solution of the Cauchy problem

$$u(0,t) = u_x(0,t) = 0 \tag{4.35}$$

for the inequality

$$|A(x,t)u_{xx} - u_t| \le B_1(x)|u_x| + B_2(x)|u|, \tag{4.36}$$

where $(x,t) \in \Pi = \{(x,t) \colon 0 \le x \le X, -T \le t \le T\}$ and $A(x,t) > 0$ for $x > 0$. We assume that the coefficients of (4.36) may have singularities of the following type as $x \to 0$:

$$A(x,t) \to 0; \quad B_1(x),\ B_2(x) \to \infty,$$

and we shall indicate conditions on the functions $A(x,t)$, $B_1(x)$, and $B_2(x)$ guaranteeing uniqueness of the solution of problem (4.35), (4.36).

THEOREM 1. *Suppose that the following conditions hold:*

1) *the function* $\alpha(x) > 0$ *is continuously differentiable on* $(0, X)$, *and* $x\alpha^{-1}(x)\alpha'(x) \to 0$ *as* $x \to 0$;

2) $A(x,t) = x^\rho \alpha(x) a(x,t)$, *where* $0 \le \rho < 2$, $a(x,t) \ge a_0 > 0$ *and* $a(x,t)$ *is continuously differentiable on* Π;

3) $B_1(x) = o(x^{\rho-1}\alpha(x))$ *and* $B_2(x) = o(x^{\rho-2}\alpha(x))$ *as* $x \to 0$;

4) *the function* $u(x,t)$ *has continuous derivatives* u_x, u_t, *and* u_{xx}, *and satisfies inequality* (4.36) *on* Π.

If, moreover, for $t \in [-T, T]$ *condition* (4.35) *is satisfied, then* $u(x,t) = 0$ *on* Π.

REMARK 1. Because of 1), for any $\mu > 0$ we have

$$\lim_{x\to 0} x^\mu \alpha(x) = \lim_{x \to 0} x^\mu \alpha^{-1}(x) = 0.$$

REMARK 2. It is not possible to replace the symbol o by O in condition 3) while preserving uniqueness: the equation

$$x^\rho u_{xx} - u_t = (3-\rho) x^{\rho-1} u_x - u$$

has the solution $u = x^{-\rho+4} e^t$ satisfying the conditions of the theorem. The condition $\rho < 2$ is also essential: the equation $x^2 u_{xx} - u_t = 0$ obviously has nontrivial analytic solutions satisfying (4.35).

The proof of the theorem is based on the following proposition.

LEMMA 1. *Suppose that the following conditions hold:*

1) *the function* $u(x,t)$ *has a continuous derivative* u_x *on* Π *and continuous derivatives* u_t *and* u_{xx} *for* $x > 0$;

2) *for some* $M > 0$ *and* $\delta > 0$

$$u^2 \le Mx^{3+\delta}, \qquad u_x^2 \le Mx^{1+\delta}, \qquad |uu_t| \le Mx^{1+\delta}, \qquad |u_x u_t| \le Mx^\delta;$$

3) *for* $(x,t) \in \Pi$

$$|L_\gamma u| = |u_{xx} + \gamma_1 x^{-1} u_x + \gamma_2 x^{-2} u - u_t| \le b(x)(x^{-1}|u_x| + x^{-2}|u|),$$

where $b(x) \to 0$ $(x \to 0)$, $\gamma_1 = 3\gamma + 6/7$, $\gamma_2 = (2\gamma + 3/7)(\gamma - 4/7)$, *and* γ *is an arbitrary nonnegative number.*

Then there exists $X_0, 0 < X_0 \le X$, *such that* $u(x,t) = 0$ *for* $x \in [0, X_0]$ *and* $t \in [-T, T]$.

PROOF OF THE LEMMA. Assuming $X < 1$, we set

$$r = r(x) = \ln(1/x), \qquad s = s(t) = (t^2 + 1/n)^{-1}, \quad n > 0,$$

and begin with the proof of the inequality

$$\int_\Pi (x^{-2}u_x^2 + x^{-4}u^2)(rs)^{2n}\,dx\,dt \le CM[n^{2n+3} r^{2n}(X) + n^2(2T^2)^{2-n}\Gamma(2n)], \tag{4.37}$$

valid for $X < \overline{X}$ and $n \geq \underline{n}$, where $\overline{X}, \underline{n}$, and C are positive constants depending on $\gamma; \Gamma(2n)$ is the gamma function.

Let $0 < \varepsilon < X$ and $\Pi_\varepsilon = \{(x,t)\colon \varepsilon \leq x \leq X, -T \leq t \leq T\}$. Introducing the auxiliary function $v(x,t) = (rs)^n u(x,t)$, we have

$$\begin{aligned}
&\int_{\Pi_\varepsilon} (rs)^{2n}(L_\gamma u)^2\, dx\, dt \\
&\quad = \int_{\Pi_\varepsilon} \{v_{xx} + x^{-1}(\gamma_1 + 2nr^{-1})v_x \\
&\qquad\qquad + [x^{-2}(\gamma_2 + (\gamma_1 - 1)nr^{-1} + n(n+1)r^{-2}) + ns^{-1}s']v - v_t\}^2\, dx\, dt \\
&\quad \geq 2\int_{\Pi_\varepsilon} x^{-1}(\gamma_1 + 2nr^{-1})v_{xx}v_x\, dx\, dt - 2\int_{\Pi_\varepsilon} v_{xx}v_t\, dx\, dt \\
&\qquad + \int_{\Pi_\varepsilon} x^{-1}(\gamma_1 + 2nr^{-1}) \\
&\qquad\qquad \times \{x^{-2}[\gamma_2 + (1-\gamma_1)nr^{-1} + n(n+1)r^{-2}] + ns^{-1}s'\}v_x v\, dx\, dt \\
&\qquad - \int_{\Pi_\varepsilon} \{x^{-2}[\gamma_2 + (1-\gamma_1)nr^{-1} + n(n+1)r^{-2}] + ns^{-1}s'\}v_t v\, dx\, dt \\
&\quad = \sum_{j=1}^{4} I_j
\end{aligned} \tag{4.38}$$

(we have used the inequality $(a+b+c+d)^2 \geq 2(ab+ad+bc+cd)$). We apply Green's formula to each of the integrals I_j to obtain integral forms containing only v_x^2 and v^2. Returning then to the function u, we obtain (we take the liberty of omitting the cumbersome but entirely trivial computations)

$$\begin{aligned}
&\int_{\Pi_\varepsilon} x^{-2}(\gamma_1 + 2nr^{-1} - 2nr^{-2})(rs)^{2n}u_x^2\, dx\, dt \\
&\qquad + \int_{\Pi_\varepsilon} \{x^{-4}[3\gamma_1\gamma_2 + (-6\gamma_1 + 3\gamma_1^2 + 6\gamma_2)nr^{-1} \\
&\qquad + [(-12 + 8\gamma_1)n^2r^{-2} + 4n^3r^{-3} - 4n^3r^{-4}](1 + O(1/n))] \\
&\qquad + x^{-2}(\gamma_1 n + 2n^2r^{-1} - 2n^2r^{-2})s^{-1}s' + n(-s^{-2}s'^2 + s^{-1}s'')\} \\
&\qquad\qquad\qquad\qquad \times (rs)^{2n}u^2\, dx\, dt \\
&\quad \leq \int_{\Pi_\varepsilon} (rs)^{2n}(L_\gamma u)^2\, dx\, dt - \sum_{j=1}^{6} i_j
\end{aligned} \tag{4.39}$$

(here and below the i_j are integrals over $\partial\Pi_\varepsilon$; all O's in the integrands are uniform with respect to $(x,t) \in \Pi$).

To prove inequality (4.37), we shall need, in addition to (4.39), the following two relations which are also easily proved by applying Green's formula to their left sides:

$$2n\int_{\Pi_\varepsilon} x^{-2}(r^{-1}-r^{-2})(rs)^{2n}uL_\gamma u\,dx\,dt$$
$$= -2n\int_{\Pi_\varepsilon} x^{-2}(r^{-1}-r^{-2})(rs)^{2n}u_x^2\,dx\,dt$$
$$+\int_{\Pi_\varepsilon}\{x^{-4}[(3\gamma_1+2\gamma_2+6)nr^{-1}$$
$$+((2\gamma_1+10)n^2r^{-2}+4n^3r^{-3}-4n^3r^{-4})(1+O(1/n))]$$
$$+2n^2x^{-2}(r^{-1}-r^{-2})s^{-1}s'\}(rs)^{2n}u^2\,dx\,dt+\sum_{j=7}^{9} i_j, \qquad (4.40)$$

$$\int_{\Pi_\varepsilon} x^{-2}(rs)^{2n}uL_\gamma u\,dx\,dt = -\int_{\Pi_\varepsilon} x^{-2}(rs)^{2n}u_x^2\,dx\,dt$$
$$+\int_{\Pi_\varepsilon}\left\{x^{-4}\left[\gamma_2+3\left(1+\frac{\gamma_1}{2}\right)+(\gamma_1+5)nr^{-1}\right.\right.$$
$$\left.\left.+2n^2r^{-2}\left(1+O\left(\frac{1}{n}\right)\right)\right]+nx^{-2}s^{-1}s'\right\}(rs)^{2n}u^2\,dx\,dt$$
$$+\sum_{j=10}^{12} i_j. \qquad (4.41)$$

We proceed to the proof of (4.37). We multiply (4.41) by $2\gamma+5/7$ and add termwise with (4.39) and (4.40). We obtain (using the definition of γ_1 and γ_2)

$$(\gamma+1/7)\int_{\Pi_\varepsilon} x^{-2}(rs)^{2n}u_x^2\,dx\,dt$$
$$+\int_{\Pi_\varepsilon}\{x^{-4}[22\gamma^3+7\gamma^2+6\gamma+2+(49\gamma^2+14\gamma+7)nr^{-1}$$
$$+((34\gamma+8)n^2r^{-2}+8n^3r^{-3}-8n^2r^{-4})(1+O(1/n))]$$
$$+x^{-2}((5\gamma+11/7)n+4n^2r^{-1}-4n^2r^{-2})s^{-1}s'$$
$$+n(-s^{-2}s'^2+s^{-1}s'')\}(rs)^{2n}u^2\,dx\,dt$$
$$\le \int_{\Pi_\varepsilon}(rs)^{2n}[(L_\gamma u)^2+x^{-2}(2\gamma+5/7+2nr^{-1}-2nr^{-2})uL_\gamma u]\,dx\,dt$$
$$-\sum_{j=1}^{9} i_j-(2\gamma+5/7)\sum_{j=10}^{12} i_j \qquad (4.42)$$

We shall obtain a lower bound for the coefficients of $x^{-2}(rs)^{2n}u_x^2$ and $(rs)^{2n}u^2$ in the first and second integrals on the left side of (4.42). We have

$$\gamma+1/7\ge 1/7 \qquad (4.43)$$

and, since $r^{-1}(x) = \ln^{-1}(1/x)$, for fixed γ for all sufficiently small x and sufficiently large n the second coefficient can be bounded below by the quantity

$$\begin{aligned} & x^{-4}(22\gamma^3 + 7\gamma^2 + 6\gamma + 2 + 7nr^{-1} + 7n^2r^{-2} + 7n^3r^{-3}) \\ & \quad - 2n^2x^{-2}\max_t |s^{-1}s'| \\ & \quad - n\max_t |-s^{-2}s'^2 + s^{-1}s''| \\ & = x^{-4}(22\gamma^3 + 7\gamma^2 + 6\gamma + 2 + 7nr^{-1} + 7n^2r^{-2} + 7n^3r^{-3}) \\ & \quad - 2n^{5/2}x^{-2} - 2n^2 \\ & > x^{-4}(22\gamma^3 + 7\gamma^2 + 6\gamma + 2 + 7nr^{-1} - 7n^2r^{-2} + 6n^3r^{-3}). \end{aligned} \tag{4.44}$$

By the Cauchy inequality, the integral over Π_ε on the right side of (4.42) can be bounded above by the quantity

$$\int_{\Pi_\varepsilon} x^{-4}\left[\frac{1}{2}(2\gamma + 5/7)^2 + n^2(r^{-1} - r^{-2})^2\right](rs)^{2n}u^2\,dx\,dt \\ + \frac{5}{2}\int_{\Pi_\varepsilon}(rs)^{2n}(L_\gamma u)^2\,dx\,dt. \tag{4.45}$$

Noting that

$$x^{-4}(22\gamma^3 + 7\gamma^2 + 6\gamma + 2 + 7nr^{-1} + 7n^2r^{-2} + 6n^3r^{-3}) \\ -x^{-4}\left[\frac{1}{2}(2\gamma + 5/7)^2 + n^2(r^{-1} - r^{-2})^2\right] > 1, \tag{4.46}$$

from (4.42)–(4.46) we obtain

$$\begin{aligned} & \int_{\Pi_\varepsilon}\left(\frac{1}{7}x^{-2}u_x^2 + x^{-4}u^2\right)(rs)^{2n}\,dx\,dt \\ & \quad \le \frac{5}{2}\int_{\Pi_\varepsilon}(rs)^{2n}(L_\gamma u)^2\,dx\,dt - \sum_{j=1}^{9} i_j - (2\gamma + 5/7)\sum_{j=10}^{12} i_j, \end{aligned}$$

and by condition 3) of Lemma 1

$$\begin{aligned} & \int_{\Pi_\varepsilon}\left(\frac{1}{8}x^{-2}u_x^2 + \frac{1}{2}x^{-4}u^2\right)(rs)^{2n}\,dx\,dt \\ & \quad \le -\sum_{j=1}^{9} i_j - (2\gamma + 5/7)\sum_{j=10}^{12} i_j, \end{aligned} \tag{4.47}$$

if X is sufficiently small and n sufficiently large. We estimate the boundary integrals by using condition 2) of the lemma:

$$
\begin{aligned}
-\sum_{j=1}^{9} i_j - (2\gamma + 5/7) \sum_{j=10}^{12} l_j \\
&\le C \int_{-T}^{T} [(nx^{-1}u_x^2 + |u_x u_t| + nx^{-1}|uu_t| + n^3x^{-3}u^2)r^{2n}(x)|_{x=\varepsilon} \\
&\qquad + |_{x=X}]s^{2n}(t)\,dt \\
&\quad + Cs^{2n}(T)\int_{\varepsilon}^{X} [(u_x^2 + n^2x^{-2}u^2)|_{t=-T} + |_{t=T}]r^{2n}(x)\,dx \\
&\le CM\left\{n^{2n+3}[\varepsilon^{\delta}r^{2n}(\varepsilon) + X^{\delta}r^{2n}(X)] + n^2(2T^2)^{-2n}\Gamma(2n)\right\} \qquad (4.48)
\end{aligned}
$$

(here and below we use the symbol C to denote various constants depending only on γ). Substituting (4.48) into (4.47) and passing to the limit as $\varepsilon \to 0$ in the resulting inequality, we obtain (4.37) for some $\overline{X} > 0$ and $\underline{n} > 0$.

The proof of the lemma can now easily be completed. Let $n > \max(\underline{n}, T^{-1})$ and let $X_0 \in (0, X)$ be such that

$$\max[2r(X)r^{-1}(X_0), 2e^{-1}T^{-2}r^{-1}(X_0)] = \theta < 1$$

(since $r(x) \to \infty$ as $x \to 0$, such an X_0 exists). From (4.37) it follows that

$$\int_0^{X_0}\int_{-1/n}^{1/n} x^{-4}(rs)^{2n}u^2\,dx\,dt \le CM[n^{2n+3}r^{2n}(X) + n^{2n+2}e^{-2n}T^{-4n}].$$

Since for $x \le X_0$ and $-1/n \le t \le 1/n$ we have $r(x) \ge r(X_0)$ and $s(t) > n/2$, from the last inequality

$$
\begin{aligned}
\int_0^{X_0}\int_{-1/n}^{1/n} u^2\,dx\,dt &\le CMX_0^4\{n^3[2r(X)r^{-1}(X_0)]^{2n} + n^2[2e^{-1}T^2r^{-1}(X_0)]^{2n}\} \\
&\le CMX_0^4(n^3 + n^2)\theta^{2n} = o(1/n),
\end{aligned}
$$

whence $\int_0^{X_0} u^2(x,0)\,dx = 0$. Since the conditions of the lemma are invariant relative to shifts in t, it follows that $u(x,t) = 0$ for $x \in [0, X_0]$ and $t \in [-T, T]$. The lemma is proved.

PROOF OF THEOREM 1. We shall show that if conditions 1)–4) are satisfied it is possible to introduce the new variables

$$\tau = t, \qquad y = y(x,t), \qquad z(y,\tau) = y^{-\nu}u(x,t) \qquad (4.49)$$

so that the function $z(y,\tau)$ satisfies the conditions of the lemma in some rectangle $\Pi(y,\tau) = \{(y,\tau)\colon 0 \le y \le Y, -T \le \tau \le T\}$.

We set

$$y(x,t) = \int_0^x A^{-1/2}(\xi,t)\,d\xi, \qquad Y = \max_t y(X,t).$$

For each t, y is a continuous and strictly increasing function of x, and $y \to 0$ as $x \to 0$. The function $x(y, \tau)$ inverse to it possesses the same properties. Using conditions 1) and 2) of the theorem and l'Hôpital's rule, we easily obtain

$$\begin{aligned} y(x,t) &= \int_0^x \xi^{-\rho/2}\alpha^{-1/2}a^{-1/2}\,d\xi \\ &= \frac{2a^{-1/2}(0,t)}{2-\rho}x^{-\rho/2+1}\alpha^{-1/2}(x)[1+o(1)](x \to 0). \end{aligned} \tag{4.50}$$

The derivatives $t_y = 0$, $t_\tau = 1$, $x_y = A^{1/2}(x,t)$ and

$$x_\tau = \frac{1}{2}A^{1/2}(x,t)\int_0^x A^{-3/2}A_t\,d\xi, \qquad x_{yy} = \frac{1}{2}A^{-1/2}(x,t)A_x(x,t)$$

are continuous on $\{0 < x \le X, -T \le t \le T\}$, and hence on $\Pi(y,\tau)$, and as $x \to 0$ we have

$$x_y = O[x^{\rho/2}\alpha^{1/2}(x)], \qquad x_\tau = O(x). \tag{4.51}$$

From condition 4) of the theorem and (4.35) it follows that

$$u(x,t) = O(x^2), \qquad u_x(x,t) = O(x), \qquad u_{xx}(x,t) = O(1), \tag{4.52}$$

and, comparing the orders of the left and right sides of (4.36), we obtain $u_t(x,t) = o[x^\rho\alpha(x)]$. Expressing u_y, u_τ, and u_{yy} in terms of u_x, u_t, u_{xx}, x_y, x_τ, and x_{yy}, from (4.51), (4.52), and the last equality we find that

$$u_y = O[x^{1+\rho/2}\alpha^{1/2}(x)], \qquad u_\tau = O[x^\rho\alpha(x)], \quad x \to 0,\ y \to 0, \tag{4.53}$$

where u_y, u_τ, and u_{yy} are continuous in y and τ for $y > 0$.

We make a change of variables in (4.36):

$$\begin{aligned} &\left|u_{yy} + \left[-\frac{1}{2}A^{-1/2}(x,t)A_x(x,t) + \frac{1}{2}A(x,t)\int_0^x A^{-3/2}A_t\,d\xi\right]u_y - u_\tau\right| \\ &\qquad \le B_1(x)A^{-1/2}(x,t)|u_y| + B_2(x)|u|. \end{aligned} \tag{4.54}$$

Proceeding from conditions 1)–3) of the theorem, it is not hard to show that

$$\begin{aligned} -\frac{1}{2}A^{-1/2}(x,t)A_x(x,t) &= -(\rho/2)x^{\rho/2-1}\alpha^{1/2}(x)a^{1/2}(0,t)[1+o(1)] \\ &= -\frac{\rho}{2-\rho}y^{-1}[1+o(1)], \\ \frac{1}{2}A(x,t)\int_0^x A^{-3/2}A_t\,d\xi &= O[x^{\rho/2+1}\alpha^{1/2}(x)] = o(y^{-1}), \\ B_1(x)A^{-1/2}(x,t) = o[x^{\rho/2-1}\alpha^{1/2}(x)] &= o(y^{-1}), \qquad B_2(x) = o(y^{-2}). \end{aligned}$$

By virtue of the last four equalities it follows from (4.54) that

$$|u_{yy} + \gamma y^{-1}u_y - u_\tau| \le b(y)(y^{-1}|u_y| + y^{-2}|u|), \tag{4.55}$$

where $\gamma = \rho/(2-\rho)$, $0 \le \gamma < \infty$, and $b(y) \to 0$ as $y \to 0$. We now set

$$z(y,\tau) = u(y,\tau)y^{-3/7-2\gamma}. \tag{4.56}$$

Using the first of equalities (4.52), equality (4.50), and Remark 1, we obtain

$$\begin{aligned} y^{-3}z^2(y,\tau) &= O(x^4 y^{-6/7-4\gamma-3}) \\ &= O[x^{4-(27/7+4\gamma)/(1+\gamma)}\alpha^{(27/7+4\gamma)/2}(x)] \\ &= O[x^{1/(7(1+\gamma))}\alpha^{(27/7+\gamma)/2}(x)] = O[x^{1/(8(1+\gamma))}] = O(y^{\delta}). \end{aligned}$$

Similarly, using (4.52), (4.53), and Remark 1, we can see that the remaining inequalities of condition 2) are satisfied. Finally, substitution of (4.56) into (4.55) leads to the inequality of condition 3) of Lemma 1 for z: the function $z(y,\tau)$ satisfies all the conditions of Lemma 1. Hence, there exists an $X_0 > 0$ such that $u(x,t) = 0$ for $0 \le x \le X_0$ and $-T \le t \le T$. The rectangle $[X_0, X] \times [-T, T]$ is exhausted by repeated application of the assertion proved. The theorem is proved.

REMARK. By analyzing the proof, it is not hard to see that condition 3) of Theorem 1 can be replaced by the following condition:

3′) $\quad |B_1(x)| < C_1 x^{\rho-1}\alpha(x), \qquad |B_2(x)| < C_2 x^{\rho-2}\alpha(x),$

where C_1 and C_2 are constants depending on ρ, a_0, and the function $\alpha(x)$.

In the next Cauchy problem we consider, the data are given on a characteristic. In constructing uniqueness classes for such problems, as a rule, it is assumed that the solution is defined in a half-space and satisfies some growth conditions. It is interesting that in our case we can give uniqueness classes for solutions defined only in a quarter-space which are reasonable from a physical viewpoint.

Thus, we consider the Cauchy problem

$$u(0,t) = f_1(t), \qquad u_x(0,t) = f_2(t) \tag{4.57}$$

for the pseudoparabolic equation

$$u_t = u_{xxt} + u_{xx}. \tag{4.58}$$

Since the line $x = 0$ is a characteristic, in the case where $u(0,t)$ and $u_x(0,t)$ are given only locally (i.e., on a compact set) the solution of this problem is not unique (this question is considered in detail in [326]). We shall assume that the solution of problem (4.57), (4.58) is defined in the quadrant $P = \{0 \le x < \infty, 0 \le t < \infty\}$, prove a uniqueness theorem in a certain class of solutions, and obtain an estimate of stability of the solution.

DEFINITION. A solution $u(x,t)$ of equation (4.58) *belongs to the class* $U(a,b)$ if the functions u, u_x, u_t, u_{xx}, u_{xt}, and u_{xxt} are continuous on P and

$$|u|, |u_x|, |u_t|, |u_{xx}|, |u_{xt}|, |u_{xxt}| \le C_u e^{ax+bt},$$

where the numbers a and b satisfy the following conditions:

a) if $a < 1$, then b is any real number;

b) if $a \geq 1$, then $b < -1 + [4a(a+\sqrt{a^2-1})]^{-1}$ (C_u is a constant depending on u).

The class $U(a,b)$ is nonempty. Indeed, suppose, for example, that $a \geq 1$. It is easy to verify that for any real α the function

$$u(x,t) = \operatorname{Re}\exp\left\{(a+i\alpha)x + \left[-1 + \frac{1-a^2+\alpha^2}{(1-a^2+\alpha^2)^2+4a^2\alpha^2} + \frac{2a\alpha}{(1-a^2+\alpha^2)^2+4a^2\alpha^2}i\right]t\right\}$$

is a solution of (4.58). If

$$\alpha \neq \pm\sqrt{a^2-1+2a\sqrt{a^2-1}},$$

then we have the inequality (see, for example, the proof of Lemma 2)

$$-1 + \frac{1-a^2+\alpha^2}{(1-a^2+\alpha^2)^2+4a^2\alpha^2} < -1 + [4a(a+\sqrt{a^2-1})]^{-1},$$

i.e.,

$$u(x,t) \in U\left[a, -1 + \frac{1-a^2+\alpha^2}{(1-a^2+\alpha^2)^2+4a^2\alpha^2}\right].$$

We note that as $\alpha \to \pm\sqrt{a^2-1+2a\sqrt{a^2-1}}$ we have

$$-1 + \frac{1-a^2+\alpha^2}{(1-a^2+\alpha^2)^2+4a^2\alpha^2} \to -1 + [4a(a+\sqrt{a^2-1})]^{-1}.$$

LEMMA 2. *Let*

$$f(p,s) = \frac{1-p^2+s^2}{(1-p^2+s^2)+4p^2s^2}.$$

Then

$$\sup_{\substack{p\geq a\\ -\infty<s<\infty}} f(p,s) = \begin{cases} [4a(a+\sqrt{a^2-1})]^{-1} & \textit{for } a \geq 1, \\ +\infty & \textit{for } a < 1. \end{cases}$$

PROOF. We shall show that the function $f(p,s)$ for $p > 0$ has no stationary points. The equalities $f_p(p,s) = f_s(p,s) = 0$ have the form

$$\frac{2p[(1-p^2+s^2)^2 - 4s^2(1+s^2)]}{[(1-p^2+s^2)+4p^2s^2]^2} = -\frac{2s[(1-p^2+s^2)^2+4p^2(1-p^2)]}{[(1-p^2+s^2)^2+4p^2s^2]^2} = 0.$$

Excluding from consideration the points of discontinuity $p = \pm 1, s = 0$ of the function $f(p,s)$, we obtain the equalities

$$(1-p^2+s^2)^2 = 4s^2(1+s^2), \tag{4.59}$$

$$(1 - p^2 + s^2)^2 = -4p^2(1 - p^2). \tag{4.60}$$

It follows from (4.60) that $p^2 > 1$ (the case $p^2 = 1$ again leads to the points of discontinuity $(\pm 1, 0)$). Suppose $1 - p^2 + s^2 \geq 0$. Then, by (4.59),

$$1 - p^2 + s^2 = 2|s|\sqrt{1 + s^2},$$

which is impossible, since

$$2|s|\sqrt{1 + s^2} \geq 2s^2 \geq s^2 > s^2 + 1 - p^2.$$

Suppose $1 - p^2 + s^2 < 0$. Then $s^2 < p^2 - 1$, $s^2 + 1 < p^2$, and hence $s^2(s^2 + 1) < p^2(p^2 - 1)$, while from (4.59) and (4.60) it follows that

$$s^2(s^2 + 1) = p^2(p^2 - 1).$$

Thus, the function $f(p, s)$ indeed has no stationary points in the half-plane $p > 0$. Since $f(-p, s) = f(p, s)$, it follows that $f(p, s)$ has no stationary points for $p < 0$ as well. It is easy to see that for sufficiently large $p^2 + s^2$ we have

$$|f(p, s)| < \frac{4}{p^2 + s^2}.$$

Thus, to determine $\sup f(p, s)$ in the half-plane $p \geq a$ it suffices for $a \geq 1$ to find $\max f(a, s)$, and for $a < 1$ to use the fact that as $p \to 1 - 0$ we have $f(p, 0) \to +\infty$. A simple computation shows that for $a \geq 1$

$$\max f(a, s) = f(a, \pm\sqrt{a^2 - 1 + 2a\sqrt{a^2 - 1}}) = [4a(a + \sqrt{a^2 - 1})]^{-1}.$$

The lemma is proved.

THEOREM 2. *Problem* (4.57), (4.58) *has at most one solution in the class* $U(a, b)$.

PROOF. We shall show that the conditions

$$f_1(t) = f_2(t) = 0 \quad (t \geq 0), \quad u \in U(a, b) \tag{4.61}$$

imply that $u(x, t) \equiv 0$. Let

$$z = p + is, \qquad \hat{u}(z, t) = \int_0^\infty e^{-zx} u(x, t)\, dx.$$

Since $u \in U(a, b)$, the function $\hat{u}(z, t)$ is analytic for $p > a$. Applying the Laplace transform to equation (4.58), we obtain

$$\hat{u}_t(z, t) = -u_{tx}(0, t) - u_x(0, t) - zu_t(0, t) - zu(0, t) + z^2[\hat{u}_t(z, t) + \hat{u}(z, t)]. \tag{4.62}$$

From this and (4.61) it follows that

$$(1 - z^2)\hat{u}_t(z, t) = z^2 \hat{u}(z, t),$$

whence for $z \neq 1$

$$\hat{u}(z,t) = C(z)\exp\left(\frac{z^2}{1-z^2}t\right).$$

We claim that $C(z) \equiv 0$. From the preceding equality for $z \neq 1$ we get

$$\begin{aligned} C(z) &= \exp\left(-\frac{z^2}{1-z^2}t\right)\hat{u}(z,t) \\ &= \exp\left\{\left[1 - f(p,s) - \frac{2ps}{(1-p^2+s^2)^2+4p^2s^2}i\right]t\right\}\int_0^\infty e^{-zx}u\,dx, \end{aligned} \tag{4.63}$$

whence

$$\begin{aligned} |C(z)| &\leq \exp\{[1-f(p,s)]t\}\int_0^\infty e^{-px}C_u e^{ax+bt}\,dx \\ &= \frac{C_u}{p-a}\exp\{[1-f(p,s)+b]t\}. \end{aligned} \tag{4.64}$$

Recalling the definition of the class $U(a,b)$ and using Lemma 2, we obtain

$$\inf_{p\geq a}[1-f(p,s)+b] = 1 - \sup_{p\geq a} f(p,s) + b < 0.$$

It is easy to see that for any a in the half-plane $p > a$ there exists a continuous curve $l \not\ni (1,0)$ on which

$$1 - f(p,s) + b < 0. \tag{4.65}$$

Indeed, examining the proof of Lemma 2, we see that for $a > 1$ it is possible to take for l any of the curves

$$s = \pm\sqrt{p^2 - 1 + 2p\sqrt{p^2-1}}, \qquad a + \delta \leq p \leq a + 2\delta;$$

if $a < 1$ it is possible to take for l the segment

$$s = 0, \qquad 1 - 2\delta \leq p \leq 1 - \delta$$

(in both cases $\delta > 0$ is sufficiently small). Letting $(p,s) \in l$ and $t \to +\infty$ in (4.64), we find that $C(z) = 0$ on l. Observing that for $u \in U(a,b)$ and $\operatorname{Re} z > a$ we have $\hat{u}(z,t) \to \hat{u}(z,0)$ as $t \to 0$ and passing to the limit as $t \to 0$ in the first of equalities (4.63), we find that $C(z) = \hat{u}(z,0)$ for $\operatorname{Re} z > a$, $z \neq 1$. Extending $C(z)$ to the point $z = 1$ (if $a < 1$) by continuity, we obtain a function analytic for $\operatorname{Re} z > a$ which vanishes on the curve l lying in this half-plane. Hence, $C(z) \equiv 0$, and so $\hat{u}(z,t) \equiv 0$. Since on the class $U(a,b)$ the Laplace transform in x is uniquely invertible, it follows that $u(x,t) \equiv 0$. The theorem is proved.

We shall show that the uniqueness class $U(a,b)$ in essence cannot be enlarged. Suppose for $a>1$

$$\begin{aligned} u(x,t) &= \frac{1}{2\pi i}\int_{a-i\infty}^{a+i\infty} e^{zx}(z-1)^{-4}\exp\left(\frac{z^2}{1-z^2}t\right)dz \\ &= \frac{1}{2\pi}e^{ax-t}\int_{-\infty}^{\infty} e^{isx}(a+is-1)^{-4} \\ &\qquad\qquad \times\exp\left[f(p,s)+\frac{2ps}{(1-p^2+s^2)^2+4p^2s^2}i\right]t\,ds. \end{aligned}$$

The integrals obtained by differentiation of the integrand twice with respect to x and once with respect to t converge uniformly; hence, the derivatives of the function $u(x,t)$ appearing in the definition of the class $U(a,b)$ exist and are continuous, while the moduli of all these derivatives, as is easily seen, are bounded above by the expression

$$C\exp[ax+(-1+\max_s f(a,s))t] = C\exp\left[ax+\left(-1+\frac{1}{4a(a+\sqrt{a^2-1})}\right)t\right].$$

The fact that $u(x,t)$ satisfies (4.58) is easily verified. Finally, the equality

$$\lim_{z\to\infty} z^2\hat{u}(z,t) = \lim_{z\to\infty} z^2(z-1)^{-4}\exp\left(\frac{z^2}{1-z^2}t\right) = 0,$$

by a familiar property of the Laplace transform, implies that

$$\lim_{x\to 0} u(x,t) = \lim_{x\to 0} u_x(x,t) = 0.$$

Thus, $u(x,t)$ is a nontrivial solution of the homogeneous Cauchy problem (4.57), (4.58). Hence, the inequality sign in the definition of the parameter b, giving together with the parameter a the class $U(a,b)$, cannot be replaced by an equality sign with preservation of uniqueness.

Solutions of problem (4.57), (4.58) (belonging to classes $U(a,b)$) are unstable relative to variations of their Cauchy data $f_1(t)$ and $f_2(t)$. Suppose, for example, that for $a>1$

$$u^{(a)}(x,t) = \frac{1}{a^2}\exp\left(ax-\frac{a^2}{a^2-1}t\right).$$

It is easy to see that each of the functions $u^{(a)}(x,t)$ is a solution of (4.58), while, since for $a>1$

$$-\frac{a^2}{a^2-1} = -1-\frac{1}{a^2-1} < -1+[4a(a+\sqrt{a^2-1})]^{-1},$$

it follows that

$$u^{(a)}(x,t)\in U\left(a,-\frac{a^2}{a^2-1}\right).$$

The Cauchy data for the solution $u^{(a)}(x,t)$ satisfy the inequality

$$|f_1(t)|, |f_1'(t)|, |f_2(t)|\,|f_2'(t)| \le \frac{a}{a^2-1}\exp\left(-\frac{a^2}{a^2-1}t\right) \to 0, \quad a \to +\infty.$$

On the contrary, the solution $u^{(a)}(x,t)$ at any fixed point with $x > 0$ grows exponentially with increasing a. This instability also occurs for solutions belonging to a fixed class $U(a,b)$. If, for example, for a natural number k

$$u^{(k)}(x,t) = \exp\left[\left(-1+\frac{1}{k^2+1}\right)t\right]\sin kx - \frac{1}{2}\exp\left[\left(-1+\frac{1}{4k^2+1}\right)t\right]\sin 2kx,$$

then $u^{(k)}(x,t)$ is a solution of (4.58) of class $U(0,-1+1/(k^2+1))$. It is easy to see that for the functions

$$f_1(t) = u(0,t) = 0,$$

$$f_2(t) = u_x(0,t) = k\left\{\exp\left[\left(-1+\frac{1}{k^2+1}\right)t\right] - \exp\left[\left(-1+\frac{1}{4k^2+1}\right)t\right]\right\},$$

we have the estimates (C is an absolute constant)

$$\|f_1(t)\|_{C^n[0,\infty)},\ \|f_2(t)\|_{C^n[0,\infty)} \le C^n k^{-1} \to 0, \quad k \to \infty.$$

At the same time, the quantity

$$u^{(k)}\left(\frac{\pi}{2k},t\right) = \exp\left[\left(-1+\frac{1}{k^2+1}\right)t\right]\sin\frac{\pi}{2} - \frac{1}{2}\exp\left[\left(-1+\frac{1}{4k^2+1}\right)t\right]\sin\pi = \exp\left[\left(-1+\frac{1}{k^2+1}\right)t\right]$$

is by no means small.

We shall show that for solutions belonging to a fixed class $U(a,b)$ and such that $C_u \le C$, where C is an absolute constant, there is continuous dependence on variations of the Cauchy data (4.57).

THEOREM 3. *Suppose that for a solution $u \in U(a,b)$ of problem* (4.57), (4.58) *the conditions*

$$C_u = 1; \qquad |f_1(t)|, |f_1'(t)|, |f_2(t)|, |f_2'(t)| \le \varepsilon e^{bt}, \quad 0 \le t < \infty, \tag{4.66}$$

are satisfied, where ε is sufficiently small. Then there exist $p_1 > a, C = C(a,b)$ and $\alpha = \alpha(a,b,p_1) > 0$ such that

$$|u(x,t)| \le C\left(\frac{\ln\ln\varepsilon^{-\alpha}}{\ln\varepsilon^{-\alpha}}\right)^{1/2} e^{p_1x+bt}.$$

PROOF. Because of linearity of the problem, the condition $C_u = 1$ is no essential restriction.

From (4.62) and (4.57) it follows that

$$(1-z^2)\hat{u}_t(z,t) - z^2\hat{u}(z,t) + z[f_1(t) + f_1'(t)] + f_2(t) + f_2'(t) = 0.$$

Integrating this equation and using for an arbitrary constant the notation $C(z)$, for $\operatorname{Re} z > a$, $z \neq 1$, we obtain

$$\hat{u}(z,t) = \exp\left(\frac{z^2}{1-z^2}t\right)\left\{C(z) - \int_0^t \left[\frac{f_1(\tau)+f_1'(\tau)}{1-z^2}z + \frac{f_2(\tau)+f_2'(\tau)}{1-z^2}\right]\exp\left(-\frac{z^2}{1-z^2}\tau\right)d\tau\right\}. \quad (4.67)$$

It turns out that under our conditions the function $C(z)$ can be expressed in terms of the given functions $f_1(t)$ and $f_2(t)$. For $z \in l$ (see the proof of Theorem 2) as $t \to \infty$ from (4.67) we find that

$$\begin{aligned}
&\left|C(z) - \int_0^t \left[\frac{f_1(\tau)+f_1'(\tau)}{1-z^2}z + \frac{f_2(\tau)+f_2'(\tau)}{1-z^2}\right]\exp\left(-\frac{z^2}{1-z^2}\tau\right)d\tau\right| \\
&\quad = \left|\exp\left(-\frac{z^2}{1-z^2}t\right)\hat{u}(z,t)\right| \\
&\quad = \left|\exp\left(-\frac{z^2}{1-z^2}t\right)\int_0^\infty e^{-zx}u(x,t)\,dx\right| \\
&\quad \le \exp\left(-\operatorname{Re}\frac{z^2}{1-z^2}t\right)\int_0^\infty e^{-px}e^{ax+bt}\,dx \\
&\quad = \frac{1}{p-a}\exp\{[1-f(p,s)+b]t\} \to 0.
\end{aligned}$$

Hence, for $z \in l$

$$C(z) = \int_0^\infty \left[\frac{f_1(\tau)+f_1'(\tau)}{1-z^2}z + \frac{f_2(\tau)+f_2'(\tau)}{1-z^2}\right]\exp\left(-\frac{z^2}{1-z^2}\tau\right)d\tau,$$

and thus, using (4.67), for $z \in l$ we obtain

$$\hat{u}(z,t) = \exp\left(\frac{z^2}{1-z^2}t\right)\int_t^\infty \left[\frac{f_1(\tau)+f_1'(\tau)}{1-z^2}z + \frac{f_2(\tau)+f_2'(\tau)}{1-z^2}\right]\exp\left(-\frac{z^2}{1-z^2}\right)d\tau. \quad (4.68)$$

The function $f(p,s)$ is continuous in some neighborhood of the curve l (we recall that l is closed and $(1,0) \notin l$) and on l satisfies (4.65). It is easy to see that there exist $\theta > 0$ and a segment L lying in the region $\operatorname{Re} z > a$ and parallel to the line $\operatorname{Re} z = a$ such that for $z \in L$

$$\operatorname{Re}\left(-\frac{z^2}{1-z^2}\right) + b = 1 - f(p,s) + b < -\theta.$$

We note that $\theta = \theta(a,b)$ and $L = L(a,b,\theta)$. (If $a > 1$, the segment L can be chosen arbitrarily close to the line $\operatorname{Re} z = a$; its length depends on a, b, and θ. If $a < 1$, then, generally speaking, the segment L must be chosen close to the point $z = 1$, and in this case its length depends on b and θ.)

For $z \in L$, using (4.66) and (4.68), we find that

$$\begin{aligned} |\hat{u}(z,t)| &\le 2\varepsilon \exp\left(\operatorname{Re}\frac{z^2}{1-z^2}t\right)\int_0^\infty \left(\left|\frac{z}{1-z^2}\right| + \left|\frac{1}{1-z^2}\right|\right) \\ &\qquad\qquad \times \exp\left[\left(-\operatorname{Re}\frac{z^2}{1-z^2}+b\right)\tau\right] d\tau \\ &\le C(a,b)\varepsilon\left(\operatorname{Re}\frac{z^2}{1-z^2}-b\right)^{-1} \exp\left[\left(\operatorname{Re}\frac{z^2}{1-z^2} - \operatorname{Re}\frac{z^2}{1-z^2}+b\right)t\right] \\ &\le C(a,b)\theta\varepsilon e^{bt}. \end{aligned}$$

Thus, for $z \in L \subset \{z\colon \operatorname{Re} z > a\}$ we have the inequality

$$|\hat{u}(z,t)| \le C(a,b)\varepsilon e^{bt}. \tag{4.69}$$

(Here and henceforth we use the notation C for various positive constants.)

We note that in the half-plane $\operatorname{Re} z > a$ for the function $\hat{u}(z,t)$ we have the estimate

$$|\hat{u}(z,t)| = \left|\int_0^\infty e^{-zx}u(x,t)\,dx\right| \le \int_0^\infty e^{-px+ax+bt}\,dx = \frac{1}{p_a}e^{bt}. \tag{4.70}$$

Let $\omega(z,L)$ be the harmonic measure of the segment L, relative to the half-plane on the boundary of which L lies, at a point z of this half-plane. For what follows we need a lower bound for the quantity $\min\omega(z,L)$ on the segment $\operatorname{Re} z = p_1, |\operatorname{Im} z| \le \lambda$ for large λ. We recall that $\omega(z,L)$ is the angle subtended by the segment L at the point z, divided by π. If $L = \{z\colon \operatorname{Re} z = p_0, |\operatorname{Im} z - s_0| \le \Delta\}$, then

$$\omega(z,L) = \frac{1}{\pi}\arctan\frac{2\Delta(\operatorname{Re} z - p_0)}{(\operatorname{Re} z - p_0)^2 + |(\operatorname{Im} z - s_0)^2 - \Delta^2|}.$$

Hence ($p_1 > p_0, \lambda$ large)

$$\begin{aligned} \min\omega(z,L) &= \frac{1}{\pi}\arctan\frac{2\Delta(p_1-p_0)}{(p_1-p_0)^2+\lambda^2-\Delta^2} \\ &> \frac{1}{\pi}\arctan\frac{2\Delta(p_1-p_0)}{(p_1-p_0)^2+\lambda^2} > \frac{\Delta(p_1-p_0)}{\pi\lambda^2}. \end{aligned} \tag{4.71}$$

Applying the theorem of R. Nevanlinna on two constants to the half-plane $\operatorname{Re} z \ge p_0$ and using (4.69)–(4.71), on the segment $\operatorname{Re} z = p_1, |\operatorname{Im} z| \le \lambda$ we obtain

$$\begin{aligned}|\hat{u}(z,t)| &\le [C(a,b)\varepsilon e^{bt}]^{\omega(z,L)}\left(\frac{e^{bt}}{p_0-a}\right)^{1-\omega(z,L)}\\ &= C^{\omega}(a,b)(p_0-a)^{-1+\omega}\varepsilon^{\omega}e^{bt}\\ &\le \frac{C(a,b)}{p_0-a}\exp\left[\frac{\Delta(p_1-p_0)}{\pi\lambda^2}\ln\varepsilon+bt\right]\end{aligned}$$

(it may be assumed that in (4.69) and (4.70) $C(a,b)>1$ and $(p_0-a)^{-1}>1$). Hence,

$$\begin{aligned}&\left|\frac{1}{2\pi i}\int_{p_1-i\lambda}^{p_1+i\lambda}e^{zx}\hat{u}(z,t)\,dt\right|\\ &\qquad\le\frac{C(a,b)}{p_0-a}\exp\left[\frac{\Delta(p_1-p_0)}{\pi\lambda^2}\ln\varepsilon+p_1x+bt\right]\lambda.\end{aligned}\tag{4.72}$$

Integrating the equality

$$\hat{u}(z,t)=\int_0^\infty e^{-zx}u(x,t)\,dx$$

by parts and using (4.57), we find that

$$\hat{u}(z,t)=\frac{1}{z}f_1(t)+\frac{1}{z^2}f_2(t)+\frac{1}{z^2}\int_0^\infty e^{-zx}u_{xx}(x,t)\,dx.\tag{4.73}$$

Using (4.66), we obtain

$$\begin{aligned}&\left|\frac{f_1(t)}{2\pi i}\left(\int_{p_1-i\infty}^{p_1-i\lambda}+\int_{p_1+i\lambda}^{p_1+i\infty}\right)e^{zx}\frac{dz}{z}\right|\\ &=\frac{|f_1(t)|}{2\pi}e^{p_1x}\left|\left(\int_{-\infty}^{-\lambda}+\int_{\lambda}^{\infty}\right)\frac{\cos sx+i\sin sx}{p_1+is}\,ds\right|\\ &=\frac{|f_1(t)|}{2\pi}e^{p_1x}\left|\left(\int_{-\infty}^{-\lambda}+\int_{\lambda}^{\infty}\right)\right.\\ &\qquad\qquad\left.\times\frac{p_1\cos sx+s\sin sx+i(-s\cos sx+p_1\sin sx)}{p_1^2+s^2}\,ds\right|\\ &=\frac{|f_1(t)|}{2\pi}e^{p_1x}\left|\left(\int_{-\infty}^{-\lambda}+\int_{\lambda}^{\infty}\right)\frac{p_1\cos sx+s\sin sx}{p_1^2+s^2}\,ds\right|\le C\varepsilon e^{p_1x+bt}\end{aligned}\tag{4.74}$$

with some absolute constant C (we recall that we assume λ to be large). Similarly,

$$\left|\frac{f_2(t)}{2\pi i}\left(\int_{p_1-i\infty}^{p_1-i\lambda}+\int_{p_1+i\lambda}^{p_1+i\infty}\right)e^{zx}\frac{dz}{z^2}\right|\le C\varepsilon e^{p_1x+bt}.\tag{4.75}$$

Further,

$$\left|\frac{1}{2\pi i}\left(\int_{p_1-i\infty}^{p_1-i\lambda}+\int_{p_1+i\lambda}^{p_1+i\infty}\right)\frac{e^{zx}}{z^2}\int_0^\infty e^{-z\xi}u_{\xi\xi}(\xi,t)\,d\xi\,dz\right|$$
$$\leq \frac{1}{\pi}\int_\lambda^\infty \frac{e^{p_1x}}{p_1^2+s^2}\int_0^\infty e^{-p_1\xi+a\xi+bt}\,d\xi\,ds=\frac{1}{\pi(p_1-a)}e^{p_1x+bt}\int_\lambda^\infty\frac{ds}{p_1^2+s^2}$$
$$<\frac{1}{\pi(p_1-a)}e^{p_1x+bt}\int_\lambda^\infty\frac{ds}{s^2}=\frac{1}{\pi(p_1-a)}\frac{1}{\lambda}e^{p_1x+bt}. \tag{4.76}$$

From (4.72)–(4.76) it follows that

$$|u(x,t)|=\left|\frac{1}{2\pi i}\int_{p_1-i\infty}^{p_1+i\infty}e^{zx}\hat{u}(z,t)\,dz\right|$$
$$\leq\left|\frac{1}{2\pi i}\int_{p_1-i\lambda}^{p_1+i\lambda}e^{zx}\hat{u}(z,t)\,dz\right|$$
$$+\left|\frac{1}{2\pi i}\left(\int_{p_1-i\infty}^{p_1-i\lambda}+\int_{p_1+i\lambda}^{p_1+\infty}\right)e^{zx}\hat{u}(z,t)\,dz\right|$$
$$<\left\{\frac{C(a,b)}{p_0-a}\exp\left[\frac{\Delta(p_1-p_0)}{\pi\lambda^2}\ln\varepsilon\right]\lambda+2C\varepsilon+\frac{1}{\pi(p_1-a)}\frac{1}{\lambda}\right\}e^{p_1x+bt}.$$

Increasing $C(a,b)$ if necessary and setting for brevity $\Delta(p_1-p_0)/\pi=\alpha$, from the foregoing inequality we obtain

$$|u(x,t)|<\frac{C(a,b)}{p_0-a}\left[\exp\left(\frac{\alpha}{\lambda^2}\ln\varepsilon\right)\lambda+\frac{1}{\lambda}\right]e^{p_1x+bt}. \tag{4.77}$$

We set here

$$\lambda=\left(\frac{\ln\varepsilon^{-\alpha}}{\ln\ln\varepsilon^{-\alpha}}\right)^{1/2}. \tag{4.78}$$

The preceding estimates were obtained only for sufficiently large λ; therefore, (4.78) can be satisfied only if ε is sufficiently small. Substituting (4.78) into (4.77), we find that

$$|u(x,t)|$$
$$<\frac{C(a,b)}{p_0-a}\left[\exp(-\ln\ln\varepsilon^{-\alpha})\left(\frac{\ln\varepsilon^{-\alpha}}{\ln\ln\varepsilon^{-\alpha}}\right)^{1/2}+\left(\frac{\ln\ln\varepsilon^{-\alpha}}{\ln\varepsilon^{-\alpha}}\right)^{1/2}\right]e^{p_1x+bt}$$
$$=\frac{C(a,b)}{p_0-a}\left[(\ln\varepsilon^{-\alpha}\ln\ln\varepsilon^{-\alpha})^{-1/2}+\left(\frac{\ln\ln\varepsilon^{-\alpha}}{\ln\varepsilon^{-\alpha}}\right)^{1/2}\right]e^{p_1x+bt}$$
$$<C\left(\frac{\ln\ln\varepsilon^{-\alpha}}{\ln\varepsilon^{-\alpha}}\right)^{1/2}e^{p_1x+bt} \tag{4.79}$$

with some new constant $C = C(a,b)$. It is obvious that $\alpha = \alpha(a,b,p_1)$. The theorem is proved.

There is an example which shows that a solution of problem (4.57), (4.58) of class $U(a,b)$ with $a = 0$ and $b < 0$ is actually only logarithmically continuous with respect to any C^m-norm of its Cauchy data.

§4. Cauchy problems with data on a timelike surface for hyperbolic and ultrahyperbolic equations

In the space R^{n+m} we consider the inequality

$$|Lu| \le A(|u| + |\nabla u|), \tag{4.80}$$

where

$$L = \left(\sum_1^n - \sum_{n+1}^{n+m}\right)\frac{\partial^2}{\partial x_i^2}, \qquad \nabla = \sum_1^{n+m} n^i \frac{\partial}{\partial x_i} \tag{4.81}$$

(n^i is the unit vector along the axis x_i). We set $|x|^2 = \sum_1^{n+m} x_i^2$ and $G(\delta) = \{x\colon |x| < \delta\}$. Let S be a surface of class C^1 passing through the point $x = 0$, and suppose that the normal to S at this point coincides with n^1. For sufficiently small δ the surface S divides $G(\delta)$ into two connected regions. Let $G^+(\delta)$ be the one of them which contains all points of the form $(\eta, 0, \dots, 0)$ with sufficiently small $\eta > 0$. Suppose in $\overline{G^+(\delta)}$ there is defined a solution $u \in C^2$ of (4.80) with the properties

$$|u(x)|, |\nabla u(x)| \le \begin{cases} \varepsilon & \text{for } x \in S \cap \overline{G(\delta)}, \\ M & \text{for } x \in \overline{G^+(\delta)}. \end{cases}$$

We shall give conditions on S under which there exists a region $W \subset G^+(\delta)$ (depending only on S and δ) such that from (4.80) and the last inequalities it follows that

$$\int_W (u^2 + |\nabla u|^2)\,dx \le C_1 \varepsilon^{C_2} M^{1-C_2}$$

with $C_1 > 0$ and $0 < C_2 < 1$ depending only on S, δ, and W. This estimate obviously implies uniqueness of the solution of the Cauchy problem for inequality (4.80) with data on S.

In what follows we use the notation

$$\sum = \sum_1^{n+m}, \qquad f_i(x) = \frac{\partial f(x)}{\partial x_i}$$

(on the other hand, x_i is always the ith coordinate of the point x).

LEMMA 1. *Let $u(x)$ and $\psi(x)$ be twice differentiable, let $l(x)$ be thrice differentiable, and let $\theta(x) = \exp l(x)$. Further, let κ_i, $i = 1, \dots, n+m$, be constants equal to ± 1, and let $L = \sum \kappa_i \partial^2/\partial x_i^2$. Then*

$$\theta^2(Lu)^2 \geq 2\theta^2 \left\{ \sum [2l_{ii} - \chi_i(Ll + \psi)]u_i^2 + 4 \sum_{1 \leq i < j \leq n+m} \kappa_i \kappa_j l_{ij} u_i u_j \right\}$$
$$+ 2\theta^2 \left[2(Ll+\psi) \sum \kappa_i l_i^2 - L\left(\sum \kappa_i l_i^2 - \frac{1}{2}\psi \right) + \sum \kappa_i \psi_i l_i - \frac{1}{2}\psi^2 \right] u^2$$
$$+ \sum_i \kappa_i \left\{ 2\theta^2 \left[l_i \sum_j \kappa_j u_j^2 - 2u_i \sum_j \kappa_j l_j u_j + \left(-2 \sum_j \kappa_j l_j^2 + \psi \right) u_i u \right.\right.$$
$$\left.\left. + \left(\left(-2l_i + \frac{\partial}{\partial x_i} \right) \sum_j \kappa_j l_j^2 - \frac{1}{2}\psi_i \right) u^2 \right] \right\}_i .$$

PROOF. Introducing the function

$$v(x) = \theta(x) u(x) \tag{4.82}$$

and using the equality

$$u_{ii} = \theta^{-1}[v_{ii} - 2l_i v_i + (l_i^2 - l_{ii})v],$$

we obtain

$$\begin{aligned}
\theta^2 \left(\sum \kappa_i u_{ii} \right)^2 &= \left\{ \sum \kappa_i [v_{ii} - 2l_i v_i + (l_i^2 - l_{ii})v] \right\}^2 \\
&\geq \left[\sum \kappa_i v_{ii} + \sum \kappa_i (l_i^2 - l_{ii})v - \psi v \right]^2 + 2\left(\sum \kappa_i v_{ii} \right) \psi v \\
&\quad - 4\left(\sum \kappa_i v_{ii} \right)\left(\sum \kappa_i l_i v_i \right) + 2\psi \sum \kappa_i (l_i^2 - l_{ii})v^2 - \psi^2 v^2 \\
&\quad - 4\left(\sum \kappa_i l_i v_i \right)\left[\sum \kappa_i (l_i^2 - l_{ii})v \right] \\
&\geq 2\left(\sum \kappa_i v_{ii} \right)\psi v - 4\left(\sum \kappa_i v_{ii} \right)\left(\sum \kappa_i l_i v_i \right) + 2\psi \sum \kappa_i (l_i^2 - l_{ii})v^2 \\
&\quad - \psi^2 v^2 - 4\left(\sum \kappa_i l_i v_i \right)\left[\sum \kappa_i (l_i^2 - l_{ii})v \right].
\end{aligned} \tag{4.83}$$

We transform each of the terms on the right side so as to obtain expressions containing (up to a divergence) only v_i^2, $v_i v_j$, and v^2. We have

$$2\left(\sum \kappa_i v_{ii} \right)\psi v = -2\psi \sum \kappa_i v_i^2 + \sum \kappa_i \psi_{ii} v^2 + 2\sum \kappa_i (\psi v_i v)_i - \sum \kappa_i (\psi_i v^2)_i .$$

From (4.82) it follows that $v_i = \theta u_i + \theta_i u$. Hence,

$$\begin{aligned}
2\left(\sum \kappa_i v_{ii} \right)\psi v = &-2\psi \sum \kappa_i (\theta u_i + \theta_i u)^2 + \theta^2 \sum \kappa_i \psi_{ii} u^2 \\
&+ \sum \kappa_i [2\psi(\theta u_i + \theta_i u)\theta u - \psi_i \theta^2 u^2]_i
\end{aligned}$$

$$
\begin{aligned}
&= -2\psi\theta^2 \sum \kappa_i u_i^2 \\
&\quad + 2\sum \kappa_i(\psi_i\theta\theta_i + \psi\theta_i^2 + \psi\theta\theta_{ii})u^2 + \sum \kappa_i(\theta^2\psi_{ii} - 2\psi\theta_i^2)u^2 \\
&\quad - 2\sum \kappa_i(\psi\theta\theta_i u^2)_i + \sum \kappa_i\{\theta^2[2\psi u_i u + (2\psi l_i - \psi_i)u^2]\}_i \\
&= -2\psi\theta^2 \sum \kappa_i u_i^2 + \theta^2 \sum \kappa_i[2\psi_i l_i + 2\psi l_i^2 + 2\psi(l_{ii} + l_i^2) \\
&\quad + \psi_{ii} - 2\psi l_i^2]u^2 + \sum \kappa_i\{\theta^2[2\psi u_i u + (2\psi l_i - \psi_i - 2\psi l_i)u^2]\}_i \\
&= -2\psi\theta^2 \sum \kappa_i u_i^2 + \theta^2 \sum \kappa_i(2\psi l_i^2 + 2\psi l_{ii} + 2\psi_i l_i + \psi_{ii})u^2 \\
&\quad + \sum \kappa_i[\theta^2(2\psi u_i u - \psi_i u^2)]_i.
\end{aligned}
$$

Further,

$$
\begin{aligned}
-4\left(\sum \kappa_i v_{ii}\right)\left(\sum \kappa_i l_i v_i\right) &= -4\sum_i\sum_j \kappa_i\kappa_j l_j v_{ii} v_j \\
&= -2\sum_i\sum_j \kappa_i\kappa_j l_{jj} v_i^2 + 4\sum_i\sum_j \kappa_i\kappa_j l_{ij} v_i v_j \\
&\quad + 2\sum_j \kappa_j\left(\sum_i \kappa_i l_j v_i^2\right)_j - 4\sum_i \kappa_i\left(\sum_j \kappa_j l_j v_i v_j\right)_i.
\end{aligned}
$$

Returning to the function $u(x)$, we obtain

$$
\begin{aligned}
-4\left(\sum \kappa_i v_{ii}\right)\left(\sum \kappa_i l_i v_i\right) &= -2\sum_i\sum_j \kappa_i\kappa_j l_{jj}(\theta u_i + \theta_i u)^2 \\
&\quad + 4\sum_i\sum_j \kappa_i\kappa_j l_{ij}(\theta u_i + \theta_i u)(\theta u_j + \theta_j u) \\
&\quad + 2\sum_i \kappa_i\left\{\sum_j \kappa_j[l_i(\theta u_j + \theta_j u)^2 - 2l_j(\theta u_i + \theta_i u)(\theta u_j + \theta_j u)]\right\}_i \\
= &-2\theta^2\sum_i\sum_j \kappa_i\kappa_j l_{jj} u_i^2 - 2\theta^2\sum_i\sum_j \kappa_i\kappa_j l_{jj} l_i^2 u^2 \\
&- 4\sum_i\sum_j \kappa_i\kappa_j l_{jj}\theta\theta_i u u_i + 4\theta^2\sum_i\sum_j \kappa_i\kappa_j l_{ij} u_i u_j \\
&+ 4\sum_i\sum_j \kappa_i\kappa_j l_{ij}\theta_j\theta u u_i + 4\sum_i\sum_j \kappa_i\kappa_j l_{ij}\theta\theta_i u u_j \\
&+ 4\theta^2\sum_i\sum_j \kappa_i\kappa_j l_{ij} l_i l_j u^2 \\
&+ 2\sum_i \kappa_i\left\{\theta^2\sum_j \kappa_j[l_i(u_j + l_j u)^2 - 2l_j(u_i + l_i u)(u_j + l_j u)]\right\}_i
\end{aligned}
$$

(*Continued*)

(Continued)

$$
\begin{aligned}
= & -2\theta^2 \sum_i \sum_j \kappa_i \kappa_j l_{jj} u_i^2 + 4\theta^2 \sum_i \sum_j \kappa_i \kappa_j l_{ij} u_i u_j \\
& + 2 \sum_i \sum_j \kappa_i \kappa_j (l_{ijj}\theta\theta_i + l_{jj}\theta_i^2 + l_{jj}\theta\theta_{ii} - l_{iij}\theta\theta_j - l_{ij}\theta_i\theta_j - l_{ij}\theta\theta_{ij} \\
& - l_{ijj}\theta\theta_i - l_{ij}\theta_i\theta_j - l_{ij}\theta\theta_{ij} - l_{jj}\theta^2 l_i^2 + 2\theta^2 l_{ij} l_i l_j) u^2 \\
& + 2 \sum_i \kappa_i \left\{ \theta^2 \sum_j \kappa_j [l_i (u_j + l_j u)^2 - 2l_j (u_i + l_i u)(u_j + l_j u)] \right\}_i \\
& - 2 \sum_i \kappa_i \left(\sum_j \kappa_j l_{jj} \theta\theta_i u^2 \right)_i + 2 \sum_i \kappa_i \left(\sum_j \kappa_j l_{ij} \theta\theta_j u^2 \right)_i \\
& + 2 \sum_j \kappa_j \left(\sum_i \kappa_i l_{ij} \theta\theta_i u^2 \right)_j \\
= & -2\theta^2 \sum_i \sum_j \kappa_i \kappa_j l_{jj} u_i^2 + 4\theta^2 \sum_i \sum_j \kappa_i \kappa_j l_{ij} u_i u_j \\
& + 2\theta^2 \sum_i \sum_j \kappa_i \kappa_j (l_{jj} l_i^2 + l_{ii} l_{jj} - l_{iij} l_j - 2l_{ij}^2 - 2l_{ij} l_i l_j) u^2 \\
& + 2 \sum_i \kappa_i \left\{ \theta^2 \sum_j \kappa_j [l_i u_j^2 - 2l_j u_i u_j - 2l_j^2 u_i u \right. \\
& \qquad \left. + (-l_i l_j^2 - l_i l_{jj} + 2l_{ij} l_j) u^2] \right\}_i .
\end{aligned}
$$

Further,

$$
2\psi \sum \kappa_i (l_i^2 - l_{ii}) v^2 - \psi^2 v^2 = 2\psi\theta^2 \sum \kappa_i (l_i^2 - l_{ii}) u^2 - \theta^2 \psi^2 u^2 .
$$

Finally,

$$
\begin{aligned}
& -4 \left(\sum \kappa_i l_i v_i \right) \left[\sum \kappa_i (l_i^2 - l_{ii}) v \right] = -4 \sum_i \sum_j \kappa_i \kappa_j l_i (l_j^2 - l_{jj}) v_i v \\
& \quad = 2 \sum_i \sum_j \kappa_i \kappa_j (l_{ii} l_j^2 - l_{ii} l_{jj} + 2l_i l_j l_{ij} - l_i l_{jji}) v^2 \\
& \qquad - 2 \sum_i \kappa_i \left[\sum_j \kappa_j l_i (l_j^2 - l_{jj}) v^2 \right]_i
\end{aligned}
$$

$$= 2\theta^2 \sum_i \sum_j \kappa_i \kappa_j (l_{ii} l_j^2 - l_{ii} l_{jj}$$

$$+ 2 l_i l_j l_{ij} - l_i l_{jji}) u^2 - 2 \sum_i \kappa_i \left[\theta^2 \sum_j \kappa_j l_i (l_j^2 - l_{jj}) u^2 \right]_i .$$

Collecting the expressions obtained for the terms of (4.83), we find that

$$\theta^2 (Lu)^2 \geq 2\theta^2 \sum_i \left(2 l_{ii} - \kappa_i \sum_j \kappa_j l_{jj} - \kappa_i \psi \right) u^2 + 4\theta^2 \sum_i \sum_{j \neq i} \kappa_i \kappa_j l_{ij} u_i u_j$$

$$+ 2\theta^2 \left[\sum_i \sum_j \kappa_i \kappa_j (2 l_{ii} l_j^2 - 2 l_{iij} l_j - 2 l_{ij}^2) \right.$$

$$\left. + \sum_i \kappa_i \left(2\psi l_i^2 + \psi_i l_i + \frac{1}{2} \psi_{ii} \right) - \frac{1}{2} \psi^2 \right] u^2$$

$$+ 2 \sum_i \kappa_i \left\{ \theta^2 \left[l_i \sum_j \kappa_j u_j^2 - 2 u_i \sum_j \kappa_j l_j u_j + \left(-2 \sum_j \kappa_j l_j^2 + \psi \right) u_i u \right. \right.$$

$$\left. \left. + \left(-2 l_i \sum_j \kappa_j l_j^2 + 2 \sum_j \kappa_j l_{ij} l_j - \frac{1}{2} \psi_i \right) u^2 \right] \right\}_i$$

$$= 2\theta^2 \sum_i [2 l_{ii} - \kappa_i (Ll + \psi)] u_i^2 + 4\theta^2 \sum_i \sum_{j \neq i} \kappa_i \kappa_j l_{ij} u_i u_j$$

$$+ 2\theta^2 \left[\sum_i \sum_j \kappa_i \kappa_j \left(2 l_{ii} l_j^2 - \frac{\partial^2}{\partial x_i^2} l_j^2 \right) \right.$$

$$\left. + \sum_i \kappa_i \left(2\psi l_i^2 + \psi_i l_i + \frac{1}{2} \psi_{ii} \right) - \frac{1}{2} \psi^2 \right] u^2$$

$$+ 2 \sum_i \kappa_i \left\{ \theta^2 \left[l_i \sum_j \kappa_j u_j^2 - 2 u_i \sum_j \kappa_j l_j u_j + \left(-2 \sum_j \kappa_j l_j^2 + \psi \right) u_i u \right. \right.$$

$$\left. \left. + \left(-2 l_i \sum_j \kappa_j l_j^2 + 2 \sum_j \kappa_j l_{ij} l_j - \frac{1}{2} \psi_i \right) u^2 \right] \right\}_i .$$

It is easy to see that the last sum coincides with the right side of the inequality to be proved. Lemma 1 is proved.

Below we use the notation

$$\sum_2 = \sum_2^{n+m}, \qquad \sum{}' = \sum_1^{n}, \qquad \sum_2{}' = \sum_2^{n}, \qquad \sum{}'' = \sum_{n+1}^{n+m}.$$

For real $\alpha_2, \dots, \alpha_{n+m}$ and ξ we set

$$\varphi(x) = (x_1 - \xi)^2 + \sum_2 \alpha_i x_i^2. \tag{4.84}$$

LEMMA 2. *If $u(x)$ is twice differentiable, then for any real λ and p the following inequality holds (L is defined in (4.81)):*

$$\begin{aligned}
e^{2\lambda\varphi}(Lu)^2 \geq 8\lambda e^{2\lambda\varphi} &\left[(1-p)u_1^2 + \sum_2{}'(\alpha_i - p)u_i^2 + \sum{}''(\alpha_i + p)u_i^2\right] \\
+ 4e^{2\lambda\varphi} &\left\{16p\lambda^3 \left[(x_1-\xi)^2 + \left(\sum_2{}' - \sum{}''\right)\alpha_i^2 x_i^2\right]\right. \\
&\left. - 4\lambda^2\left(1 + \sum_2 \alpha_i^2\right) - \lambda^2\left[1 + \left(\sum_2{}' - \sum{}''\right)\alpha_i - 2p\right]^2\right\} u^2 \\
+ \left(\sum_i{}' - \sum_i{}''\right) &\left\{4e^{2\lambda\varphi}\left[\left(\sum_j{}' - \sum_j{}''\right)(\lambda\alpha_i x_i u_j^2 - 2\lambda\alpha_j x_j u_j u_i \right.\right. \\
&\qquad - 4\lambda^2\alpha_j^2 x_j^2 u_i u - \lambda\alpha_j u_i u - 8\lambda^3\alpha_i\alpha_j^2 x_i x_j^2 u^2) \\
&\qquad \left.\left. - 2\lambda p u_i u + 4\lambda^2\alpha_i x_i u^2\right]\right\}_i,
\end{aligned}$$

where the expressions α_i, α_j and x_i, x_j in the divergence term for $i = 1$ or $j = 1$ must be replaced by the expressions 1 and $x_1 - \xi$ respectively.

PROOF. Setting in Lemma 1

$$\kappa_1 = \dots = \kappa_n = 1, \qquad \kappa_{n+1} = \dots = \kappa_{n+m} = -1,$$

$$l(x) = \lambda\varphi(x), \qquad \psi = -2\lambda\left(1 + \sum_2{}' \alpha_i - \sum{}'' \alpha_i\right) + 4\lambda p,$$

we arrive at the inequality to be proved.

Below we denote by C constants depending only on $n + m$ (generally speaking, these constants are distinct). If it is necessary to emphasize the dependence of a constant on some parameters, we indicate this dependence explicitly.

THEOREM 1. *Let T, R, z, and ς be numbers such that*

$$0 < T \le R, \qquad 0 < \varsigma \le z < R - \sqrt{R^2 - T^2}.$$

Set

$$\omega^{(1)}(z,\varsigma) = \left\{ x \colon (x_1 - R)^2 + \sum_2{}' x_i^2 = R^2, \qquad \sum{}'' x_i^2 \le T^2 \frac{2R\varsigma - \varsigma^2}{2Rz - z^2} \right\},$$

$$\omega^{(2)}(z,\varsigma) = \left\{ x \colon (x_1 - R)^2 + \sum_2{}' x_i^2 - \frac{2Rz - z^2}{T^2} \sum{}'' x_i^2 = (R - \varsigma)^2 \right\}$$

and denote by $\Omega(z,\varsigma)$ the region of the space R^{n+m} bounded by the surfaces $\omega^{(1)}(z,\varsigma)$ and $\omega^{(2)}(z,\varsigma)$. Suppose in the region $\Omega(z,z)$ there is defined a solution $u(x) \in C^1(\overline{\Omega}(z,z)) \cap C^2(\Omega(z,z))$ of inequality (4.80) *such that*

$$|u(x)|, |\nabla u(x)| \le \begin{cases} \varepsilon & \text{for } x \in \omega^{(1)}(z,z), \\ 1 & \text{for } x \in \overline{\Omega}(z,z). \end{cases} \tag{4.85}$$

Then there exist $C_1 = C_1(R,T,z,A) > 0$ and $C_2 = C_2(R,T,z) > 0$ such that for any ε with $\ln(1/\varepsilon) \ge C_1$ and any ς, $0 < \varsigma < z$,

$$\int_{\Omega(z,\varsigma)} (|\nabla u|^2 + \ln^2 \varepsilon u^2)\, dx \le C_2 \ln^2 \varepsilon \exp\left[2 \left(1 - \frac{2R\varsigma - \varsigma^2}{2Rz - z^2} \right) \ln \varepsilon \right].$$

PROOF. The surface $\omega^{(1)}(z,\varsigma)$ is a circular polycylinder with base of radius R and height

$$2T\sqrt{(2R\varsigma - \varsigma^2)/(2Rz - z^2)};$$

$\omega^{(2)}(z,\varsigma)$ is a hyperboloid. The surfaces $\omega^{(1)}(z,\varsigma)$ and $\omega^{(2)}(z,\varsigma)$ intersect along spheres whose points satisfy the relations

$$(x_1 - R)^2 + \sum_2{}' x_i^2 = R^2, \qquad \sum{}'' x_i^2 = T^2 \frac{2R\varsigma - \varsigma^2}{2Rz - z^2}.$$

We assume that the Cauchy data for inequality (4.80) are given on the polycylinder

$$\omega^{(1)}(z,z) = \left\{ x \colon (x_1 - R)^2 + \sum_2{}' x_i^2 = R^2, \qquad \sum{}'' x_i^2 \le T^2 \right\}$$

and admit the first of the estimates (4.85) on it; we estimate the solution in the region $\Omega(z,\varsigma)$. It is here assumed that a solution exists in the region $\Omega(z,z) \supset \Omega(z,\varsigma)$ and admits there the second of the estimates (4.85).

In the inequality of Lemma 2 we set

$$\xi = R, \qquad \alpha_2 = \cdots = \alpha_n = 1, \qquad \alpha_{n+1} = \cdots = \alpha_{n+m} = -T^{-2}(2Rz - z^2),$$
$$p = (2T^2)^{-1}(T^2 + 2Rz - z^2).$$

In this case

$$\phi(x) = (x_1 - R)^2 + \sum_2' x_i^2 - \frac{2Rz - z^2}{T^2} \sum'' x_i^2,$$

and hence the surface $\omega^{(2)}(z, \varsigma)$ is a level surface of the function $\varphi(x)$ and thus also of the function $\exp[2\lambda\varphi(x)]$.

We shall show that for $i = 2, \ldots, n$ and $j = n+1, \ldots, n+m$

$$0 < p, 1-p, \qquad \alpha_i - p, -\alpha_j, \alpha_j + p < 1. \tag{4.86}$$

From the inequality $z < R - \sqrt{R^2 - T^2}$ it follows that

$$\sqrt{R^2 - T^2} < R - z, \qquad R^2 - T^2 < (R-z)^2, \qquad T^2 + z^2 - 2Rz > 0, \tag{4.87}$$

whence

$$T^2 + 2Rz - z^2 < 2T^2. \tag{4.88}$$

Since $z < R$, it follows that $z^2 - 2Rz < 2Rz - z^2$. From this inequality and the last inequality of (4.87) we have $T^2 + 2Rz - z^2 > 0$. The inequalities $0 < p < 1$ and analogous inequalities for $1 - p$ follow from this and (4.88). Since $\alpha_i - p = \alpha_j + p = 1 - p$, the same inequalities hold also for $\alpha_i - p$ and $\alpha_j + p$.

Further, $-\alpha_j = (2Rz - z^2)/T^2$. Since $z < R - \sqrt{R^2 - T^2} \leq R$, it follows that $-\alpha_j > 0$, and from (4.87) it follows that $-\alpha_j < 1$. The inequalities (4.86) have been justified. We apply the inequality of Lemma 2 to the region $\Omega(z, z)$ with the ξ, α_i, and p chosen above. By (4.86)

$$-4\left(1 + \sum_2 \alpha_i^2\right) - \left[1 + \left(\sum_2' - \sum''\right)\alpha_i - 2p\right]^2 \geq -C,$$

$$\begin{aligned}(x_1 - \xi)^2 &+ \left(\sum_2' - \sum''\right)\alpha_i^2 x_i^2 \\ &= (x_1 - R)^2 + \sum_2' x_i^2 - T^{-2}(2Rz - z^2)^2 \sum'' x_i^2 \\ &\geq \inf_{\Omega(z,z)} \left[(x_1 - R)^2 + \sum_2' x_i^2 - T^{-2}(2Rz - z^2)^2 \sum'' x_i^2\right] = (R - z)^2.\end{aligned}$$

From this and Lemma 2 it follows that

$$\begin{aligned}e^{2\lambda\varphi}(Lu)^2 &\geq 8\lambda(1-p)e^{2\lambda\varphi}|\nabla u|^2 \\ &\quad + 64\lambda^3 p e^{2\lambda\varphi}\left[(R-z)^2 - \frac{C}{16\lambda p}\right]u^2 + \operatorname{div} e^{2\lambda\varphi}U,\end{aligned}$$

where the modulus of the vector-valued function U, as is easily seen, for $\lambda \geq 1$ admits the estimate

$$|U(x)| \leq C\lambda^3(1 + R^3 + T^3)[|\nabla u(x)|^2 + u^2(x)].$$

Thus,

$$e^{2\lambda\varphi}\left\{8\lambda(1-p)|\nabla u|^2+64\lambda^3 p\left[(R-z)^2-\frac{C}{16\lambda p}\right]u^2-(Lu)^2\right\}$$
$$\le -\operatorname{div} e^{2\lambda\varphi}U. \quad (4.89)$$

From (4.80) we obtain

$$(Lu)^2 \le A^2[|\nabla u|^2+u^2+2|\nabla u|\,|u|] \le 2A^2(|\nabla u|^2+u^2).$$

From this and (4.89) we find that

$$e^{2\lambda\varphi}\left\{[8\lambda(1-p)-2A^2]|\nabla u|^2\right.$$
$$\left.+\left[64\lambda^3 p\left((R-z)^2-\frac{C}{16\lambda p}\right)-2A^2\right]u^2\right\} \le -\operatorname{div} e^{2\lambda\varphi}U. \quad (4.90)$$

Let λ be so large that

$$8\lambda(1-p)-2A^2 \ge 4\lambda(1-p), \quad (4.91)$$

$$64\lambda^3 p\left[(R-z)^2-\frac{C}{16\lambda p}\right]-2A^2 \ge 32\lambda^3 p(R-z)^2. \quad (4.92)$$

It is easy to see that (4.91) and (4.92) are satisfied if, for example,

$$\lambda \ge C(R,T,z,A), \quad (4.93)$$

where

$$C(R,T,z,A)=\max\left\{1,\ \frac{A^2}{2(1-p)},\ \frac{C}{2p(R-z)^2},\ \left[\frac{A^2}{12p(R-z)^2}\right]^{1/3}\right\} \quad (4.94)$$

(we recall that we restrict our attention to λ such that $\lambda \ge 1$). For such λ (4.90) gives

$$e^{2\lambda\varphi}[4\lambda(1-p)|\nabla u|^2+32\lambda^3 p(R-z)^2u^2] \le -\operatorname{div} e^{2\lambda\varphi}U.$$

We integrate this inequality over $\Omega(z,z)$. We have

$$\int_{\Omega(z,z)} e^{2\lambda\varphi}[4\lambda(1-p)|\nabla u|^2+32\lambda^3 p(R-z)^2u^2]\,dx$$
$$\le \left(\int_{\omega^{(1)}(z,z)}+\int_{\omega^{(2)}(z,z)}\right)e^{2\lambda\varphi}|U|\,ds.$$

Noting that for $x\in\omega^{(2)}(z,z)$ we have $\varphi(x)=(R-z)^2$ while for $x\in\omega^{(1)}(z,z)$ we have $\max\varphi(x)=R^2$, from this and the inclusion $\Omega(z,\varsigma)\subset\Omega(z,z)$ we obtain

$$\int_{\Omega(z,\varsigma)} e^{2\lambda\varphi}[4\lambda(1-p)|\nabla u|^2+32\lambda^3 p(R-z)^2u^2]\,dx$$
$$\le \exp(2\lambda R^2)\int_{\omega^{(1)}(z,z)}|U|\,ds+\exp[2\lambda(R-z)^2]\int_{\omega^{(2)}(z,z)}|U|\,ds. \quad (4.95)$$

Using the estimate for the vector U and (4.85), we have

$$|U(x)| \leq \begin{cases} C\lambda^3(1+R^3+T^3)\epsilon^2 & \text{for } x \in \omega^{(1)}(z,z), \\ C\lambda^3(1+R^3+T^3) & \text{for } x \in \omega^{(2)}(z,z). \end{cases}$$

Noting that

$$\operatorname{meas}\omega^{(1)}(z,z),\ \operatorname{meas}\omega^{(2)}(z,z) \leq CR^{n-1}T^m,$$

from (4.95) we obtain

$$\int_{\Omega(z,\varsigma)} e^{2\lambda\varphi}[4\lambda(1-p)|\nabla u|^2 + 32\lambda^3 p(R-z)^2u^2]\,dx \\ \leq C(1+R^3+T^3)R^{n-1}T^m\lambda^3\{\exp(2\lambda R^2)\varepsilon^2 + \exp[2\lambda(R-z)^2]\}.$$

Since $\varphi(x) > (R-\varsigma)^2$ for $x \in \Omega(z,\varsigma)$, from this we get

$$\int_{\Omega(z,\varsigma)} [4(1-p)|\nabla u|^2 + 32\lambda^2 p(R-z)^2u^2]\,dx \\ \leq C(1+R^3+T^3)R^{n-1}T^m\lambda^2\left\{\exp[2\lambda(R^2-(R-\varsigma)^2)]\varepsilon^2 \right. \\ \left. + \exp[-2\lambda((R-\varsigma)^2-(R-z)^2)]\right\}. \quad (4.96)$$

Let ε be so small that

$$\ln(1/\varepsilon) \geq [R^2-(R-z)^2]C,$$

where C is defined in (4.94). Setting in (4.96)

$$\lambda = [R^2-(R-z)^2]^{-1}\ln(1/\varepsilon),$$

we then obtain

$$\int_{\Omega(z,\varsigma)} [4(1-p)|\nabla u|^2 + 32p(R-z)^2(2Rz-z^2)^{-2}\ln^2\varepsilon u^2]\,dx \\ \leq C(1+R^3+T^3)R^{n-1}T^m(2Rz-z^2)^{-2}\ln^2\varepsilon \\ \times \exp\left[2\left(1-\frac{2R\varsigma-\varsigma^2}{2Rz-z^2}\right)\ln\varepsilon\right].$$

Setting

$$C_1(R,T,z,A) = (2Rz-z^2)C(R,T,z,A),$$
$$C_2(R,T,z) = C(1+R^3+T^3)R^{n-1}T^m(2Rz-z^2)^{-2} \\ \times \min{}^{-1}[4(1-p), 32p(R-z)^2(2Rz-z^2)^{-2}],$$

we arrive at the assertion of the theorem.

COROLLARY. *If a solution of* (4.80) *of class* $C^1(\overline{\Omega}(z,z)) \cap C^2(\Omega(z,z))$ *is defined in the region* $\Omega(z,z)$ *and vanishes on* $\omega^{(1)}(z,z)$ *together with its gradient, then this solution also vanishes in* $\Omega(z,z)$.

Indeed, it follows from Theorem 1 that in this case $u(x) = 0$ in any region $\Omega(z,\varsigma)$ for $\varsigma < z$, whence our assertion follows. This assertion is valid for any $z < R - \sqrt{R^2 - T^2}$. If the function $u(x)$ and its gradient vanish on $\omega^{(1)}(R - \sqrt{R^2 - T^2}, R - \sqrt{R^2 - T^2})$ and a solution of this class exists in $\Omega(R - \sqrt{R^2 - T^2}, R - \sqrt{R^2 - T^2})$, then this solution obviously vanishes in $\Omega(R - \sqrt{R^2 - T^2}, R - \sqrt{R^2 - T^2})$. We note that the region $\Omega(R - \sqrt{R^2 - T^2}, R - \sqrt{R^2 - T^2})$ does not exhaust the entire region bounded by $\omega^{(1)}(R - \sqrt{R^2 - T^2}, R - \sqrt{R^2 - T^2})$ and the corresponding characteristic conoids. At the same time, it should be expected that the domain of definition for $\omega^{(1)}(R - \sqrt{R^2 - T^2}, R - \sqrt{R^2 - T^2})$ is precisely the latter region. Theorem 1 confirms this assumption only in the case $R = T$. Indeed, the conditions $u(x) = \nabla u(x) = 0$ on $\omega^{(1)}(R,R)$ in this case imply that $u(x) = 0$ in the region $\Omega(R,R)$ bounded by $\omega^{(1)}(R,R)$ and $\omega^{(2)}(R,R)$. However, the equation of $\omega^{(2)}(R,R)$ is

$$(x_1 - R)^2 + \sum_2{}' x_i^2 - \sum{}'' x_i^2 = 0,$$

i.e., $\omega^{(2)}(R,R)$ is the characteristic conoid with vertex at the point $(R, 0, \dots, 0)$.

We have considered the question of uniqueness and stability of a solution of inequality (4.80) in the case where the Cauchy data are given on the entire lateral surface of the cylinder

$$(x_1 - R)^2 + \sum_2{}' x_i^2 = R^2, \qquad \sum{}'' x_i^2 \le T^2.$$

Study of this problem in the case where the Cauchy data are given only on part of this surface is also of interest. As previously, the consideration of this question will be based on Lemma 2.

Suppose there is given a collection of positive quantities R, T, r with the condition $r < 2R$. We shall assume that the Cauchy data for inequality (4.80) are given on the surface

$$\omega(R,T,r) = \left\{ x\colon (x_1 - R)^2 + \sum_2{}' x_i^2 = R^2,\ \sum{}' x_i^2 \le r^2,\ \sum{}'' x_i^2 \le T^2 \right\}$$

In subsequent considerations there are no distinguished variables among those contained in each of the groups $x_2, \dots, x_n$ and $x_{n+1}, \dots, x_{n+m}$. Therefore, it is natural to choose the coefficients α_i appearing in the definition of the function $\varphi(x)$ (see Lemma 2) so that $\alpha_2 = \cdots = \alpha_n = \alpha$ and $\alpha_{n+1} = \cdots = \alpha_{n+m} = \beta$. Thus,

$$\varphi(x) = (x_1 - \xi)^2 + \alpha \sum_2{}' x_i^2 + \beta \sum{}'' x_i^2,$$

and the level surfaces of the weight $e^{2\lambda\varphi}$ have the form

$$\varphi(x) \equiv (x_1 - \xi)^2 + \alpha \sum_2{}' x_i^2 + \beta \sum{}'' x_i^2 = c.$$

Use of Lemma 2 makes it possible to prove uniqueness of a solution of inequality (4.80) in the region $\Omega \subset \{x\colon x_1 > 0\}$ bounded by the surface $\omega(R, T, r)$ and the surface $\varphi(x) = c$ (more precisely, by parts of these surfaces). Here it is natural to require that the region Ω for given R, T, and r be maximal. If the point of intersection of the surface $\varphi(x) = c$ with the line $x_1 = s$, $x_2 = \cdots = x_{n+m} = 0$ is $(a, 0, \ldots, 0)$, then maximality of the region Ω corresponds to maximality of a. In this case the level surface of the function $\varphi(x)$ bounding the region Ω has the form

$$(x_1 - \xi)^2 + \alpha \sum_2{}' x_i^2 + \beta \sum{}'' x_i^2 = (a - \xi)^2. \tag{4.97}$$

The intersection of the support $\omega(R, T, r)$ of the Cauchy data with the plane $x_{n+1} = \cdots = x_{n+m} = 0$ is

$$\left\{ x\colon (x_1 - R)^2 + \sum_2{}' x_i^2 = R^2, \quad \sum{}' x_i^2 \le r^2, \quad x_{n+1} = \cdots = x_{n+m} = 0 \right\},$$

and the boundary of this intersection is the sphere of dimension $n - 2$ defined by the equations

$$x_1 = \frac{r^2}{2R}, \qquad \sum_2{}' x_i^2 = r^2 \left(1 - \frac{r^2}{4R^2}\right), \qquad x_{n+1} = \cdots = x_{n+m} = 0.$$

We shall require that this boundary belong to the level surface (4.97). For this it is necessary that

$$\left(\frac{r^2}{2R} - \xi\right)^2 + \alpha r^2 \left(1 - \frac{r^2}{4R^2}\right) = (a - \xi)^2$$

whence

$$\alpha = r^{-2} \left(1 - \frac{r^2}{4R^2}\right)^{-1} \left(a - \frac{r^2}{2R}\right) \left(a - 2\xi + \frac{r^2}{2R}\right). \tag{4.98}$$

We also require that the surface (4.97) contain the boundary of the intersection of $\omega(R, T, r)$ with the plane $x_2 = \cdots = x_n = 0$, i.e., the sphere

$x_1 = \cdots = x_n = 0, \sum'' x_i^2 = T^2$. For this it is necessary that $\xi^2 + \beta T^2 = (a-\xi)^2$, whence

$$\beta = -\frac{a(2\xi - a)}{T^2}. \tag{4.99}$$

The surface (4.97) in which the parameters α and β are defined by (4.98) and (4.99) we denote by $\gamma(R, T, r, a, \xi)$. The part cut out on $\omega(R, T, r)$ by the surface $\gamma(R, T, r, a, \xi)$ and containing the point $(0, \ldots 0)$ we denote by $\omega^{(1)}(R, T, r, a, \xi)$. Equations of the boundary of $\omega^{(1)}(R, T, r, a, \xi)$ can be obtained by jointly solving the equation

$$(x_1 - R)^2 + \sum_2' x_i^2 = R^2$$

and equation (4.97) and using the fact that for $x \in \omega(R, T, r)$ we have $0 \le x_1 \le r^2/2R$. After executing this procedure, we obtain

$$\partial\omega^{(1)}(R, T, r, a, \xi) = \left\{ x : 0 \le x_1 \le \frac{r^2}{2R}, \sum_2' x_i^2 = -x_1^2 + 2Rx_1, \right.$$
$$\left. \sum'' x_i^2 = \frac{1}{\beta}[-(1-\alpha)x_1^2 + 2(\xi - \alpha R)x_1] + T^2 \right\}. \tag{4.100}$$

That part of the surface $\gamma(R, T, r, a, \xi)$ which is bounded by the surface (4.100) and contains the point $(a, 0, \ldots, 0)$ we denote by $\omega^{(2)}(R, T, r, a, \xi)$. The region bounded by the surfaces $\omega^{(1)}(R, T, r, a, \xi)$ and $\omega^{(2)}(R, T, r, a, \xi)$ we denote by $\Omega(R, T, r, a, \xi)$. Thus,

$$\partial\Omega = \omega^{(1)} \cup \omega^{(2)}$$

(in the sequel we shall also omit all or part of the indices R, T, r, a, ξ where it causes no confusion). The character of Lemma 2 shows that for uniqueness of a solution of (4.80) in Ω it suffices to give the Cauchy data on the surface $\omega^{(1)}$.

It follows from Lemma 2 that to be able to obtain an estimate guaranteeing uniqueness in Ω the quantities α, β, ξ, and a defining the function $\varphi(x)$ should satisfy the following conditions:

a) $d(x_1 - \xi)^2/dx_1 < 0$ for $0 \le x_1 \le a$; $\beta < 0$;

b) there exists $p, 0 < p < 1$, such that $\alpha - p$, $\beta + p > 0$;

c) $(x_1 - \xi)^2 + \alpha^2 \sum_2' x_i^2 - \beta^2 \sum'' x_i^2 \ge \delta > 0$ for $x \in \Omega$.

Conditions a) and b) are equivalent to the conditions

$$\xi > a, \ \beta < 0; \qquad \alpha > -\beta, \ -\beta < 1,$$

while condition c) follows from a) and b). Indeed, since by a) and b) $\alpha > 0$, $\alpha > -\beta$, and $0 < -\beta < 1$, it follows that for $x \in \overline{\Omega}$

$$(x_1-\xi)^2+\alpha^2\sum_2{}' x_i^2-\beta^2\sum{}'' x_i^2 > \left[(x_1-\xi)^2+\alpha\sum_2{}' x_i^2+\beta\sum{}'' x_i^2\right](-\beta)$$
$$\geq -\beta\min_{\overline{\Omega}}\varphi(x) = -\beta(a-\xi)^2 > 0. \quad (4.101)$$

Thus, to obtain an estimate guaranteeing uniqueness in Ω it suffices to require that the quantities a, ξ and α, β (defined by (4.98) and (4.99)) satisfy conditions a) and b).

The second of conditions a) follows from the first by (4.99). Thus, conditions a) and b) lead to the following three conditions on the quantities a and ξ:

$$\xi > a, \qquad \frac{a(2\xi-a)}{T^2} < 1,$$
$$r^{-2}\left(1-\frac{r^2}{4R^2}\right)^{-1}\left(a-\frac{r^2}{2R}\right)\left(a-2\xi+\frac{r^2}{2R}\right) > \frac{a(2\xi-a)}{T^2}. \quad (4.102)$$

We rewrite these inequalities as follows:

$$\xi > a, \qquad \xi < \frac{a}{2}+\frac{T^2}{2a},$$
$$2\left[a-\frac{r^2/2R^2}{1+(r^2/T^2)(1-r^2/4R^2)}\right]\xi < a^2-\frac{r^4/4R^2}{[1+(r^2/T^2)(1-r^2/4R^2)]^2}$$
$$-\frac{r^4/4R^2}{1+(r^2/T^2)(1-r^2/4R^2)}+\frac{r^4/4R^2}{[1+(r^2/T^2)(1-r^2/4R^2)]^2}.$$

Thus, conditions (4.102) are equivalent to the following conditions:

$$\xi > a, \qquad \xi < \frac{a}{2}+\frac{T^2}{2a},$$
$$\xi \lessgtr \frac{1}{2}\left\{a+\frac{r^2/2R}{1+(r^2/T^2)(1-r^2/4R^2)}\right.$$
$$-\left[\frac{r^4/4R^2}{1+(r^2/T^2)(1-r^2/4R^2)}-\frac{r^4/4R^2}{[1+(r^2/T^2)(1-r^2/4R^2)]^2}\right]$$
$$\left.\times\left[a-\frac{r^2/2R}{1+(r^2/T^2)(1-r^2/4R^2)}\right]^{-1}\right\}$$
$$\left(a \gtrless \frac{r^2/2R}{1+(r^2/T^2)(1-r^2/4R^2)}\right). \quad (4.103)$$

The (open) region $\sigma(R,T,r)$ of variation of (a,ξ) satisfying these conditions is bounded by the segment $[r^2/4R,\infty)$ of the ξ-axis and the two segments of the hyperbolas

$$\xi = \frac{a}{2}+\frac{T^2}{2a}, \quad (4.104)$$

$$\xi = \frac{1}{2}\left\{a + \frac{r^2/2R}{1+(r^2/T^2)(1-r^2/4R^2)} - \left[\frac{r^4/4R^2}{1+(r^2/T^2)(1-r^2/4R^2)} - \frac{r^4/4R^2}{[1+(r^2/T^2)(1-r^2/4R^2)]^2}\right] \times \left[a - \frac{r^2/2R}{1+(r^2/T^2)(1-r^2/4R^2)}\right]^{-1}\right\}. \quad (4.105)$$

The segments of the hyperbolas intersect at the point with coordinates

$$\bar{a} = \frac{T^2 + r^2 - \sqrt{(T^2+r^2)^2 - T^2 r^4/R^2}}{r^2/R}, \qquad \bar{\xi} = R(1 + T^2/r^2). \quad (4.106)$$

We note that

$$\bar{a} = \sup_{(a,\xi)\in\sigma} a.$$

Thus, if for given R, T, and r ($r < 2R$) the pair (a, ξ) is such that $(a,\xi) \in \sigma(R,T,r)$ and the quantities α and β are defined by (4.98) and (4.99), then conditions a)–c) are satisfied for the collection (ξ, α, β).

After these preliminary considerations we proceed to the formulation of theorems on estimating a solution of the Cauchy problem of interest to us.

THEOREM 2. *Suppose there are given positive numbers R, T, and r ($r < 2R$), and the numbers a and ξ are such that $(a,\xi) \in \sigma(R,T,r)$. Suppose that on $\overline{\Omega}(R,T,r,a,\xi)$ there is defined a solution $u(x)$ of class $C^1(\overline{\Omega}) \cap C^2(\Omega)$ of inequality* (4.80) *with the conditions*

$$|u(x)|, |\nabla u(x)| \le \begin{cases} \varepsilon & \text{for } x \in \omega^{(1)}(R,T,r,a,\xi), \\ 1 & \text{for } x \in \omega^{(2)}(R,T,r,a,\xi), \end{cases} \quad (4.107)$$

where $\ln(1/\varepsilon) > C_1(R,T,r,a,\xi,A)$. *Let*

$$\Omega(R,T,r,a,\xi|a') = \Omega(R,T,r,a,\xi) \cap \{x\colon \varphi(x) > (a'-\xi)^2\}.$$

Then, for any region $\Omega(R,T,r,a,\xi|a')$ *with* $0 < a' < a$,

$$\int_{\Omega(\cdots|a')} (|\nabla u|^2 + \ln^2 \varepsilon u^2)\,dx \le C_2(R,T,r,a,\xi)\ln^2\varepsilon \exp\left[2\left(1 - \frac{2a'\xi - a'^2}{2a\xi - a^2}\right)\ln\varepsilon\right]. \quad (4.108)$$

PROOF. Since $(a,\xi) \in \sigma(R,T,r)$, conditions (4.102) are satisfied. Defining α and β by (4.98) and (4.99) and setting $p = (\alpha - \beta)/2$ if $\alpha < 1$ and $p = (1-\beta)/2$ if $\alpha > 1$, we satisfy all conditions a)–c) guaranteeing strict positivity of the leading terms on the right side of the inequality of Lemma 2 and the "regular" character of the decay of the function $\varphi(x)$. Indeed, suppose that $\alpha < 1$. Since $0 < -\beta < 1$, it follows that

$$0 < p = \frac{\alpha - \beta}{2} < 1;$$
$$\alpha_i - p = \alpha - p = \frac{\alpha + \beta}{2} > 0, \quad i = 2, \dots, n;$$
$$\alpha_i + p = \beta + p = \frac{\alpha + \beta}{2} > 0, \quad i = n+1, \dots, n+m$$

($\alpha + \beta > 0$ because of b). If $\alpha > 1$, then

$$0 < p = \frac{1 - \beta}{2} < 1;$$
$$\alpha_i - p = \alpha - p = \alpha - \frac{1-\beta}{2} > 1 - \frac{1-\beta}{2} = \frac{1+\beta}{2} > 0, \quad i = 2, \dots, n;$$
$$\alpha_i + p = \beta + p = \frac{1+\beta}{2} > 0, \quad i = n+1, \dots, n+m.$$

The leading coefficient of u^2 on the right side of the inequality of Lemma 2 by (4.101) admits the estimate

$$64p\lambda^3[(x_1 - \xi)^2 + \sum_2' \alpha_i^2 x_i^2 - \sum{}'' \alpha_i^2 x_i^2]e^{2\lambda\varphi} \geq 64\beta p\lambda^3(a - \xi)^2 e^{2\lambda\varphi}$$

in the region $\Omega(R, T, r, a, \xi)$. Since

$$\frac{d}{dx_1}(x_1 - \xi)^2 < 0 \quad \text{for } 0 \leq x_1 \leq a; \quad \alpha > 0, \ \beta < 0,$$

the function $\varphi(x)$ achieves a maximum on $\overline{\Omega}(R, T, r, a, \xi)$ at a point of the support of the Cauchy data $\omega^{(1)}(R, T, r, a, \xi)$: $\max \varphi(x) = \varphi(0) = \xi^2$, and for $x \in \omega^{(2)}(R, T, r, a, \xi)$ we have $\varphi(x) = (a - \xi)^2 < \xi^2$.

Proceeding just as in the proof of Theorem 1, for $\lambda \geq C_1(R, T, r, a, \xi, A)$ we obtain

$$\int_{\Omega(R,t,r,a,\xi)} e^{2\lambda\varphi}(|\nabla u|^2 + \lambda^2 u^2)\, dx \leq C(R, T, r, a, \xi)\lambda^2\{\exp(2\xi^2\lambda)\varepsilon^2 + \exp[2(a - \xi)^2\lambda]\}. \quad (4.109)$$

Let $0 < a' < a$. It is easy to see that then

$$\Omega(R, T, r, a, \xi | a') \subset \Omega(R, T, r, a, \xi)$$

and for $x \in \overline{\Omega}(R, T, r, a, \xi | a')$

$$\min \varphi(x) = (a' - \xi)^2 > (a - \xi)^2.$$

Because of this, from (4.109) we obtain

$$\int_{\Omega(\cdots|a')} (|\nabla u|^2 + \lambda^2 u^2)\,dx$$
$$\leq C(R,T,r,a,\xi)\lambda^2 \left\{\exp[2(\xi^2 - (a'-\xi)^2)\lambda]\varepsilon^2 + \exp[-2((a'-\xi)^2 - (a-\xi)^2)\lambda]\right\}.$$

In this inequality we set

$$\lambda = \frac{1}{\xi^2 - (a-\xi)^2}\ln(1/\varepsilon),$$

which is possible if

$$\frac{1}{\xi^2 - (a-\xi)^2}\ln(1/\varepsilon) \geq C_1(R,T,r,a,\xi,A),$$

i.e.,

$$\ln(1/\varepsilon) \geq [\xi^2 - (a-\xi)^2]C_1(R,T,r,a,\xi,A).$$

We obtain

$$\int_{\Omega(\cdots|a')} \left\{|\nabla u|^2 + \frac{1}{[\xi^2 - (a-\xi)^2]^2}\ln^2\varepsilon u^2\right\} dx$$
$$\leq \frac{2C(R,T,r,a,\xi)}{[\xi^2 - (a-\xi)^2]^2}\ln^2\varepsilon \exp\left\{2\left[1 - \frac{\xi^2 - (a'-\xi)^2}{\xi^2 - (a-\xi)^2}\right]\ln\varepsilon\right\}.$$

From this we obtain the assertion of the theorem with some $C_1(R,T,r,a,\xi,A)$ and $C_2(R,T,r,a,\xi)$.

THEOREM 3. *Suppose there are given positive numbers R, T, and r ($r < 2R$), and the numbers $\bar{a}$ and $\bar{\xi}$ are defined by* (4.106). *Suppose in the region $\Omega(R,T,r,\bar{a},\bar{\xi})$ a solution $u(x) \in C^1(\overline{\Omega}) \cap C^2(\Omega)$ of inequality* (4.80) *is defined such that*

$$u(x), \nabla u(x) = 0 \quad \textit{for } x \in \omega^{(1)}(R,T,r,\bar{a},\bar{\xi}). \tag{4.110}$$

Then

$$u(x) = 0 \quad \textit{for } x \in \Omega(R,T,r,\bar{a},\xi). \tag{4.111}$$

PROOF. The point with coordinates $\bar{a}, \bar{\xi}$ lies on the intersection of the hyperbolas (4.104) and (4.105), i.e., $(\bar{a},\bar{\xi}) \in \partial\sigma(R,T,r)$. We recall that

$$\bar{a} = \sup_{(a,\xi)\in\sigma} a.$$

If α and β are defined by (4.98) and (4.99), then $\beta = -1$ and $\alpha = 1$, and hence the surface $\gamma(R,T,r,\bar{a},\bar{\xi})$ (defined by (4.97)) has in this case the form

$$(x_1 - \xi)^2 + \sum_2' x_i^2 - \sum'' x_i^2 = (\bar{a} - \bar{\xi})^2,$$

while the boundary of the surface $\omega^{(1)}(R,T,r,\bar{a},\bar{\xi})$ (cut out on the cylinder $(x_1-R)^2+\sum'_2 x_i^2=R^2$ by the surface $\gamma(R,T,r,\bar{a},\bar{\xi})$; see (4.100)) is given by the equation

$$\partial\omega^{(1)}(R,T,r,\bar{a},\bar{\xi})=\left\{x\colon 0\le x_1\le\frac{r^2}{2R},\ \sum_2{}' x_i^2=-x_1^2+2Rx_1,\right.$$
$$\left.\sum{}'' x_i^2=-2(\bar{\xi}-R)x_1+T^2\right\}.$$

Suppose the point (a,ξ) is such that $(a,\xi)\in\sigma(R,T,r)$ and

$$(a-\bar{a})^2+(\xi-\bar{\xi})^2<\delta^2 \tag{4.112}$$

with sufficiently small $\delta>0$. Using the method of Theorem 2, it is then easy to obtain from (4.110) the equality

$$u(x)=0\quad\text{for } x\in\Omega(R,T,r,\bar{a},\bar{\xi})\cap\Omega(R,T,r,a,\xi).$$

Points of the regions $\Omega(R,T,r,\bar{a},\bar{\xi})$ and $\Omega(R,T,r,a,\xi)$ are precisely those points for which

$$(x_1-R)^2+\sum_2{}' x_i^2<R^2,\qquad (x_1-\bar{\xi})^2+\sum_2{}' x_i^2-\sum{}'' x_i^2>(\bar{a}-\bar{\xi})^2$$

and

$$(x_1-R)^2+\sum_2{}' x_i^2<R^2,\qquad (x_1-\xi)^2+\alpha\sum_2{}' x_i^2+\beta\sum{}'' x_i^2>(a-\xi)^2$$

respectively. From (4.98) and (4.99) it follows easily that if $(a,\xi)\to(\bar{a},\bar{\xi})$, then $\alpha(a,\xi)\to 1$ and $\beta(a,\xi)\to -1$. Therefore, for all $x\in\Omega(R,T,r,\bar{a},\bar{\xi})$, there exists a $\delta>0$ such that from (4.112) it follows that

$$x\in\Omega(R,T,r,\bar{a},\bar{\xi})\cap\Omega(R,T,r,a,\xi),$$

whence we obtain (4.111).

THEOREM 4. *Let*

$$\Omega(R)=\left\{x\colon (x_1-R^2)+\sum_2{}' x_i^2<R^2,(x_1-2R)^2+\sum_2{}' x_i^2-\sum{}'' x_i^2>0\right\},$$
$$\omega^{(1)}(R)=\overline{\Omega}(R)\cap\left\{x\colon (x_1-R)^2+\sum_2{}' x_i^2=R^2\right\}.$$

Suppose that a solution $u(x)\in C^1(\overline{\Omega})\cap C^2(\Omega)$ of inequality (4.80) is defined in the region $\Omega(R)$ and satisfies the condition $u(x),\nabla u(x)=0$ for $x\in\omega^{(1)}(R)$. Then $u(x)=0$ for $x\in\Omega(R)$.

PROOF. $\Omega(R)$ is a region of the type $\Omega(R, T, r, \bar{a}, \bar{\xi})$ of Theorem 3 for which $T = r = 2R$. The boundary of the support of the Cauchy data $\omega^{(1)}(R)$ is described as follows:

$$\partial\omega^{(1)}(R) = \left\{x\colon 0 \le x_1 \le 2R,\ \sum_2{}' x_i^2 = -x_1^2 + 2Rx_1, \right.$$
$$\left. \sum{}'' x_i^2 = -2Rx_1 + 4R^2 \right\}.$$

The region $\Omega(R)$ is the part of the interior of the cylinder

$$(x_1 - R)^2 + \sum_2{}' x_i^2 = R^2,$$

cut out by the characteristic conoid

$$(x_1 - 2R)^2 + \sum_2{}' x_i^2 - \sum{}'' x_i^2 = 0.$$

Let r' be such that $0 < 2R - r' < \delta$. Applying the method of Theorem 2, we prove that $u(x) = 0$ in any region of the form

$$\Omega(R) \cap \Omega(R, T, r', a, \xi),$$

where $(a, \xi) \in \sigma(R, T, r')$ and $(a - \bar{a})^2 + (\xi - \bar{\xi})^2 \le \delta^2$ (the quantities $\bar{a}$ and $\bar{\xi}$ are determined by (4.106) in terms of R, T, and r'). Since

$$\Omega(R, T, r', a, \xi) \to \Omega(R, T, r', \bar{a}, \bar{\xi})$$

as $(a, \xi) \to (\bar{a}, \bar{\xi})$, and

$$\Omega(R, T, r', \bar{a}, \bar{\xi}) \to \Omega(R)$$

as $r' \to 2R$, the equality $u(x) = 0$ holds for any point $x \in \Omega(R)$. We note that the region $\Omega(R)$ contains the entire ball

$$(x_1 - R)^2 + \sum_2{}' x_i^2 \le R^2, \qquad \sum{}'' x_i^2 = 0.$$

It is easy to see that if the set N is compactly imbedded in $\Omega(R, T, r, \bar{a}, \bar{\xi})$ or $\Omega(R)$, then the $L^2(N)$-norm of a smooth solution of the Cauchy problem in question for inequality (4.80) with an a priori bounded C^1-norm depends in a Hölder-continuous fashion on the C^1-norm of the Cauchy data. (Indeed, if $N \Subset \Omega$, then there exists a pair (a, ξ) such that $(a, \xi) \in \sigma(R, T, r)$ and $N \Subset \Omega(R, T, r, a, \xi)$. It remains to apply the method of Theorem 2 to the intersection $\Omega(R, T, r, a, \xi) \cap \Omega(R, T, r, \bar{a}, \bar{\xi})$ or $\Omega(R, T, r, a, \xi) \cap \Omega(R)$.)

It follows from our results that a nonhyperbolic Cauchy problem for an inequality containing the wave operator possesses the property of Hölder stability if we restrict ourselves to a suitable class of solutions. The Cauchy data are here given on the surface of a cylinder with generators parallel to the time axis, and the solution is estimated inside the cylinder. It is natural to call such a problem an interior problem. As the following example due to F. John shows, the analogous problem in the case where the solution is sought outside the cylinder (it is natural to call it an exterior problem) possesses essentially different properties. Namely, a solution of this problem depends on the Cauchy data logarithmically even in the case where the solution is an entire analytic function of all variables.

EXAMPLE. We consider the equation

$$\Box u \equiv u_{tt} - u_{rr} - \frac{1}{r}u_r - \frac{1}{r^2}u_{\varphi\varphi} = 0. \tag{4.113}$$

It is obvious that for each natural number n the function

$$u_n = J_n(nr)\exp[in(t+\varphi)], \tag{4.114}$$

where $J_n(\xi)$ is the Bessel function of order n, is a solution of (4.113). Clearly u_n is an entire analytic function of the variables t, r, and φ. For $0 < r < 1$

$$|J_n(nr)| \leq \left| \frac{re^{\sqrt{1-r^2}}}{1+\sqrt{1-r^2}} \right|^n \tag{4.115}$$

([73], §8.7, formula (1)), and for such r the modulus on the right is less than one. From this it follows that u_n decays exponentially with increasing n at any point of the cylinder $r < 1$, and this decay is uniform in any smaller cylinder. On the contrary, for $r = 1$

$$|u_n| = |J_n(n)| = \Gamma(1/3)2^{-2/3}3^{-1/6}\pi^{-1}n^{-1/3} + O(n^{-5/3}), \tag{4.116}$$

while for $r > 1$

$$J_n(nr) = \sqrt{\frac{2}{\pi n}}(r^2-1)^{-1/4}\cos\left[-\frac{\pi}{4} + n\left(\sqrt{r^2-1} - \arccos\frac{1}{r}\right)\right] + o(n^{-1/2})$$

uniformly on each compact subinterval of the ray $r > 1$. Thus, for $r \geq 1, u_n$ decays like a negative power of n. Analogous estimates hold also for the derivatives of u_n. For derivatives with respect to t and φ this follows from (4.114)–(4.116), while the derivatives with respect to r can be estimated by using the formula $2J_n' = J_{n-1} + J_{n+1}$. Proceeding in this way, it can be shown that the derivatives of u_n of order s can be estimated by a quantity of order $n^{s-1/3}$ uniformly with respect to r, t, and φ.

We now consider the problem of extending a solution u from a cylinder S_0 of radius $r_0 < 1$ to a cylinder S_1 of radius 1, and we shall consider only solutions that are entire analytic functions in t, r, and φ. By Holmgren's principle, a solution u is defined for all t, r, and φ by giving its Cauchy data on S_r. We introduce the norms $(-\infty < t < \infty,\ 0 \le \varphi \le 2\pi)$

$$\|u\| = \sup_{\alpha \le s} |D^\alpha u| \quad (0 < r < \infty),$$

$$\|u\|_1 = \sup |u| \quad (r = 1), \qquad \|u\|_0 = \sup_{\alpha \le 1} |D^\alpha u| \quad (r = r_0).$$

By the foregoing we have

$$\|u_n\|_1 \ge a n^{-1/3}, \qquad \|u_n\| < b n^{s-1/3}, \qquad \|u_n\|_0 \le q^n, \qquad a, b > 0; \quad q < 1.$$

From this it follows that there exists a constant $C > 0$ such that for $u = u_n$ and sufficiently large n

$$\frac{\|u\|_1}{\|u\|} \ge C \left(\ln \frac{\|u\|}{\|u\|_0} \right)^{-s}.$$

This shows that u on S_1 is at best logarithmically continuous with respect to the norm of its Cauchy data defined on S_0.

It is of interest to construct classes of solutions for which the exterior nonhyperbolic problem possesses the property of Hölder-continuous dependence of solutions on the initial data. We shall show that such a class is formed by functions $u(t, r, \varphi)$ twice differentiable with respect to t and r which admit analytic continuation in φ to a fixed complex domain containing the segment $[0, 2\pi]$ and which are bounded in this domain by a fixed constant. For this we need some preliminary results.

LEMMA 3. *Suppose that the functions $u(t, r)$ and $\psi(t, r)$ are twice differentiable with respect to t and r, $l(t, r)$ is differentiable three times with respect to these variables, and $\theta(t, r) = \exp l(t, r)$. Then*

$$\begin{aligned}
\theta^2 (u_{tt} - u_{rr})^2 \ge{} & 2\theta^2 [(\Delta l - \psi) u_t^2 + (\Delta l + \psi) u_r^2 - 4 l_{tr} u_t u_r] \\
& + 2\theta^2 \Big[2 \Box l (l_t^2 - l_r^2) - \Box (l_t^2 - l_r^2) + 2\psi (l_t^2 - l_r^2) \\
& \qquad + \psi_t l_t - \psi_r l_r + \frac{1}{2} \Box \psi - \frac{1}{2} \psi^2 \Big] u^2 \\
& + \Big\{ 2\theta \big[l_t (u_t^2 - u_r^2) - 2 u_t (l_t u_t - l_r u_r) + (2(l_r^2 - l_t^2) + \psi) u_t u \\
& \qquad - \Big(\Big(2 l_t - \frac{\partial}{\partial t} \Big) (l_t^2 - l_r^2) + \frac{1}{2} \psi_t \Big) u^2 \big] \Big\}_t \\
& + \Big\{ 2\theta^2 \big[l_r (u_r^2 - u_t^2) - 2 u_r (l_r u_r - l_t u_t) - (2(l_r^2 - l_t^2) + \psi) u_r u \\
& \qquad + \Big(\Big(2 l_r - \frac{\partial}{\partial r} \Big) (l_t^2 - l_r^2) + \frac{1}{2} \psi_r \Big) u^2 \big] \Big\}_r
\end{aligned}$$

$$\left(\Delta = (\partial^2/\partial t^2) + (\partial^2/\partial r^2), \qquad \Box = (\partial^2/\partial t^2) - (\partial^2/\partial r^2) \right).$$

PROOF. Lemma 3 follows from Lemma 1 if we set $\kappa_1 = 1, \kappa_2 = -1, x_1 = t$ and $x_2 = r$.

LEMMA 4. *Suppose $u(t,r)$, $l(t,r)$, and $\theta(t,r)$ satisfy the conditions of Lemma 3. Then for any real n*

$$\theta^2\left(u_{tt} - u_{rr} - \frac{1}{r}u_r + \frac{n^2}{r^2}u\right)^2$$
$$\geq 2\theta^2\left[\left(\Delta l - \frac{1}{r}l_r + \frac{1}{2r^2}\right)u_t^2 + \left(\Delta l - \frac{1}{r}l_r + \frac{1}{r^2}\right)u_r^2 + \left(-4l_{tr} + \frac{2}{r}l_t\right)u_tu_r\right]$$
$$+ 2\theta^2\left[2\Box l(l_t^2 - l_r^2) + \Box l\frac{n^2}{r^2} + 3l_r\frac{n^2}{r^3} - \Box(l_t^2 - l_r^2) - \frac{3}{2}\frac{n^2}{r^4}\right]u^2$$
$$+ \Bigg\{2\theta^2\Big[l_t(u_t^2 - u_r^2) - 2u_t(l_tu_t - l_ru_r)$$
$$- \frac{1}{r}u_tu_r - 2(l_t^2 - l_r^2)u_tu - \left(\left(2l_t - \frac{\partial}{\partial t}\right)(l_t^2 - l_r^2) + l_t\frac{n^2}{r^2}\right)u^2\Big]\Bigg\}_t$$
$$+ \Bigg\{2\theta^2\Big[l_r(u_r^2 - u_t^2) + 2u_r(l_tu_t - l_ru_r) + \frac{1}{2r}(u_t^2 + u_r^2)$$
$$+ 2(l_t^2 - l_r^2)u_ru + \left(\left(2l_r - \frac{\partial}{\partial r}\right)(l_t^2 - l_r^2) + l_r\frac{n^2}{r^2} - \frac{1}{2}\frac{n^2}{r^3}\right)u^2\Big]\Bigg\}_r.$$

PROOF. We have

$$\theta^2\left(u_{tt} - u_{rr} - \frac{1}{r}u_r + \frac{n^2}{r^2}u\right)^2 = \theta^2(u_{tt} - u_{rr})^2 + \theta^2\frac{1}{r^2}u_r^2 + \theta^2\frac{n^4}{r^4}u^2$$
$$-2\theta^2\frac{1}{r}(u_{tt} - u_{rr})u_r + 2\theta^2\frac{n^2}{r^2}(u_{tt} - u_{rr})u - 2\theta^2\frac{n^2}{r^3}uu_r. \quad (4.117)$$

We transform the last three terms to obtain expressions containing (up to a divergence) only u_t^2, u_r^2, u_tu_r, and u^2. We have

$$-2\theta^2\frac{1}{r}(u_{tt} - u_{rr})u_r = 2\theta^2\frac{1}{r}u_tu_{tr} + 4\theta\theta_t\frac{1}{r}u_tu_r$$
$$- \left(2\theta\theta_r\frac{1}{r} - \theta^2\frac{1}{r^2}\right)u_r^2 - \left(2\theta^2\frac{1}{r}u_tu_r\right)_t + \left(\theta^2\frac{1}{r}u_r^2\right)_r$$
$$= -\left(2\theta\theta_r\frac{1}{r} - \theta^2\frac{1}{r^2}\right)(u_t^2 + u_r^2) + 4\theta\theta_t\frac{1}{r}u_tu_r - \left(2\theta^2\frac{1}{r}u_tu_r\right)_t$$
$$+ \left[\theta^2\frac{1}{r}(u_t^2 + u_r^2)\right]_r$$
$$= 2\theta^2\left(-\frac{1}{r}l_r + \frac{1}{2r^2}\right)(u_t^2 + u_r^2)$$
$$+ 4\theta^2\frac{1}{r}l_tu_tu_r - \left(2\theta^2\frac{1}{r}u_tu_r\right)_t + \left[\theta^2\frac{1}{r}(u_t^2 + u_r^2)\right]_r,$$

$$
\begin{aligned}
&2\theta^2\frac{n^2}{r^2}(u_{tt}-u_{rr})u\\
&\quad= -2\theta^2\frac{n^2}{r^2}u_t^2 + 2\theta^2\frac{n^2}{r^2}u_r^2 - 4\theta\theta_t\frac{n^2}{r^2}uu_t\\
&\qquad + 2\left(2\theta\theta_r\frac{n^2}{r^2} - 2\theta^2\frac{n^2}{r^3}\right)uu_r + \left(2\theta^2\frac{n^2}{r^2}u_tu\right)_t - \left(2\theta^2\frac{n^2}{r^2}u_ru\right)_r\\
&\quad= -2\theta^2\frac{n^2}{r^2}(u_t^2-u_r^2) + 2(\theta_t^2+\theta\theta_{tt})\frac{n^2}{r^2}u^2\\
&\qquad - \left(2\theta_r^2\frac{n^2}{r^2} + 2\theta\theta_{rr}\frac{n^2}{r^2} - 8\theta\theta_r\frac{n^2}{r^3} + 6\theta^2\frac{n^2}{r^4}\right)u^2\\
&\qquad + \left(2\theta^2\frac{n^2}{r^2}u_tu - 2\theta\theta_t\frac{n^2}{r^2}u^2\right)_t\\
&\qquad + \left[-2\theta^2\frac{n^2}{r^2}u_ru\left(2\theta\theta_r\frac{n^2}{r^2} - 2\theta^2\frac{n^2}{r^3}\right)u^2\right]_r.
\end{aligned}
$$

Since $\theta_{rr}/\theta = l_{rr} + l_r^2$, it follows that

$$
\begin{aligned}
&2\theta^2\frac{n^2}{r^2}(u_{tt}-u_{rr})u\\
&\quad= -2\theta^2\frac{n^2}{r^2}(u_t^2-u_r^2) + 2\theta^2\frac{n^2}{r^2}\left[2(l_t^2-l_r^2) + \Box l + \frac{4}{r}l_r - \frac{3}{r^2}\right]u^2\\
&\qquad + \left[2\theta^2\frac{n^2}{r^2}(u_tu - l_tu^2)\right]_t + \left\{2\theta\frac{n^2}{r^2}\left[-u_ru + \left(l_r - \frac{1}{r}\right)u^2\right]\right\}_r.
\end{aligned}
$$

Further,

$$
\begin{aligned}
-2\theta^2\frac{n^2}{r^3}uu_r &= \left(2\theta\theta_r\frac{n^2}{r^3} - 3\theta^2\frac{n^2}{r^4}\right)u^2 + \left(\theta^2\frac{n^2}{r^3}u^2\right)_r\\
&= 2\theta^2\left(l_r\frac{n^2}{r^3} - \frac{3}{2}\frac{n^2}{r^4}\right)u^2 - \left(\theta^2\frac{n^2}{r^3}u^2\right)_r.
\end{aligned}
$$

Using Lemma 3 with $\psi = -n^2r^{-2}$ to estimate $\theta^2(u_{tt}-u_{rr})^2$ and substituting the result into (4.117), we obtain the lemma.

LEMMA 5. *Suppose that for* $0<\alpha<1$, $R\geq 1$, *and* $T\geq 1$

$$
\begin{aligned}
\theta(t,r) &= \exp l(t,r) = \exp\{\lambda[-\alpha t^2 + (R+T-r)^2]\},\\
\Omega_\alpha &= \{(t,r)\colon -T<t<T, R<r<R+T-\sqrt{(1-\alpha)T^2+\alpha t^2}\}.
\end{aligned}
$$

Suppose, moreover, that

$$
\lambda \geq 2(1-\alpha)^{-1}, \qquad N = \frac{1}{2}(1-\alpha)^{1/2}\lambda, \qquad |n|\leq N.
$$

Then for any point $(t,r)\in\Omega_\alpha$

$$(1-\alpha)\lambda\theta^2(u_t^2+u_r^2)+(1-\alpha)\lambda^3\theta^2u^2 \le \theta^2\left(u_{tt}-u_{rr}-\frac{1}{r}u_r+\frac{n^2}{r^2}u\right)^2+\operatorname{div}U,$$

where

$$|U|\le C\lambda^3T^3\theta^2(u_t^2+u_r^2+u^2)$$

with some absolute constant C.

PROOF. We note that Ω_α is the region of the half-plane $r>R$ bounded by the level line of the function $l(t,r)$ passing through the points (T,R) and $(-T,R)$. If π is the triangle with vertices (T,R), $(0,R+T)$, and $(-T,R)$, then $\Omega_\alpha\subset\pi$ and $\Omega_\alpha\to\pi$ as $\alpha\to 1$. The inequality of Lemma 4 in the present case has the form

$$\begin{aligned}\theta^2&\left(u_{tt}-u_{rr}-\frac{1}{r}u_r+\frac{n^2}{r^2}u\right)^2\\ &\ge 2\theta^2\left\{\left[2\lambda\left(\frac{R+T}{r}-\alpha\right)+\frac{1}{2r^2}\right]u_t^2\right.\\ &\qquad\left.+\left[2\lambda\left(\frac{R+T}{r}-\alpha\right)+\frac{1}{r^2}\right]u_r^2-4\alpha\lambda\frac{t}{r}u_tu_r\right\}\\ &\quad+2\theta^2\left\{16(1+\alpha)\lambda^3[(R+T-r)^2-\alpha^2t^2]-2(1+\alpha)\lambda\frac{n^2}{r^2}\right.\\ &\qquad\left.-6\lambda(R+T-r)\frac{n^2}{r^3}-8(1+\alpha^2)\lambda^2-\frac{3}{2}\frac{n^2}{r^4}\right\}u^2+\operatorname{div}U \qquad (4.118)\end{aligned}$$

(we shall write out the components of the vector U later).

Since

$$-4\alpha\lambda\frac{t}{r}u_tu_r\ge-2\alpha\lambda\frac{|t|}{r}(u_t^2+u_r^2),$$

the first term on the right side of (4.118) can be bounded below by the quantity

$$4\lambda\theta^2\left(\frac{R+T}{r}-\alpha\frac{|t|}{r}-\alpha\right)(u_t^2+u_r^2).$$

In the domain Ω_α, the function

$$f(t,r)=\frac{R+T}{r}-\alpha\frac{|t|}{r}-\alpha$$

has no extremal points ($f_t=-\alpha\operatorname{sgn}t\cdot r^{-1}\ne 0$, and $f_r(0,r)=-(R+T)/r^2\ne 0$). Therefore, to bound $f(t,r)$ below on $\overline{\Omega}_\alpha$ we consider its behavior on $\partial\Omega_\alpha$.

Let

$$\Gamma'_\alpha = \{(t,r)\colon\ -T \le t \le T,\ r = R\},$$
$$\Gamma''_\alpha = \{(t,r)\colon\ -T \le t \le T,\ -\alpha t^2 + (R+T-r)^2 = (1-\alpha)T^2\}.$$

Then $\Gamma'_\alpha \cup \Gamma''_\alpha = \partial\Omega_\alpha$. We have

$$f(t,r)|_{\Gamma'_\alpha} = \frac{R+T-\alpha|t|}{R} - \alpha \ge \frac{R+T-\alpha T}{R} - \alpha$$
$$= (1-\alpha)\frac{R+T}{R} > 1-\alpha.$$

For $(t,r) \in \Gamma''_\alpha$ we have

$$t^2 = \alpha^{-1}[(R+T-r)^2 - (1-\alpha)T^2],$$

whence

$$|t| = \alpha^{-1/2}\sqrt{(R+T-r)^2 - (1-\alpha)T^2} < \alpha^{-1/2}(R+T-r),$$

and so

$$f(t,r)|_{\Gamma''_\alpha} > \frac{R+T}{r} - \alpha^{1/2}\frac{R+T-r}{r} - \alpha = (1-\alpha^{1/2})\frac{R+T}{r} + \alpha^{1/2} - \alpha$$
$$> (1-\alpha^{1/2})\frac{R+T}{r} > 1-\alpha^{1/2} > \frac{1}{2}(1-\alpha).$$

Thus, $f(t,r) > \frac{1}{2}(1-\alpha)$ for $(t,r) \in \overline{\Omega}_\alpha$, and hence the first term on the right side of (4.118) can be bounded below by the quantity

$$2(1-\alpha)\lambda\theta^2(u_t^2 + u_r^2).$$

The second term on the right side of (4.118) on the basis of the hypotheses of the lemma ($T, R \ge 1, |n| \le N$) can be bounded below by the expression

$$2\theta^2\left\{16\lambda^3[(R+T-r)^2 - \alpha t^2] - 4\lambda N^2R^{-2} - 6\lambda N^2TR^{-3}\right.$$
$$\left. -16\lambda^2 - \frac{3}{2}N^2R^{-4}\right\}u^2$$
$$\ge 2\theta^2\left[16(1-\alpha)\lambda^3T^2 - N^2\left(4\lambda + 6\lambda T + \frac{3}{2}\right) - 16\lambda^2\right]u^2$$
$$\ge 2\theta^2[16(1-\alpha)\lambda^3T^2 - 12\lambda TN^2 - 16\lambda^2]u^2$$
$$\ge 2\theta^2\left\{8\lambda T\left[(1-\alpha)\lambda^2 - \frac{3}{2}N^2\right] + 8\lambda^2[(1-\alpha)\lambda - 2]\right\}u^2.$$

If $\lambda \ge 2/(1-\alpha)$ and $N = (\lambda/2)\sqrt{1-\alpha}$, the last expression is not less than

$$8(1-\alpha)\lambda^3\theta^2u^2.$$

We shall bound (this time above) the modulus of U in (4.118). We have

$$
\begin{aligned}
U_1 = 2\theta^2 \{&2\alpha\lambda t(u_r^2 - u_t^2) + 4\lambda u_t[\alpha t u_t - (R+T-r)u_r] \\
&- (1/r)u_t u_r + 8\lambda^2[(R+T-r)^2 - \alpha^2 t^2]u_t u \\
&+ \left[-16\alpha\lambda^3 t((R+T-r)^2 - \alpha^2 t^2) + 8\alpha^2\lambda^2 t + 2\alpha\lambda n^2(t/r^2)\right] u^2\},
\end{aligned}
$$

whence we easily obtain

$$|U_1| \le C\lambda^3 T^3 \theta^2 (u_t^2 + u_r^2 + u^2),$$

and, since

$$
\begin{aligned}
U_2 = 2\theta^2 \Bigg\{&2\lambda(R+T-r)(u_t^2 - u_r^2) + 4\lambda u_r[-\alpha t u_t + (R+T-r)u_r] \\
&+ \frac{1}{2r}(u_t^2 + u_r^2) - 8\lambda^2[(R+T-r)^2 - \alpha^2 t^2]u_r u \\
&+ \Bigg[16\lambda^3(R+T-r)((R+T-r)^2 - \alpha^2 t^2) \\
&\quad - 8\lambda^2(R+T-r) - 2\lambda n^2(R+T-r)\frac{1}{r^2} - \frac{1}{2}\frac{n^2}{r^3}\Bigg] u^2\Bigg\},
\end{aligned}
$$

it follows that $|U_2|$ admits the analogous estimate $|U_2| \le C\lambda^3 T^3 \theta^2 (u_t^2 + u_r^2 + u^2)$. The lemma is proved.

Below, for functions of the form $f(t,r,\varphi)$, we use the notation

$$\|f(t,r)\| = \left[\int_0^{2\pi} |f(t,r,\varphi)|^2\, d\varphi\right]^{1/2}.$$

THEOREM 5. *Suppose the coefficients and solution of the equation*

$$
\begin{aligned}
&u_{tt} - u_{rr} - (1/r)u_r - (1/r^2)u_{\varphi\varphi} \\
&\qquad = a(t,r,\varphi)u_t + b(t,r,\varphi)u_r + c(t,r,\varphi)u_\varphi + d(t,r,\varphi)u
\end{aligned}
$$

are defined for $(t,r) \in \overline{\Omega}_\alpha(T,R)$ and $0 \le \varphi \le 2\pi$, where $T, R > 1$, and satisfy the following conditions:

1) $|a|$, $|b|$, $|c|$, $|d| \le A$;

2) u, u_t, u_r, u_{tt}, *and* u_{rr} *for* $(t,r) \in \overline{\Omega}_\alpha(T,R)$ *are analytic in* φ *in some (open) region* Φ *of the complex* φ*-plane containing the real segment* $[0, 2\pi]$, *where for* $(t,r) \in \overline{\Omega}_\alpha(T,R)$ *and* $\varphi \in \Phi$

$$\sup(|u|, |u_t|, |u_r|, |u_\varphi|) \le 1;$$

3) *for* $t \in [-T, T]$

$$\|u(t,R)\|,\ \|u_t(t,R)\|,\ \|u_r(t,R)\| \le \varepsilon.$$

Then there exist positive constants $\kappa(\alpha, A, \Phi)$, $C(\alpha, T, A)$, and $\beta(\alpha, T, \Phi)$ $(< \alpha)$, such that for $\varepsilon^{-1} \geq \kappa(\alpha, A, \Phi)$,

$$\int_{\omega(\beta)} \|u\|^2 \, dt \, dr \leq C(\alpha, T, A)\varepsilon$$

$$(\omega(\beta) = \{(t, r)\colon -\sqrt{\beta/\alpha}\, T \leq t \leq \sqrt{\beta/\alpha}\, T,$$
$$R \leq r \leq R + T - \sqrt{(1-\beta)T^2 + \alpha t^2}\}).$$

PROOF. Let

$$u_0 = \frac{1}{\sqrt{2\pi}} \int_0^{2\pi} u \, d\varphi, \qquad u_{\pm n} = \frac{1}{\sqrt{\pi}} \int_0^{2\pi} u_{\sin}^{\cos} n\varphi \, d\varphi, \quad n = 1, 2, \ldots .$$

From condition 2) of the theorem it follows that there exists $\sigma > 0$ such that

$$|u_n|, |u_{nt}|, |u_{nr}|, |nu_n| \leq Ce^{-\sigma n} \tag{4.119}$$

with some absolute constant C. Exponential decay holds for u_{ntt} and u_{nrr} as well.

Because of condition 1) of the theorem, we have

$$\|au_t + bu_r + cu_\varphi + du\|^2 \leq 4A^2(\|u_t\|^2 + \|u_r\|^2 + \|u_\varphi\|^2 + \|u\|^2)$$
$$= 4A^2 \sum_{-\infty}^{\infty} [u_{nt}^2 + u_{nr}^2 + (n^2+1)u_n^2]. \tag{4.120}$$

Suppose the function $\theta(t, r)$, the region Ω_α, and the numbers n, N, and λ satisfy the conditions of Lemma 5. Integrating (4.120) over Ω_α and applying Lemma 5 to $u_n(t, r)$ for $|n| \leq N$, we obtain

$$\int_{\Omega_\alpha} \theta^2 \|au_t + bu_r + cu_\varphi + du\|^2 \, dt \, dr$$
$$\leq 4A^2 \sum_{|n|\leq N} \int_{\Omega_\alpha} \theta^2 (u_{nt}^2 + u_{nr}^2) \, dt \, dr + 4A^2(N^2+1) \sum_{|n|\leq N} \int_{\Omega_\alpha} \theta^2 u_n^2 \, dt \, dr$$
$$+ 4A^2 \int_{\Omega_\alpha} \theta^2 \sum_{|n|>N} [u_{nt}^2 + u_{nr}^2 + (n^2+1)u_n^2] \, dt \, dr$$
$$\leq 4A^2(1-\alpha)^{-1}\lambda^{-1} \sum_{|n|\leq N} \int_{\Omega_\alpha} \theta^2 \left(u_{ntt} - u_{nrr} - \frac{1}{r}u_{nr} + (n^2/r^2)u_n \right)^2 dt \, dr$$
$$+ 4A^2(1-\alpha)^{-1} C\lambda^2 T^3 \sum_{|n|\leq N} \int_{\partial\Omega_\alpha} \theta^2 (u_{nt}^2 + u_{nr}^2 + u_n^2) \, ds$$
$$+ 4A^2(1-\alpha)^{-1}\lambda^{-3}(N^2+1) \sum_{|n|\leq N} \int_{\Omega_\alpha} \theta^2 (u_{ntt} - u_{nrr} - (1/r)u_{nr}$$
$$+ (n^2/r^2)u_n)^2 \, dt \, dr$$

(*Continued*)

(*Continued*)

$$+ 4A^2(1-\alpha)^{-1}CT^3(N^2+1)\sum_{|n|\le N}\int_{\partial\Omega_\alpha}\theta^2(u_{nt}^2+u_{nr}^2+u_n^2)\,ds$$

$$+ 4A^2\int_{\Omega_\alpha}\theta^2\sum_{|n|>N}[u_{nt}^2+u_{nr}^2+(n^2+1)u_n^2]\,dt\,dr$$

$$= 4A^2(1-\alpha)^{-1}\lambda^{-3}(\lambda^2+N^2+1)\sum_{|n|\le N}\int_{\Omega_\alpha}\theta^2(u_{ntt}-u_{nrr}$$

$$-(1/r)u_{nr}+(n^2/r^2)u_n)^2\,dt\,dr$$

$$+ 4A^2(1-\alpha)^{-1}CT^3(\lambda^2+N^2+1)\sum_{|n|\le N}\int_{\partial\Omega_\alpha}\theta^2(u_{nt}^2+u_{nr}^2+u_n^2)\,ds$$

$$+ 4A^2\int_{\Omega_\alpha}\theta^2\sum_{|n|>N}[u_{nt}^2+u_{nr}^2+(n^2+1)u_n^2]\,dt\,dr$$

$$\le 4A^2(1-\alpha)^{-1}\lambda^{-1}\left(1+\frac{1-\alpha}{4}+\lambda^{-2}\right)\int_{\Omega_\alpha}\sum_{-\infty}^{\infty}\theta^2(u_{ntt}-u_{nrr}-(1/r)u_{nr}$$

$$+(n^2/r^2)u_n)^2\,dt\,dr$$

$$+ 4A^2(1-\alpha)^{-1}C\lambda^2T^3\left(1+\frac{1-\alpha}{4}+\lambda^{-2}\right)\int_{\partial\Omega_\alpha}\sum_{-\infty}^{\infty}\theta^2(u_{nt}^2+u_{nr}^2+u_n^2)\,ds$$

$$+ 4A^2\int_{\Omega_\alpha}\theta\sum_{|n|>N}[u_{nt}^2+u_{nr}^2+(n^2+1)u_n^2]\,dt\,dr.$$

Since

$$\left\|u_{tt}-u_{rr}-\frac{1}{r}u_r-\frac{1}{r^2}u_{\varphi\varphi}\right\|^2$$

$$=\left\|\left(\frac{\partial^2}{\partial t^2}-\frac{\partial^2}{\partial r^2}-\frac{1}{r}\frac{\partial}{\partial r}-\frac{1}{r^2}\frac{\partial^2}{\partial\varphi^2}\right)\right.$$

$$\left.\times\left[\frac{1}{\sqrt{2\pi}}u_0+\frac{1}{\sqrt{\pi}}\sum_{n=1}^{\infty}(u_n\cos n\varphi+u_{-n}\sin n\varphi)\right]\right\|^2$$

$$=\sum_{-\infty}^{\infty}\left(u_{ntt}-u_{nrr}-\frac{1}{r}u_{nr}+\frac{n^2}{r^2}u_n\right)^2,$$

$$\|u_t\|^2+\|u_r\|^2+\|u\|^2=\sum_{-\infty}^{\infty}(u_{nt}^2+u_{nr}^2+u_n^2),$$

$$1+\frac{1-\alpha}{4}+\lambda^{-2}\le 1+\frac{1-\alpha}{4}+\frac{(1-\alpha)^2}{4}<2,$$

it follows that

$$\int_{\Omega_\alpha} \theta^2 \|au_t + bu_r + cu_\varphi + du\|^2 \, dt\, dr$$
$$\leq 8A^2(1-\alpha)^{-1}\lambda^{-1} \int_{\Omega_\alpha} \theta^2 \left\| u_{tt} - u_{rr} - \frac{1}{r}u_r - \frac{1}{r^2}u_{\varphi\varphi} \right\|^2 dt\, dr$$
$$+ 8A^2(1-\alpha)^{-1}C\lambda^2T^3 \int_{\Omega_\alpha} \theta^2(\|u_t\|^2 + \|u_r\|^2 + \|u\|^2)\, ds$$
$$+ 4A^2 \int_{\Omega_\alpha} \sum_{|n|>N} [u_{nt}^2 + u_{nr}^2 + (n^2+1)u_n^2]\, dt\, dr.$$

Since

$$\|u_{tt} - u_{rr} - (1/r)u_r - (1/r^2)u_{\varphi\varphi}\| = \|au_t + bu_r + cu_\varphi + du\|,$$

from the last inequality it follows that

$$\left[1 - 8A^2(1-\alpha)^{-1}\lambda^{-1}\right] \int_{\Omega_\alpha} \theta^2 \|au_t + bu_r + cu_\varphi + du\|^2 \, dt\, dr$$
$$\leq 8A^2(1-\alpha)^{-1}C\lambda^2T^3 \int_{\partial\Omega_\alpha} \theta^2(\|u_t\|^2 + \|u_r\|^2 + \|u\|^2)\, ds$$
$$+ 4A^2 \int_{\Omega_\alpha} \theta^2 \sum_{|n|>N} [u_{nt}^2 + u_{nr}^2 + (n^2+1)u_n^2]\, dt\, dr. \tag{4.121}$$

Let

$$\lambda \geq 16A^2(1-\alpha)^{-1}. \tag{4.122}$$

Under this condition $1 - 8A^2(1-\alpha)^{-1}\lambda^{-1} \geq 1/2$, and from (4.121) we obtain

$$\int_{\Omega_\alpha} \theta^2 \|au_t + bu_r + cu_\varphi + du\|^2 \, dt\, dr$$
$$\leq CA^2(1-\alpha)^{-1}\lambda^2T^3 \int_{\partial\Omega_\alpha} \theta^2(\|u_t\|^2 + \|u_r\|^2 + \|u\|^2)\, ds$$
$$+ 8A^2 \int_{\Omega_\alpha} \theta^2 \sum_{|n|>N} [u_{nt}^2 + u_{nr}^2 + (n^2+1)u_n^2]\, dt\, dr \tag{4.123}$$

with some new constant C. We have estimated the right side of the original equation. We shall use this estimate to obtain an analogous estimate of the solution $u(t, r, \varphi)$. Again applying Lemma 5, we have

$$\int_{\Omega_\alpha} \theta^2 \|u\|^2 \, dt\, dr = \sum_{|n|\leq N} \int_{\Omega_\alpha} \theta^2 u_n^2 \, dt\, dr + \int_{\Omega_\alpha} \theta^2 \sum_{|n|>N} u_n^2 \, dt\, dr$$
$$\leq (1-\alpha)^{-1}\lambda^{-3} \sum_{|n|\leq N} \int_{\Omega_\alpha} \theta^2 \left(u_{ntt} - u_{nrr} - (1/r)u_{nr} + (n^2/r^2)u_n\right)^2 dt\, dr$$

(Continued)

(*Continued*)

$$
\begin{aligned}
&+C(1-\alpha)^{-1}T^3 \sum_{|n|\le N} \int_{\partial\Omega_\alpha} \theta^2 (u_{nt}^2 + u_{nr}^2 + u_n^2)\, ds \\
&+\int_{\Omega_\alpha} \theta^2 \sum_{|n|>N} u_n^2 \, dt\, dr \\
\le\ & (1-\alpha)^{-1}\lambda^{-3} \int_{\Omega_\alpha} \theta^2 \left\| u_{tt} - u_{rr} - \frac{1}{r}u_r - \frac{1}{r^2}u_{\varphi\varphi} \right\|^2 dt\, dr \\
&+C(1-\alpha)^{-1}T^3 \int_{\partial\Omega_\alpha} \theta^2 (\|u_t\|^2 + \|u_r\|^2 + \|u\|^2)\, ds \\
&+\int_{\Omega_\alpha} \theta^2 \sum_{|n|>N} u_n^2 \, dt\, dr \\
=\ & (1-\alpha)^{-1}\lambda^{-3} \int_{\Omega_\alpha} \theta^2 \|au_t + bu_r + cu_\varphi + du\|^2 \, dt\, dr \\
&+C(1-\alpha)^{-1}T^3 \int_{\partial\Omega_\alpha} \theta^2 (\|u_t\|^2 + \|u_r\|^2 + \|u\|^2)\, ds + \int_{\Omega_\alpha} \theta^2 \sum_{|n|>N} u_n^2 \, dt\, dr.
\end{aligned}
$$

If λ satisfies (4.122), it follows from the last inequality and (4.123) that

$$
\begin{aligned}
\int_{\Omega_\alpha} \theta^2 \|u\|^2 \, dt\, dr \le\ & [CA^2(1-\alpha)^{-2}\lambda^{-1}T^3 + C(1-\alpha)^{-1}T^3] \\
&\times \int_{\partial\Omega_\alpha} \theta^2 (\|u_t\|^2 + \|u_r\|^2 + \|u\|^2)\, ds \\
&+ 8A^2(1-\alpha)^{-1}\lambda^{-3} \int_{\Omega_\alpha} \theta^2 \sum_{|n|>N} [u_{nt}^2 + u_{nr}^2 + (n^2+1)u_n^2]\, dt\, dr \\
&+ \int_{\Omega_\alpha} \theta^2 \sum_{|n|>N} u_n^2 \, dt\, dr.
\end{aligned}
\tag{4.124}
$$

By (4.122),

$$
\begin{aligned}
&CA^2(1-\alpha)^{-2}\lambda^{-1}T^3 + C(1-\alpha)^{-1}T^3 \\
&\quad \le CA^2(1-\alpha)^{-2}\frac{1-\alpha}{16}A^{-2}T^3 + C(1-\alpha)^{-1}T^3 \le C(1-\alpha)^{-1}T^3,
\end{aligned}
$$

and, since λ satisfies the conditions of Lemma 5,

$$
\begin{aligned}
&8A^2(1-\alpha)^{-1}\lambda^{-3} \sum_{|n|>N} [u_{nt}^2 + u_{nr}^2 + (n^2+1)u_n^2] + \sum_{|n|>N} u_n^2 \\
&\quad \le 8A^2(1-\alpha)^{-1}\frac{(1-\alpha)^3}{8} \sum_{|n|>N} [u_{nt}^2 + u_{nr}^2 + (n^2+1)u_n^2] + \sum_{|n|>N} u_n^2 \\
&\quad \le (A^2+1) \sum_{|n|>N} [u_{nt}^2 + u_{nr}^2 + n^2 u_n^2].
\end{aligned}
$$

From this and (4.124) we obtain

$$\int_{\Omega_\alpha} \theta^2 \|u\|^2 \, dt \, dr \leq C(1-\alpha)^{-1} T^3 \int_{\partial\Omega_\alpha} \theta^2 (\|u_t\|^2 + \|u_r\|^2 + \|u\|^2) \, ds$$
$$+(A^2+1) \int_{\Omega_\alpha} \theta^2 \sum_{|n|>N} [u_{nt}^2 + u_{nr}^2 + n^2 u_n^2] \, dt \, dr. \quad (4.125)$$

Suppose $0 < \beta < \alpha$ and let $\omega(\beta)$ be the region lying in the half-plane $r > R$ and bounded by the line $r = R$ and the left branch of the hyperbola

$$-\alpha t^2 + (R + T - r)^2 = (1-\beta)T^2$$

(this branch intersects the line $r = R$ at the points $(-\sqrt{\beta/\alpha}T, R), (\sqrt{\beta/\alpha}T, R)$ and the line $t = 0$ at the point $(0, R + (1 - \sqrt{1-\beta})T))$. It is obvious that $\omega(\beta) \subset \Omega_\alpha$. For the regions Ω_α and $\omega(\beta)$ we have

$$\max_{\overline{\Omega}_\alpha} \theta^2 = \max_{\bar{\omega}(\beta)} \theta^2 = \theta^2(0, R) = \exp(2\lambda T^2),$$
$$\min_{\bar{\omega}(\beta)} \theta^2 = \theta^2(\pm\sqrt{\beta/\alpha}T, R) = \exp(2\lambda(1-\beta)T^2),$$
$$\max_{\Gamma'_\alpha} \theta^2 = \theta^2(0, R) = \exp(2\lambda T^2),$$
$$\max_{\Gamma''_\alpha} \theta^2 = \theta^2(\pm T, R) = \exp[2\lambda(1-\alpha)T^2].$$

Further, by the hypotheses of the theorem

$$\|u_t\|^2 + \|u_r\|^2 + \|u\|^2 \leq \begin{cases} 3\varepsilon^2 & \text{for } (t,r) \in \Gamma'_\alpha, \\ 3 & \text{for } (t,r) \in \Gamma''_\alpha. \end{cases}$$

Noting finally that

$$\operatorname{meas} \Gamma'_\alpha = 2T, \qquad \operatorname{meas} \Gamma''_\alpha \leq 4T, \qquad \operatorname{meas} \Omega_\alpha \leq 2T^2,$$

from (4.125) we obtain

$$\exp[2\lambda(1-\beta)T^2] \int_{\omega(\beta)} \|u\|^2 \, dt \, dr$$
$$\leq C(1-\alpha)^{-1} T^4 \varepsilon^2 \exp(2\lambda T^2) + C(1-\alpha)^{-1} T^4 \exp[2\lambda(1-\alpha)T^2]$$
$$+ C(A^2+1)T^2 \exp(2\lambda T^2) \max_{\overline{\Omega}_\alpha} \sum_{|n|>N} [u_{nt}^2 + u_{nr}^2 + n^2 u_n^2]. \quad (4.126)$$

It follows from (4.119) that

$$\sum_{|n|>N} [u_{nt}^2 + u_{nr}^2 + n^2 u_n^2] \leq C \exp(-2\sigma N).$$

From this and (4.126) we get

$$\begin{aligned}\int_{\omega(\beta)} \|u\|^2\,dt\,dr &\le C(1-\alpha)^{-1}T^4\varepsilon^2\exp(2\lambda\beta T^2) + C(1-\alpha)^{-1}T^4\exp[-2\lambda(\alpha-\beta)T^2] \\ &\quad + C(A^2+1)T^2\exp(2\lambda\beta T^2 - 2\sigma N) \\ &\le C(1-\alpha)^{-1}(A^2+1)T^4\left\{\varepsilon^2\exp(2\lambda\beta T^2) + \exp[-2\lambda(\alpha-\beta)T^2]\right. \\ &\quad \left. + \exp\left[-2\lambda\left((1-\alpha)^{1/2}\frac{\sigma}{2} - \beta T^2\right)\right]\right\}. \qquad (4.127)\end{aligned}$$

Let

$$\gamma = \min\left(\alpha T^2, (1-\alpha)^{1/2}\frac{\sigma}{2}\right).$$

We set $\beta = \frac{1}{2}\gamma T^{-2}$. Then

$$2\lambda(\alpha-\beta)T^2, \qquad 2\lambda\left[(1-\alpha)^{1/2}\frac{\sigma}{2} - \beta T^2\right] \ge 2\lambda(\gamma - \beta T^2) = \lambda\gamma$$

and from (4.127) we obtain

$$\int_{\omega(\beta)} \|u\|^2\,dt\,dr \le C(1-\alpha)^{-1}(A^2+1)T^4[\varepsilon^2\exp(\lambda\gamma) + \exp(-\lambda\gamma)]. \qquad (4.128)$$

Let ε^{-1} be so large that

$$\ln(1/\varepsilon) \ge \gamma\max[2(1-\alpha)^{-1}, 16A^2(1-\alpha)^{-1}].$$

Setting $\lambda\gamma = \ln(1/\varepsilon)$ in (4.128), we then obtain

$$\int_{\omega(\beta)} \|u\|^2\,dt\,dr \le C(1-\alpha)^{-1}(A^2+1)T^4\varepsilon.$$

The theorem is proved.

Uniqueness of the solution of the Cauchy problem in question in the class of functions analytic in the variable φ follows from the last theorem.

CHAPTER V

Volterra Equations

The question of "Volterra" regularization of a Volterra equation of the first kind is considered in §1 of this chapter; §2 is devoted to the operator Volterra equations introduced by M. M. Lavrent'ev [149], [152] in connection with problems of integral geometry. The results of this chapter are due to V. O. Sergeev [245] and A. L. Bukhgeĭm [59]. The reader will find other results connected with the material of this chapter in [66], [113], [230], and [266].

§1. Regularization of a Volterra equation of the first kind

Suppose that the kernel $K(x,t)$, $x,t \in [0,x_0]$, of the Volterra integral equation of the first kind

$$\int_0^x K(x,t)\varphi(t)\,dt = f(x) \tag{5.1}$$

vanishes on the diagonal together with its derivatives with respect to x through order $n-2$, the function $K_n(x,t) = D_x^n K(x,t)$ is continuous in x and t, $K_{n-1}(x,x) = 1$, and $|K_n(x,t)| \leq K_n$.

After n-fold differentiation of both sides of (5.1) we obtain

$$\varphi(x) + \int_0^x K_n(x,t)\varphi(t)\,dt = f^{(n)}(x). \tag{5.2}$$

If the function $f \in C_0^n[0,x_0] = \{v(x)\colon v \in C^n[0,x_0], v^{(i)}(0) = 0, i = 0,1,\dots,n-1\}$, then equation (5.2) has a unique solution $\varphi(x) \in C[0,x_0]$. It is supposed a priori that $\varphi \in C^m[0,x_0]$, $m > 0$.

We consider the case where the function $f(x)$ is known with some error which takes the right side of (5.1) out of the space $C_0^n[0,x_0]$. Then by a solution one should understand either a quasisolution or a solution of a regularized equation. Methods of solving such problems presume passage to a selfadjoint operator which for (5.1) means that it ceases to be a Volterra equation. We shall give a method of regularization which preserves this property of the original equation.

We suppose first that the values $D_x^k \varphi(x)|_{x=0} = \varphi_k$, $k = 0, 1, \ldots, m-1$, are known. We consider the function

$$W_\varepsilon(x) = \sum_{k=0}^{n+1} \frac{1}{\varepsilon^k} C_{m+k-1}^{m-1} e^{-x/\varepsilon} \frac{x^{m+k-1}}{(m+k-1)!}.$$

Using the Leibniz formula, it is not hard to verify that

$$W_\varepsilon^{(j)}(0) = 0 \text{ for } 0 \le j \le m+n, \quad j \ne m-1; \qquad W_\varepsilon^{(m-1)}(0) = 1.$$

We define the operator $B_\varepsilon \in \mathcal{L}(C^0[0, x_0] \to C_0^n[0, x_0])$ by $B_\varepsilon u = y$,

$$y(x) = u(x) - D_x^m \int_0^x W_\varepsilon(x-t) u(t)\, dt = -\int_0^x W_\varepsilon^{(m)}(x-t) u(t)\, dt.$$

For $\varepsilon > 0$ the operator B_ε has an inverse. Indeed, the kernel $W^{(m)}(x-t)$ of this integral operator vanishes on the diagonal together with its derivatives with respect to x through order n, while $W^{(m+n+1)}(x-t)$ for $\varepsilon > 0$ is bounded and $W^{(m+n+1)}(0) = -\varepsilon^{-n-2} C_{m+n+1}^{n+2}$. Thus, for $\varepsilon > 0$ (5.2) and the equation

$$B_\varepsilon K_n \varphi = B_\varepsilon f^{(n)} - B_\varepsilon \varphi \tag{5.3}$$

are equivalent. Let $f_0(x)$ be a function of boundary-layer type:

$$f_0(x) = \sum_{k=0}^{m-1} \varphi_k W_\varepsilon^{(m-k-1)}(x).$$

Integrating by parts, we find that

$$\varphi - f_0 - B_\varepsilon \varphi = -f_0(x) + \int_0^x W_\varepsilon^{(m-1)}(x-t) \varphi'(t)\, dt$$
$$+ \varphi_0 W_\varepsilon^{(m-1)}(x) = \cdots = \int_0^x W_\varepsilon(x-t) \varphi^{(m)}(t)\, dt.$$

Setting $\max_{[0,x_0]} |\varphi^{(m)}(x)| = \Phi_m$, we then obtain the estimate

$$\|\varphi - f_0 - B_\varepsilon \varphi\|_{C^0} \le \Phi_m \varepsilon^m, \qquad A = C_{m+n+1}^m. \tag{5.4}$$

We next consider the kernel of the integral operator $B_\varepsilon K_n$:

$$u = B_\varepsilon K_n \varphi, \qquad u(x) = -\int_0^x \varphi(t) \int_t^x W_\varepsilon^{(m)}(x-\tau) K_n(\tau, t)\, d\tau\, dt.$$

The modulus of the kernel can be estimated as follows:

$$\left| \int_t^x W_\varepsilon^{(m)}(x-\tau) K_n(\tau, t)\, d\tau \right| \le K_n \int_0^\infty |W_\varepsilon^{(m)}(t)|\, dt \le 2^m K_n A = B. \tag{5.5}$$

The norm of $B_\varepsilon f^{(n)}$ can be estimated similarly:

$$\|B_\varepsilon f^{(n)}\|_{C^0} \le \max_{[0,x_0]} |f^{(n)}(x)| 2^m A. \tag{5.6}$$

We consider the equation

$$\varphi_\varepsilon + B_\varepsilon K_n \varphi_\varepsilon = B_\varepsilon f^{(n)} + f_0. \tag{5.7}$$

For $\varepsilon > 0$ the right side of this equation is bounded, and by virtue of (5.5) we obtain existence and uniqueness of its solution $\varphi_\varepsilon(x)$. The difference $w_\varepsilon(x) = \varphi(x) - \varphi_\varepsilon(x)$ satisfies the equation

$$w_\varepsilon + B_\varepsilon K_n w_\varepsilon = \varphi - B_\varepsilon \varphi - f_0,$$

and hence, according to the estimate (5.4),

$$\|\varphi - \varphi_\varepsilon\|_{C^0} \le \varepsilon^m A \Phi_m e^{Bx_0}. \tag{5.8}$$

Thus, as $\varepsilon \to 0$, $\varphi_\varepsilon \to \varphi$ uniformly on $[0, x_0]$. Integrating by parts and using the fact that $f^{(i)}(0) = 0$, $i = 0, 1, \dots, n-1$, we obtain for the function $B_\varepsilon f^{(n)}$ the representation

$$-\int_0^x W_\varepsilon^{(m)}(x-t) f^{(n)}(t)\,dt = -\int_0^x W_\varepsilon^{(n+m)}(x-t) f(t)\,dt. \tag{5.9}$$

We extend the domain of the right sides of (5.7) according to (5.9). We estimate $\|B_\varepsilon f^{(n)}\|_{C^0}$ in terms of $\|f\|_{C^0}$:

$$\|B_\varepsilon f^{(n)}\|_{C^0} \le \frac{2^{n+m} A}{\varepsilon^n} \|f\|_{C^0}.$$

Suppose now that in place of $f(x)$ on the right side of (5.7) there is given a function $f^\delta(x) = f(x) + \delta(x)$, where $\delta(x)$ is continuous and $\|\delta(x)\| \le \delta$. We consider the equation

$$\varphi(x) - \int_0^x \varphi(t) \int_t^x W_\varepsilon^{(m)}(x-\tau) K_n(\tau, t)\,d\tau\,dt = f_0(x) - \int_0^x W_\varepsilon^{(n+m)}(x-t) f^\delta(t)\,dt, \tag{5.10}$$

whose solution we denote by $\varphi_\varepsilon^\delta(x)$. The difference $w_\varepsilon^\delta(x) = \varphi_\varepsilon(x) - \varphi_\varepsilon^\delta(x)$ then satisfies the equation

$$w_\varepsilon^\delta(x) - \int_0^x w_\varepsilon^\delta(t) \int_t^x W_\varepsilon^{(m)}(x-\tau) K_n(\tau, t)\,d\tau\,dt = -\int_0^x W_\varepsilon^{(m+n)}(x-t)\delta(t)\,dt,$$

and hence

$$\|\varphi_\varepsilon(x) - \varphi_\varepsilon^\delta(x)\|_{C^0} \le \frac{2^{n+m}}{\varepsilon^n} e^{Bx_0} \delta. \tag{5.11}$$

Combining (5.8) and (5.11), we obtain

$$\|\varphi - \varphi_\varepsilon^\delta\|_{C^0} \le \varepsilon^m A \Phi_m e^{Bx_0} + \frac{\delta}{\varepsilon^n} 2^{n+m} A e^{Bx_0}.$$

A minimum of the right side of the inequality is achieved for

$$\varepsilon(\delta) = 2\left(\frac{n}{m\Phi_m}\right)^{\frac{1}{n+m}} \delta^{\frac{1}{n+m}}. \tag{5.12}$$

We thus have the following result.

THEOREM 1. *Suppose the exact solution φ of equation (5.1) is in C^m, $m > 0$, and the values $\varphi^{(i)}(x)$, $i = 0, 1, \ldots, m-1$, are known at $x = 0$. If the norm of the error $\delta(x)$ of the right side tends to zero, then the solution of the regularized equation (5.10) with ε defined by (5.12) tends uniformly to the exact solution, and*

$$\|\varphi - \varphi_\varepsilon^\delta\| \leq 2^m A e^{Bx_0}(1 - m/n)\Phi_m^{\frac{n}{n+m}}(n\delta/m)^{\frac{m}{n+m}}. \tag{5.13}$$

We shall compare the order with respect to δ of this estimate with the maximal attainable order of magnitude of the error under our assumptions regarding the error of the right side of (5.1) and the smoothness of the exact solution.

We shall consider functions $\delta(x) \in C^{n+m}$ whose derivatives up to order $n+m$ vanish at $x = 0$. Passing to equation (5.2) with right side $f^{(n)}(x) + \delta^{(n)}(x)$, for its solution $\tilde{\varphi}_\delta(x)$ we obtain the estimate

$$\|\varphi - \tilde{\varphi}_\delta\|_{C^0} \leq e^{K_n x_0}\|\delta^{(n)}\|_{C^0},$$

and, since $\delta^{(n)} \in C_0^m$,

$$\|\delta^{(n)}(x)\|_{C^0} \leq C\|\delta(x)\|_{C^0}^{\frac{m}{n+m}} \leq C\delta^{\frac{m}{n+m}}$$

(see, for example, [137]).

Thus, the order of the estimate obtained in regularization coincides with the maximal attainable order of error of the solution under our assumptions.

If the values φ_k, $k = 0, 1, \ldots, m-1$, are unknown we consider equation (5.10) with right side

$$-\int_0^x W_\varepsilon^{(m+n)}(x-t)f^\delta(t)\,dt;$$

the difference $\tilde{w}_\varepsilon(x)$ between the solution $\tilde{\varphi}_\varepsilon(x)$ of this equation and the solution $\varphi_\varepsilon^\delta(x)$ then satisfies

$$\tilde{w}_\varepsilon(x) - \int_0^x \tilde{w}_\varepsilon(t)\int_t^x W_\varepsilon^{(m)}(x-\tau)K_n(\tau,t)\,d\tau\,dt = f_0(x),$$

and can be represented in the form of the Neumann series

$$\tilde{w}_\varepsilon(x) = \sum_{k=0}^{m-1}\varphi_k W_\varepsilon^{(m-k-1)}(x)$$
$$+\sum_{k=0}^{m-1}\varphi_k\int_0^x W_\varepsilon^{(m-k-1)}(t)\int_t^x W_\varepsilon^{(m)}(x-\tau)K_n(\tau,t)\,d\tau\,dt + \cdots$$

However,

$$\left|\int_0^x W_\varepsilon^{(m-k-1)}(t)\int_t^x W_\varepsilon^{(m)}(x-\tau)K_n(\tau,t)\,d\tau\,dt\right| \leq 2^{m-k-1}A\varepsilon^{k+1},$$
$$k = 0, 1, \ldots, m-1.$$

Thus, $\tilde{w}_\varepsilon(x)$ is the sum of a function of boundary-layer type and a function which decreases together with ε.

§2. Operator Volterra equations of the first kind

Let X be a complex Banach space, and let $\mathcal{L}(X,X)$ be the algebra of bounded linear operators acting in X.

We recall that a *spectral measure* is a set function $E(\delta)$ defined on the Borel system of sets B of the complex plane with values in $\mathcal{L}(X,X)$ and such that the following conditions are satisfied:

1) $E(\delta_1 \cap \delta_2) = E(\delta_1)E(\delta_2)$, $\delta_1, \delta_2 \in B$; $E(\phi) = 0$ and $E(p) = I$ (p is the entire complex plane, and I is the identity operator);

2) $\|E(\delta)\| \leq M$, $\delta \in B$, and M does not depend on δ;

3) $E(\delta)$ is countably additive in the strong operator topology, i.e., for each sequence of disjoint Borel sets

$$E\left(\bigcup_{n=1}^{\infty} \delta_n\right) x = \sum_{n=1}^{\infty} E(\delta_n)x, \quad x \in X.$$

An operator $T \in \mathcal{L}(X,X)$ is called *spectral* if there exists a spectral measure $E(\delta)$ such that $TE(\delta) = E(\delta)T$ for $\delta \in B$, and $\sigma(T_\delta) \subseteq \bar{\delta}$; here T_δ is the restriction of T to the subspace $E(\delta)X$, and $\sigma(T_\delta)$ is the spectrum of the operator T_δ.

The function $E(\delta)$ is called a *spectral decomposition* of the operator T.

An operator S having a spectral decomposition for which $S = \int \lambda E(d\lambda)$ is called a *scalar operator*. In particular, a normal operator in a Hilbert space is scalar.

We shall need the following two theorems.

THEOREM 1 (DUNFORD [303]). *An operator T is spectral if and only if $T = S + N$, where S is a scalar operator and N is a quasinilpotent operator commuting with S. Moreover, this representation is unique, and the spectra and spectral decompositions of the operators S and N coincide.*

The operator S is called the *scalar part* of the operator T, and the operator N is called the *radical part* (an operator N is said to be *quasinilpotent* if $\lim_{n\to\infty} \|N^n\|^{1/n} = 0$).

THEOREM 2 (FOGUEL [304]). *Let R be a right (left) ideal of the algebra $\mathcal{L}(X,X)$ which is closed in the uniform operator topology. If $T \in R$, then S, N, and $E(\delta)$ $(0 \notin \delta)$ also belong to R.*

COROLLARY. *If $Tu = 0$, then $Su = Nu = E(\delta)u = 0$ $(0 \notin \bar{\delta})$.*

Let $L^p(0,T;X)$ be the Banach space of functions defined on $(0,T)$, taking values in X, and summable in norm to pth power:

$$\|u\|_{L^p(0,T;X)} = \left\{ \int_0^T \|u(t)\|_X^p \, dt \right\}^{1/p}, \quad 1 \leq p < \infty.$$

Let $N(t,\tau)$ be a two-parameter family of operators, $N(t,\tau) \in \mathcal{L}(X,X)$, $(t,\tau) \in \Omega = \{(t,\tau)\colon 0 < \tau < t < T\}$. We assume that the formula

$$(Nu)(t) = \int_0^t N(t,\tau)u(\tau)\, d\tau \tag{5.14}$$

(the integral is understood in the Bochner sense) defines a quasinilpotent operator acting in $L^p(0,T;X)$. For this, for example, it suffices to require that $\|N(t,\tau)\| \leq M$, $(t,\tau) \in \Omega$, where M does not depend on (t,τ).

REMARK. If it is additionally assumed that the operators $N(t,\tau)$ are compact and X is a Hilbert space, then the operator (5.14) for $p = 2$ is a Volterra operator in the terminology of [95].

It is natural to begin the study of equations containing such operators from an operator Volterra equation of the second kind by which we mean an equation

$$Su(t) + (Nu)(t) = f(t), \tag{5.15}$$

where S is a scalar operator not depending on t:

$$S \in \mathcal{L}(X,X), \qquad \ker S = 0, \tag{5.16}$$

and N is the quasinilpotent operator (5.14).

LEMMA 1. *If operators $A_n \in \mathcal{L}(X,X)$ converge strongly to an operator A, then the operators $\overline{A}_n$ acting in $L^p(0,T;X)$ according to the formula $(\overline{A}_n x)(t) = A_n x(t)$, $x(t) \in L^p(0,T;X)$, converge strongly to the operator $\overline{A}$; that is, $(\overline{A}x)(t) = Ax(t)$.*

The proof is trivial.

LEMMA 2. *If $SN(t,\tau) = N(t,\tau)S$ for $(t,\tau) \in \Omega$, then equation (5.15) has at most one solution $u(t) \in L^p(0,T;X)$.*

PROOF. As above, we denote by $\overline{S}$ the extension of the operator S to the space $L^p(0,T;X)$. Since S is a scalar operator, $\overline{S}$ is also a scalar operator generated by the spectral measure $\overline{E}(\delta)$, where $E(\delta)$ is the spectral decomposition of S. Indeed, conditions 1) and 2) for the operators $\overline{E}(\delta)$ are obvious, while condition 3) follows from Lemma 1. Further, since $SN(t,\tau) = N(t,\tau)S$ for any $(t,\tau) \in \Omega$, it follows that $\overline{S}N = N\overline{S}$. The operator N is quasinilpotent; it thus follows from Dunford's theorem that the operator $T = \overline{S} + N$

is spectral. From this it follows that if $(\overline{S}u)(t) + (Nu)(t) = 0$, then (by the corollary of Foguel's theorem) $(\overline{S}u)(t) = 0$. But by (5.16) $\ker \overline{S} = 0$, i.e., $u(t) = 0$ almost everywhere. The lemma is proved.

REMARK. If S has no bounded inverse (for example, if S is compact), then problem (5.15), generally speaking, is ill-posed in the Hadamard sense.

We now consider an operator Volterra equation of the first kind:

$$(Ku)(t) = \int_0^t \frac{K(t,\tau)}{(t-\tau)^\alpha} u(\tau)\,d\tau = f(t), \tag{5.17}$$

$$K(t,\tau) \in \mathcal{L}(X,X), \quad (t,\tau) \in \overline{\Omega};\ 0 \le \alpha < 1.$$

We assume that the integral in (5.17) exists for any $u(t) \in L^p(0,T;X)$ for some $p \in [1,\infty)$.

As a rule, in cases interesting for applications the derivative of the operator $K(t,\tau)$ with respect to t is already not a bounded operator in X. It is nevertheless often possible to choose a Banach space Y imbedded in X such that $K(t,\tau)$ is strongly continuously differentiable with respect to t from Y to X in Ω. The formula

$$K_t(t,\tau)y = [K(t,\tau)y]_t$$

then defines an operator $K_t(t,\tau) \in \mathcal{L}(Y,X)$ strongly continuous in $\overline{\Omega}$.

THEOREM 3. *Suppose the operator $K(t,\tau)$ is strongly continuously differentiable with respect to t in $\overline{\Omega}$ from Y to X, where Y is a Banach space imbedded in X. Suppose the operator $K(t,t)$ has a bounded inverse, $\|K^{-1}(t,t)\| \le M$, where M does not depend on t. If there exists an operator S such that*

1) $S \in \mathcal{L}(X,Y)$, $SK(t,\tau) = K(t,\tau)S$, $(t,\tau) \in \Omega$,

2) *S as an operator $X \to X$ is a scalar operator, and*

3) $\ker S = 0$,

then (5.17) *has at most one solution $\alpha(t) \in L^p(0,T;X)$.*

PROOF. Suppose first that $\alpha = 0$. Applying operator $R = K^{-1}(t,t)(d/dt)S$ to both sides of (5.17) and using condition 1, we obtain the equation

$$Su(t) + \int_0^t N(t,\tau)u(\tau)\,d\tau = (Rf)(t),$$

where $N(t,\tau) = K^{-1}(t,t)K_t(t,\tau)S$. Since $S \in \mathcal{L}(X,Y)$, $K_t(t,\tau) \in \mathcal{L}(Y,X)$, and $K^{-1}(t,t) \in \mathcal{L}(X,X)$, it follows that $N(t,\tau) \in \mathcal{L}(X,X)$. From the conditions imposed on $K(t,\tau)$ it follows that $\|N(t,\tau)\|$ is uniformly bounded, i.e., the operator

$$(Nu)(t) = \int_0^t N(t,\tau)u(\tau)\,d\tau$$

is quasinilpotent. It remains to appeal to Lemma 2. In the case $\alpha > 0$ we set $R = K^{-1}(t,t)(d^{1-\alpha}/dt^{1-\alpha})S$ and again use Lemma 2. Here $d^{1-\alpha}/dt^{1-\alpha}$ is the operator of differentiation of order $1-\alpha$ (see, for example, [86]).

REMARK 1. The condition of strong continuous differentiability in $\overline{\Omega}$ can be relaxed. In particular, $K_t(t,\tau)$ may have a singularity on the boundary of Ω. If this singularity is weak, i.e.,

$$\|K_t(t,\tau)\|_{\mathcal{L}(Y,X)} \leq M(t-\tau)^{-\beta}, \quad 0 \leq \beta < 1,$$

then quasinilpotence of the operator N is preserved. Uniqueness is thus also preserved.

REMARK 2. If $Y = X$, then for S it is possible to take the identity operator. In this case we arrive at an operator version of Volterra's classical theorem. If $Y \subset X$ and $Y \neq X$, then an operator S satisfying conditions 1)–3) exists for far from all operators $K(t,\tau)$. It is therefore important to distinguish a class of operators $K(t,\tau)$ for which such an operator certainly exists.

Let H be a separable Hilbert space, and let $G = \{U_h\}$ be a strongly continuous group of unitary operators in H:

$$U_{h_1+h_2} = U_{h_1}U_{h_2}, \qquad h_1, h_2 \in (-\infty,\infty); \qquad U_0 = I.$$

By Stone's theorem the group G has a generator A, and $i^{-1}A$ is a selfadjoint operator. We introduce the Hilbert space H^l with norm

$$\|u\|_l = \|(I - A^2)^{l/2}u\|_H, \quad l > 0.$$

It is easy to see that $H^{l_1} \subset H^{l_2}$ $(l_1 > l_2)$, $H^0 = H$.

We recall that an operator $B \in \mathcal{L}(H,H)$ is called *invariant* relative to G if $U_hB = BU_h$, $h \in (-\infty,\infty)$.

THEOREM 4. *Suppose a family of operators $K(t,\tau)$ invariant relative to the group G is strongly continuously differentiable with respect to t from H^l to H in $\overline{\Omega}$ and*

$$\|K^{-1}(t,t)\|_{\mathcal{L}(H,H)} \leq M,$$

where M does not depend on t. Then equation (5.17) *has at most one solution $u(t) \in L^p(0,T;H)$.*

PROOF. It suffices to establish the existence of an operator S satisfying conditions 1)–3) of Theorem 3. We shall verify that $S = (I-A^2)^{-l/2}$ possesses these properties. Since

$$\|Su\|_l = \|u\|_H, \tag{5.18}$$

it follows that $S \in \mathcal{L}(H,H^l)$. Further, because of the invariance of $K(t,\tau)$ relative to G, $K(t,\tau)A = AK(t,\tau)$, and thus the operators $K(t,\tau)$ commute with S. Therefore, condition 1) is satisfied. The operator $i^{-1}A$ is selfadjoint, and hence the operator S is Hermitian (and a fortiori scalar). Thus, condition 2) is also satisfied. Condition 3) follows from (5.18). The theorem is proved.

We consider the question of stability of problem (5.17). Let X be a normed space. We consider the problem of finding $u \in D(A)$ from the equation

$$Au = f, \tag{5.19}$$

where A is a linear, generally speaking, unbounded operator $X \to X$. Problem (5.19) is often ill-posed in the Hadamard sense. We define a set of well-posedness $\mathfrak{M}$ by means of a nonnegative, homogeneous functional l with domain $D(l) \equiv D(A)$:

$$\mathfrak{M} = \{u \in X \colon l(u) \leq 1\},$$

and we call problem (5.19) *conditionally well-posed* if there exists a nondecreasing function $\omega(\xi) \in C[0, \infty)$, $\omega(0) = 0$, such that

$$\forall u \in \mathfrak{M} \colon \|u\| \leq \omega(\|Au\|). \tag{5.20}$$

If, for example, $\mathfrak{M}$ is compact and A is a continuous operator or an operator with closed graph, the existence of such a function $\omega(\xi)$ follows from general topological considerations.

We are interested in the problem of constructive determination of the function $\omega(\xi)$. For this in a number of cases it is convenient to alter the above definition in the following manner: problem (5.19) is called *conditionally well-posed* if there exists a positive, nonincreasing function $\varphi(\varepsilon) \in C(0, \alpha)$, $\alpha > 0$, such that

$$\forall u \in X, \qquad \forall \varepsilon \in (0, \alpha) \qquad \|u\| \leq \varepsilon l(u) + \varphi(\varepsilon)\|Au\|. \tag{5.21}$$

REMARK 1. If $u \notin D(A)$, then by definition we set the right side of (5.21) equal to infinity.

REMARK 2. If problem (5.19) is ill-posed in the Hadamard sense, then

$$\lim_{\varepsilon \to 0} \varphi(\varepsilon) = \infty.$$

We shall compare these definitions. Suppose (5.21) holds. Then for $u \in \mathfrak{M}$ from (5.21) we have $\|u\| \leq \varepsilon + \varphi(\varepsilon)\|Au\|$. Setting

$$\omega(\xi) = \min_{\varepsilon}[\varepsilon + \varphi(\varepsilon)\xi],$$

we obtain (5.20), where $\omega(\xi)$ possesses the required properties. Suppose now that (5.20) is satisfied. It may be assumed with no loss of generality that

$$\omega(\xi) \in C^2(0, a), \qquad \omega''(\xi) < 0, \qquad 0 < \xi < a. \tag{5.22}$$

We set

$$p = \omega'(\xi), \qquad H(p) = \omega(\xi) - p\xi, \tag{5.23}$$

where ξ is found from (5.23). Up to sign the function $H(p)$ coincides with the Legendre transform of the function $\omega(\xi)$. It is well known (see [85], Chapter IV, §15.1) that if we consider the expression $H(p) + p\xi$ as a function of two independent variables which for $p = \omega'(\xi)$ coincides with the function $\omega(\xi)$, then

$$\omega(\xi) = \min_p [H(p) + p\xi],$$

and hence

$$\omega(\xi) \le H(p) + p\xi. \tag{5.24}$$

The function $H(p)$ is called the *Young dual* of the function $\omega(\xi)$. We set $\varepsilon = H(p)$. Because of conditions (5.22), this equation defines in some interval $(0, \alpha)$ a continuous, nonincreasing function $\varphi(\varepsilon) = p$. Inequality (5.24) can then be rewritten as follows:

$$\omega(\xi) \le \varepsilon + \varphi(\varepsilon)\xi. \tag{5.25}$$

Let $u \in D(l)$, $l(u) \neq 0$. Then $u/l(u) \in \mathfrak{M}$, and from (5.20) and (5.25) we have

$$\left\| \frac{u}{l(u)} \right\| \le \omega\left(\left\| A\frac{u}{l(u)} \right\| \right) \varepsilon + \varphi(\varepsilon) \left\| A\frac{u}{l(u)} \right\|,$$

whence we obtain (5.21). If $l(u) = 0$, then $\lambda u \in \mathfrak{M}$ for all $\lambda \ge 0$, and from (5.20) and (5.25) we obtain

$$\lambda\|u\| \le \omega(\|A\lambda u\|) \le \varepsilon + \varphi(\varepsilon)\|A\lambda u\|.$$

Dividing both sides by λ and letting $\lambda \to \infty$, we again arrive at (5.21). Thus, the definitions (5.20) and (5.21) are equivalent, and the functions ω and $-\varphi^{-1}$ are dual in the Young sense.

An analogous assertion holds also for a nonlinear, homogeneous operator A.

We now proceed to obtain estimates of stability for spectral operators.

THEOREM 5. *Let $T = S + N$ be the canonical decomposition of the spectral operator T in the Banach space X. Then*

$$\|u\| \le \varepsilon\|S^{-1}u\| + \varphi(\varepsilon)\|Tu\|, \quad \forall\varepsilon > 0, \tag{5.26}$$

where

$$\varphi(\varepsilon) = M \sum_{n=0}^{\infty} \left(\frac{16M^2}{\varepsilon} \right)^{n+1} \|N^n\|$$

(M is the constant appearing in the definition of the spectral measure).

REMARK. If $u \notin R(S)$, we set $\|S^{-1}u\| = \infty$.

PROOF. Let $u \in R(S)$, and let $E(\alpha)$ be the spectral decomposition of T. We set

$$E_\varepsilon = E(\alpha_\varepsilon), \qquad \alpha_\varepsilon = \{\lambda : |\lambda| \le \varepsilon\}, \qquad E_{1-\varepsilon} = I - E_\varepsilon.$$

Then any vector u can be uniquely decomposed into a sum $u = E_\varepsilon u + E_{1-\varepsilon} u = u_\varepsilon + u_{1-\varepsilon}$. From the triangle inequality we get

$$\|u\| \le \|u_\varepsilon\| + \|u_{1-\varepsilon}\|. \tag{5.27}$$

We shall estimate $\|u_\varepsilon\|$. Since $u \in R(S)$, it follows that $u = Sv$, $v \in X$. Therefore,

$$u_\varepsilon = E_\varepsilon u = E_\varepsilon S v, \qquad E_\varepsilon S = E_\varepsilon \int \lambda E(d\lambda) = \int_{|\lambda| \le \varepsilon} \lambda E(d\lambda).$$

Since [303] for any bounded, measurable function $f(\lambda)$

$$\left\| \int f(\lambda) E(d\lambda) \right\| \le 4M \sup |f(\lambda)|, \tag{5.28}$$

it follows that $\|E_\varepsilon S\| \le \varepsilon 4M$. Thus,

$$\|u_\varepsilon\| \le \varepsilon \cdot 4M \|v\| = \varepsilon \cdot 4M \|S^{-1} u\|. \tag{5.29}$$

We now estimate $\|u_{1-\varepsilon}\|$. We denote by $T_{1-\varepsilon}$ the restriction of the operator T to $E_{1-\varepsilon}X$. By property 2) of a spectral operator,

$$\sigma(T_{1-\varepsilon}) \subseteq \overline{\alpha}_{1-\varepsilon} = \{\lambda : |\lambda| \ge \varepsilon\},$$

and hence $T_{1-\varepsilon}$ has a bounded inverse. We have identically

$$u_{1-\varepsilon} = E_{1-\varepsilon} u = T_{1-\varepsilon}^{-1} T_{1-\varepsilon} E_{1-\varepsilon} u = T_{1-\varepsilon}^{-1} T E_{1-\varepsilon} u = T_{1-\varepsilon}^{-1} E_{1-\varepsilon} T u.$$

Therefore,

$$\|u_{1-\varepsilon}\| \le \|T_{1-\varepsilon}^{-1} E_{1-\varepsilon}\| \, \|Tu\|,$$

$$T_{1-\varepsilon}^{-1} = (S_{1-\varepsilon} + N_{1-\varepsilon})^{-1} = S_{1-\varepsilon}^{-1} \sum_{n=0}^{\infty} (-S_{1-\varepsilon}^{-1} N_{1-\varepsilon})^n, \tag{5.30}$$

where $S_{1-\varepsilon}$ and $N_{1-\varepsilon}$ are the restrictions of the operators S and N to the space $E_{1-\varepsilon}X$. (By Dunford's theorem $\sigma(S_{1-\varepsilon}) = \sigma(T_{1-\varepsilon})$, i.e., $S_{1-\varepsilon}$ is invertible.) Further,

$$S_{1-\varepsilon}^{-1} = \int_{|\lambda| > \varepsilon} \lambda^{-1} E(d\lambda),$$

whence by (5.28)

$$\|S_{1-\varepsilon}^{-1}\| \le \frac{1}{\varepsilon} 4M. \tag{5.31}$$

By property 2) of the definition of a spectral measure we have

$$\|E_{1-\varepsilon}\| \le M. \tag{5.32}$$

The identity (5.30) and the estimates (5.31) and (5.32) give

$$\|T_{1-\varepsilon}^{-1}E_{1-\varepsilon}\| \le M\|T_{1-\varepsilon}^{-1}\| \le M\sum_{n=0}^{\infty}\left(\frac{4M}{\varepsilon}\right)^{n+1}\|N_{1-\varepsilon}^{n}\|$$
$$\le M\sum_{n=0}^{\infty}\left(\frac{4M}{\varepsilon}\right)^{n+1}\|N^{n}\| = \psi(\varepsilon). \tag{5.33}$$

Since N is quasinilpotent, $\psi(\varepsilon)$ is an entire function of $1/\varepsilon$. From (5.27), (5.29), and (5.33) we obtain

$$\|u\| \le \varepsilon\cdot 4M\|S^{-1}u\| + \psi(\varepsilon)\|Tu\|,$$

which up to notation coincides with (5.26). The theorem is proved.

In Hilbert space it is possible to obtain a more elegant estimate. Namely, the next result is proved in a manner similar to Theorem 5.

THEOREM 6. *Let $X = H$ be Hilbert space, and let $T = S + N$ be the canonical decomposition of the spectral operator T, where S is a nonnegative Hermitian operator, i.e., $S^* = S$ and $(SU, u) \ge 0$ for all $u \in H$. Then*

$$\forall\varepsilon > 0,\ \delta > 0, \qquad \|u\| \le \varepsilon\|S^{-\delta}u\| + \varphi_\delta(\varepsilon)\|Tu\|,$$
$$\varphi_\delta(\varepsilon) = \sum_{n=0}^{\infty}\varepsilon^{-\frac{n+1}{\delta}}\|N^n\|$$

(if $u \notin R(S^\delta)$, then we set $\|S^{-\delta}u\| = \infty$).

EXAMPLE. Let $T = RK$, where K is the operator of (5.17) satisfying the conditions of Theorem 2, and

$$R = K^{-1}(t,t)\frac{d^{1-\alpha}}{dt^{1-\alpha}}S,$$

Since in this case $S = (I - A^2)^{-l/2}$, it follows that

$$\|S^{-\delta}u\|_{L^2(0,T;H)} = \|u\|_{L^2(0,T;H^{\delta l})}.$$

Therefore, the estimate of Theorem 4 gives us

$$\|u\|_{L^2(0,T;H)} \le \varepsilon\|u\|_{L^2(0,T;H^{\delta l})} + \varphi_\delta(\varepsilon)\|RKu\|_{L^2(0,T;H)}.$$

Hence, if it is known a priori that the solution of our problem belongs to the set

$$\mathfrak{M}_\delta = \{u\colon \|u\|_{L^2(0,T;H^{\delta l})} \le M\},$$

then problem (5.17) becomes stable in the usual sense.

Just as for scalar Volterra equations, it is easy to show that $\|N^n\| \leq C^n[(n-1)!]^{-1}$; hence,

$$\varphi_\delta(\varepsilon) \leq \varepsilon^{-1/\delta}[1+\exp(C\varepsilon^{-1/\delta})] \leq 2\varepsilon^{-1/\delta}\exp(C\varepsilon^{-1/\delta}).$$

Here C depends on the number T and the maximum of the norm of the operators $K^{-1}(t,t)K_t(t,\tau)S$.

Chapter VI is devoted to a systematic examination of problems of integral geometry. Now, however, we shall use one of the results of that chapter to illustrate possible applications of the theorems proved above.

In §2 of Chapter VI it is shown that the problem of integral geometry of recovering a function $u(x,t)$, $t \in (0,T)$, $x \in R^n$, from its integrals over a family of surfaces of "cap" type invariant with respect to translation along the hyperplane $t=0$ under certain regularity conditions reduces to the solution of the following integral equation of the first kind:

$$\int_0^t \int_{|\nu|=1} K\left(t,\sqrt{t-\tau},\nu\right) u\left[x+g\left(t,\sqrt{t-\tau},\nu\right),\tau\right] d\omega_\nu\, d\tau = f(x,t). \quad (5.34)$$

Here $g(t,p,\nu) = (g_1,\dots,g_n)$ is a given smooth vector-valued function of the variables t, p, and ν ($\nu \in R^n$, $|\nu|=1$), $d\omega_\nu$ is the area element of the unit sphere in R^n, and $K(t,p,\nu)$ is a known weight function. We suppose that $g(t,p,\nu)$ and $K(t,p,\nu)$ are continuous and bounded together with their first derivatives with respect to t and p, and that

$$g(t,0,\nu)=0, \qquad \int_{|\nu|=1} K(t,0,\nu)\, d\omega_\nu \geq 1/M.$$

We set $X = L^2(R^n)$ and define the two-parameter family of operators $K(t,\tau) \in \mathcal{L}(X,X)$:

$$K(t,\tau)v(x) = \int_{|\nu|=1} K\left(t,\sqrt{t-\tau},\nu\right) v\left[x+g\left(t,\sqrt{t-\tau},\nu\right)\right] d\omega_\nu, \quad v \in L^2(R^n).$$

Equation (5.34) can then be considered the operator Volterra equation (5.17) for $\alpha=0$. We set $Y = W_2^1(R^n)$ and $S = (I-\Delta)^{1/2}$ (Δ is the Laplace operator in R^n). Then it is easy to see that

$$\|K^{-1}(t,t)\|_{\mathcal{L}(X,X)} \leq M, \qquad \|K_t(t,\tau)\|_{\mathcal{L}(Y,X)} \leq C(t-\tau)^{-1/2}$$

and $SK(t,\tau) = K(t,\tau)S$. Therefore, by Theorem 1 (taking account of Remark 1) the solution of (5.34) is unique in the space $L^2[(0,T)\times R^n]$, and it is easy to see that the function $\varphi(\varepsilon)$ occurring in the estimate of stability in Theorem 5 admits an estimate of the form $\varphi(\varepsilon) \leq \exp(C\varepsilon^{-2})$.

CHAPTER VI

Integral Geometry

We shall consider the following problem [83]. Let $u(x)$ be a sufficiently smooth function defined in n-dimensional space $x = (x_1, \dots, x_n)$, and let $\{\mathfrak{M}(\lambda)\}$ be a family of smooth manifolds in this space depending on a parameter $\lambda = (\lambda_1, \dots, \lambda_k)$. Suppose, further, that about the function $u(x)$ we know the integrals

$$\int_{\mathfrak{M}(\lambda)} u(x)\, d\sigma = v(\lambda), \tag{6.1}$$

where $d\sigma$ defines the element of measure on $\mathfrak{M}(\lambda)$. On the basis of the function $v(\lambda)$ it is required to find the function $u(x)$.

The main questions that arise in the study of this problem are the following. The first question and the most important in principle is: does the function $v(\lambda)$ uniquely determine the function $u(x)$? The next question is: how do we find $u(x)$ from $v(\lambda)$? Here, of course, it is desirable to obtain an analytic formula expressing $u(x)$ in terms of $v(\lambda)$. We observe, by the way, that this is possible only in exceptional cases. Finally, there is a question related in a natural way to the existence theorem for the problem: what are the necessary and sufficient conditions that $v(\lambda)$ belong to the class of functions representable in the form (6.1)?

As is evident from (6.1), the problem of integral geometry is the problem of solving a special integral equation of the first kind. As we shall see below (see, for example, §1 of this chapter), among the problems of integral geometry there are problems which are strongly unstable. In connection with this an estimate of conditional stability of the problem assumes major importance. On the other hand, the question of constructing a function $u(x)$ from the function $v(\lambda)$ in the general case requires the creation of special computational algorithms based on the general theory of ill-posed problems. In connection with this it is expedient, for example, to pose the question of constructing regularizers for a problem of integral geometry which take into account its special features.

Problems of integral geometry in the formulation (6.1) were first considered at the beginning of our century. The Radon transform became widely known in mathematics. It assigns to a function $u(x)$ its integrals $v(\lambda)$ over all possible hyperplanes. Finding the inverse transformation is the problem (6.1) of integral geometry for the case where $\{\mathfrak{M}(\lambda)\}$ is the family of all possible hyperplanes in n-dimensional space. This problem was first solved by Radon [325]; it was then considered in various aspects by F. John [308], [309], A. A. Khachaturov [280], P. O. Kostelyanets and Yu. G. Reshentnyak [133], I. M. Gel'fand and M. I. Graev [80]–[82], and a number of other authors [79], [84].

The problem of integral geometry on linear manifolds turned out to be closely connected with the theory of representations of Lie groups (see [78]). This circumstance to a considerable extent explains the interest in this problem. Among other problems of integral geometry the greatest popularity was enjoyed by problems of integral geometry on surfaces of second order: spheres and ellipsoids. For the case where the family $\{\mathfrak{M}(\lambda)\}$ is a family of spheres of arbitrary radius with centers which run through the set of points of a fixed hyperplane, problem (6.1) was considered in Courant's book [139], while for a family of spheres of fixed radius it was considered in F. John's book [120]. Various formulations of the problem of integral geometry on ellipsoids are considered by F. John ([120, Chapter V]), G. I. Plaksin [201], [202], V. G. Romanov [209], S. V. Uspenskiĭ [100], [270]–[274], and M. V. Klibanov [128]–[130].

For the case where $\mathfrak{M}(\lambda)$ is a more complex geometric object, problems of integral geometry have come to be studied relatively recently. The stimulus for this study was the connection found between problems of integral geometry and multidimensional inverse problems for differential equations. M. M. Lavrent'ev and V. G. Romanov in [158] first focused attention on this connection. Subsequently, Romanov constructed a number of examples in which the investigation of inverse problems for equations of hyperbolic type was reduced to problems of integral geometry (see, for example, [159], [217], [223], and also the next chapter). The manifolds which arise here are connected in a natural way with the original differential equation. These are either sections of characteristic conoids or projections of the bicharacteristics onto the space orthogonal to the time coordinate. For an equation with variable coefficients these are rather complex geometric objects.

Romanov [210] obtained the first results in integral geometry for a rather general family of curves $\mathfrak{M}(\lambda)$ in two-dimensional space which is invariant under the rotation group. This result was then generalized [217] to the case of families of curves and hypersurfaces in n-dimensional space which are invariant under parallel translations of these objects along a fixed plane. The theory of problems of integral geometry was further developed in the works of Yu. E. Anikonov [15], [20], [22], M. M. Lavrent'ev and A. L. Bukhgeim [60]–[65], [153], [154], V. G. Romanov [223], [224], [230], [236], R. G. Mukhometov [189], [190], and other authors.

§1. The problem of finding a function from its spherical means

With the example of this simplest problem we shall show that among problems of integral geometry there are problems which are strongly unstable, completely analogous in character of instability to the Cauchy problem for the Laplace equation. The exposition of the material of this section is based on results presented in Courant's book [139], Chapter VI, §17.

1. In n-dimensional space $x_1, \dots, x_n$ $(n \geq 2)$ we consider the problem of finding a function $u(x_1, \dots, x_n)$ in terms of its mean values over spheres (over circles for $n = 2$) of arbitrary radius r $(0 < r < \infty)$ whose centers run through the set of points of a fixed plane. For convenience we take this plane to be the coordinate plane $x_n = 0$, and we denote an arbitrary point of the space $x_1, \dots, x_n$ by (x, y), where $x = (x_1, \dots, x_{n-1})$ and $y = x_n$. Thus, $u = u(x, y)$. The sphere of radius r with center at the point $(x, 0)$ we denote by $S(x, r)$. In this notation our problem consists in finding a function $u(x, r)$ from its integrals

$$\frac{1}{\omega_n} \int_{S(x,r)} u(\xi, \eta)\, d\omega = v(x, r). \tag{6.2}$$

Here (ξ, η) is a variable point on the sphere $S(x, r)$; $\xi = (\xi_1, \dots, \xi_{n-1})$; ω_n is the area of the unit sphere in n-dimensional space: $\omega_n = 2\pi^{n/2}/\Gamma(n/2)$; $d\omega$ is the element of solid angle connected with the element dS of surface area by the formula $dS = r^{n-1}\, d\omega$.

It is obvious that any function $u(x, y)$ odd in the variable y is a solution of the homogeneous equation (6.2), i.e., of the equation in which $v(x, r) = 0$. In connection with this it is reasonable to pose the problem of determining from $v(x, r)$ only the even part of the function $u(x, y)$, i.e., the function $u_1(x, y) = \frac{1}{2}[u(x, y) + u(x, -y)]$. This is equivalent to considering a class of functions $u(x, y)$ even in y. There is the following uniqueness theorem.

THEOREM 1. *Any function $u(x, y)$ continuous in the region $D(x_0, r_0) = \{(x, y)\colon |x - x_0|^2 + y^2 < r_0^2\}$ and even in y is uniquely determined in this region by giving the function $v(x, r)$ in the region $G_\varepsilon(x_0, r_0) = \{(x, r)\colon |x - x_0| < \varepsilon,$ $0 < r < r_0 - |x - x_0|\}$, where ε is any fixed positive number such that $\varepsilon \leq r_0$.*

The proof is based on calculating all moments of the function $u(x, t)$ corresponding to the variable x on each fixed sphere $S(x, r)$ such that $(x, r) \in G_\varepsilon(x_0, r_0)$. The region $G_\varepsilon(x_0, r_0)$ is reminiscent of a sharpened pencil (Figure

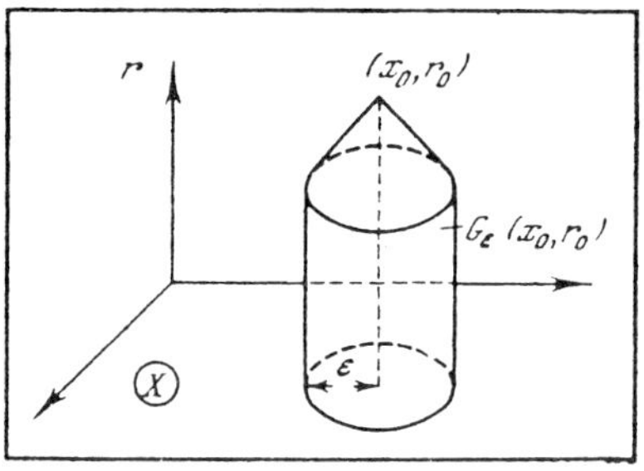

Figure 6

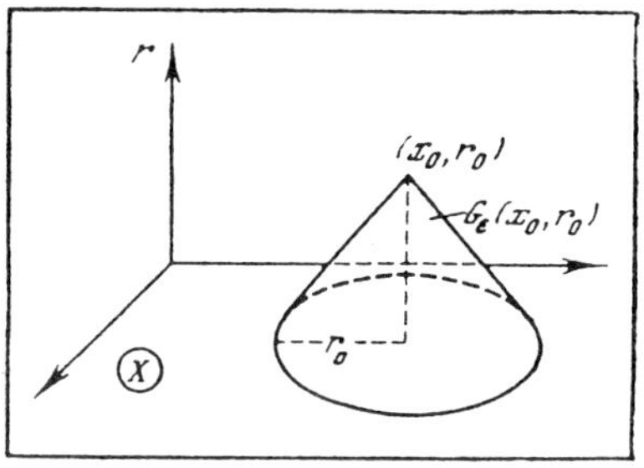

Figure 7

6). For $\varepsilon = r_0$ this "pencil" degenerates into a piece of a cone (Figure 7). For any point $(x, r) \in G_\varepsilon(x_0, r_0)$, $0 < \varepsilon \leq r_0$, the sphere $S(x, r)$ lies entirely inside the region $D(x_0, r_0)$.

We consider the result of applying the operator

$$A_i v = \frac{\partial}{\partial x_i} \int_0^r \tau^{n-1} v(x, \tau)\, d\tau, \quad i = 1, 2, \ldots, n-1,$$

to equation (6.2). Carrying out a number of computations in succession, we find

$$\begin{aligned} A_i v &= \frac{1}{\omega_n} \frac{\partial}{\partial x_i} \int_0^r \tau^{n-1} \left[\int_{S(x,\tau)} u(\xi, \eta)\, d\omega \right] d\tau \\ &= \frac{1}{\omega_n} \frac{\partial}{\partial x_i} \int_{|\xi - x|_i^2 + \eta^2 \leq r^2} \left[\int_{x_i - \sqrt{r^2 - \eta^2 - |\xi - x|_i^2}}^{x_i + \sqrt{r^2 - \eta^2 - |\xi - x|_i^2}} u(\xi, \eta)\, d\xi_i \right] \\ &\qquad\qquad \times d\eta\, d\xi_1 \ldots d\xi_{i-1}\, d\xi_{i+1} \cdots d\xi_{n-1} \\ &= \frac{1}{\omega_n} \int_{S(x,r)} u(\xi, \eta) \cos(n, \xi_i)\, dS \\ &= \frac{r^{n-2}}{\omega_n} \int_{S(x,r)} u(\xi, \eta)(\xi_i - x_i)\, d\omega. \end{aligned}$$

Here $|\xi - x|_i$ denotes the expression $\sqrt{|\xi - x|^2 - |\xi_i - x_i|^2}$ which, in the intermediate computations, coincides numerically with the length of the projection of the segment of the line joining the points $(\xi, 0)$ and $(x, 0)$ onto the plane $x_i = 0$, $y = 0$; n denotes the unit vector of the outer normal to $S(x, r)$.

Denoting now by L_i the operator defined by

$$L_i v = \frac{1}{r^{n-2}} A_i v + x_i v(x, r), \quad i = 1, 2, \ldots, n-1,$$

from the formula we have obtained we find that the result of its application to (6.2) is equivalent to computing the spherical mean of the function $u_i(x, y) = x_i u(x, y)$. Indeed,

$$L_i v = \frac{1}{\omega_n} \int_{S(x,r)} u(\xi, \eta) \xi_i \, d\omega.$$

It is clear that the result of applying the composition of operators $L_j L_i$ to the function $v(x, r)$ is equivalent to computing the spherical mean of the function $u(x, y) x_i x_j$. Generally, if $P_m(x)$ is a polynomial of degree m with constant coefficients, then the result of applying the operator $P(L)$, $L = (L_1, \ldots, L_{n-1})$, to equation (6.2) leads to the equality

$$P_m(L) v = \frac{1}{\omega_n} \int_{S(x,r)} u(\xi, \eta) P_m(\xi) \, d\omega.$$

We now fix x and r. Let $D_0(x, r) = \{\xi \colon |\xi - x| \le r\}$. On the sphere $S(x, r)$ our equality can be represented in the form

$$P_m(L) v = \int_{D_0(x,r)} \varphi(\xi, x, r) P_m(\xi) \, d\xi, \tag{6.3}$$

where the function $\varphi(\xi, x, r)$ (x and r are fixed!) is connected with $u(x, y)$ by the formula

$$\varphi(\xi, x, r) = \frac{2}{\omega_n r^{n-2}} \frac{u(\xi, \sqrt{r^2 - |x - \xi|^2})}{\sqrt{r^2 - |\xi - x|^2}}.$$

Here the parity of $u(x, y)$ in the variable y has been taken into account.

The function $\varphi(\xi, x, r)$, and hence also $u(x, y)$, is uniquely determined by the system of equalities (6.3) corresponding to distinct polynomials $P_m(\xi)$. The theorem is thus proved.

For a constructive determination of $u(x, y)$ from $v(x, r)$ it is possible to take for $P_m(\xi)$ a system of polynomials orthogonal in $D_0(x, r)$. The function $\varphi(\xi, x, r)$ can then be written out explicitly in the form of a Fourier series. This requires, however, more stringent conditions on $u(x, y)$ in order to guarantee convergence of the Fourier series.

COROLLARY. *If the function $u(x, y)$ is continuous in the region $D(x_0, r_0)$ together with its partial derivative $u_y(x, y)$, then it is uniquely determined in $D(x_0, r_0)$ by giving the spherical means of the function $u(x, y)$ itself and of its partial derivative $u_y(x, y)$ over all possible spheres $S(x, r)$ such that $(x, r) \in G_\varepsilon(x_0, r_0)$, $0 < \varepsilon \le r_0$.*

Indeed, in this case the part of the function $u(x, y)$ even in the variable y and the even part of the partial derivative $u_y(x, y)$ are uniquely determined. But any function of the class considered can be represented in the form

$$u(x, y) = \frac{1}{2}[u(x, y) + u(x, -y)] + \frac{1}{2} \int_0^y [u_y(x, y) + u_y(x, -y)] \, dy$$

and it is thus possible to determine it uniquely.

From the theorem just proved it is evident that giving the function $v(x,r)$ in the region $G_\varepsilon(x_0,r_0)$ for arbitrarily small positive ε determines $u(x,y)$ in $D(x_0,r_0)$. But, knowing $u(x,y)$ in $D(x_0,r_0)$, we can compute the integrals over all possible spheres lying entirely inside $D(x_0,r_0)$ and thus find $v(x,r)$ for $(x,r)\in G_\varepsilon(x_0,r_0)$ with $\varepsilon=r_0$. Thus, giving $v(x,r)$ in $G_\varepsilon(x_0,r_0)$ determines it uniquely in the larger region $G_{r_0}(x_0,r_0)\supset G_\varepsilon(x_0,r_0)$.

This bears witness to the fact that the function $v(x,r)$ cannot be prescribed arbitrarily. Moreover, any function $v(x,r)$ representable in the form (6.2) possesses a property analogous to a property of analytic functions: it is uniquely determined in $G_{r_0}(x_0,r_0)$ by its values in an arbitrarily "thin" region $G_\varepsilon(x_0,r_0)$. At the same time, it is clear that to nonanalytic functions $u(x,y)$ there correspond functions $v(x,r)$ which are also nonanalytic. In connection with what has been said above, a constructive description of the class of functions $v(x,r)$ representable in the form (6.2) constitutes a rather difficult problem.

2. We shall now show that the problem of solving equation (6.2) is classically ill-posed, namely, it is strongly unstable relative to small changes of the function $v(x,r)$. This is most conveniently done for three-dimensional space; therefore, we henceforth assume that $n=3$. Moreover, we assume here that the function $u(x,y)$ is twice continuously differentiable in the entire space x,y and is even in the variable y. We introduce the function

$$w(x,y,r)=\frac{r}{\omega_3}\int_{|\xi-x|^2+(\eta-y)^2=r^2} u(\xi,\eta)\,d\omega, \tag{6.4}$$

which up to the factor r is the spherical mean over a sphere of radius r with center at the point (x,y). From a course in equations of mathematical physics (see, for example, [199], [246] or [248]) it is known that the function $w(x,y,r)$ in the half-space $r>0$ satisfies the wave equation

$$w_{rr}=\Delta_{x,y}w. \tag{6.5}$$

Here Δ_{xy} is the Laplace operator in the variables x and y. It is evident directly from (6.4) that

$$w(x,y,0)=0. \tag{6.6}$$

From (6.2) we obtain

$$w(x,y,r)|_{y=0}=rv(x,r). \tag{6.7}$$

Finally, the condition that $u(x,y)$ be even in y leads to the equality

$$w_y(x,y,r)|_{y=0}=0. \tag{6.8}$$

The problem of finding a function $w(x,y,r)$ satisfying (6.5) and conditions (6.6)–(6.8) is a boundary value problem which is obviously equivalent to the

Cauchy problem (6.5), (6.7), (6.8) in the entire space x, y, r if $v(x, r)$ is extended to the region $r < 0$ as an even function.

THEOREM 2. *Problem* (6.5)–(6.8) *is equivalent to the problem of solving equation* (6.2) *in the class of functions even in* y.

Indeed, if $u(x, y)$ is a solution of (6.2) such that $u(x, y) = u(x, -y)$, then the function $w(x, y, r)$ computed by (6.4) is a solution of problem (6.5)–(6.8). This solution is unique, since the homogeneous problem (6.5)–(6.8) has only the zero solution. Thus, to each function $u(x, y)$ there corresponds one and only one solution of (6.5)–(6.8).

We shall now show that the converse assertion holds. Namely, each solution of problem (6.5)–(6.8) generates one and only one solution of (6.2) such that $u(x, -y) = u(x, y)$. Let $w(x, y, r)$ be a solution of (6.5)–(6.8). We set

$$w_r(x, y, 0) = u(x, y). \tag{6.9}$$

The function $w(x, y, r)$ in the region $r \geq 0$ can then be uniquely represented in the form (6.4) as a solution of the Cauchy problem with data (6.6), (6.9). Comparing formula (6.4) for $y = 0$ with (6.7), we find that $u(x, y)$ is a solution of (6.2). Computing the derivatives with respect to y of both sides of (6.4), setting $y = 0$, and using (6.8), we find that

$$\int_{S(x,r)} u_\eta(\xi, \eta)\, d\omega = 0. \tag{6.10}$$

From this and Theorem 1 it follows that the even part of $u_y(x, y)$ is equal to zero:

$$u_y(x, -y) + u_y(x, y) = 0,$$

i.e., $u(x, -y) = u(x, y)$.

Thus, the function $u(x, y)$ computed from the solution of problem (6.5)–(6.8) by means of (6.9) possesses the property that it is even in the variable y. By Theorem 1, to each solution of problem (6.5)–(6.8) there corresponds only one such function.

We have thus proved that there is a one-to-one correspondence between solutions of (6.5)–(6.8) and solutions of (6.2) under the condition $u(x, -y) = u(x, y)$. This constitutes the equivalence theorem.

We shall now illustrate the instability of problem (6.5)–(6.8). Let

$$v(x, r) = \frac{1}{k^s} \frac{\sin kr}{r} \sin kx_1 \sin kx_2.$$

For sufficiently large k and s the function $v(x, r)$ is small together with any finite number of derivatives. At the same time, it is easy to verify directly that the solution of (6.5)–(6.8) in this case is given by

$$w(x, y, r) = \frac{1}{k^s} \sin kr \sin kx_1 \sin kx_2 \cosh ky.$$

But this solution for any fixed y tends to infinity as $k \to \infty$. The function $u(x, y)$ computed according to (6.9) and providing the solution of (6.2) has the same property:

$$u(x, y) = \frac{1}{k^{s-1}} \sin kx_1 \sin kx_2 \cosh ky,$$

and this means strong instability of the solution of problem (6.5)–(6.8) and of the equivalent problem of solving equation (6.2).

The problem of solving (6.2) nevertheless becomes conditionally well-posed if the class of solutions considered is restricted. Namely, we consider a set U of functions $u(x, y)$ even in the variable y which is compact in the space C (for example, a bounded set of functions with first derivatives bounded by a given constant), and we let V be the set of images $v(x, r)$ of the functions $u(x, y) \in U$. Under the condition that a solution of (6.2) exists and belongs to U it will then be stable because of the familiar theorem from functional analysis regarding the continuity of the inverse mapping (see Chapter II, §1, Theorem 1).

§2. Problems of integral geometry on a family of manifolds which is invariant under a group of transformations of the space

1. We shall consider the problem of integral geometry on a family of manifolds $\mathfrak{M}(\xi)$ of the following type. Suppose in m-dimensional space $t = (t_1, \dots, t_m)$, $m \geq 1$, there is given an n-parameter family of regions $\Delta(\xi)$ with smooth boundary $\gamma(\xi)$, $\xi = (\xi_1, \dots, \xi_n)$, and let $\mathfrak{M}(\xi)$ be the image of $\Delta(\xi)$ in n-dimensional space $x = (x_1, \dots, x_n)$, $n > m$, under a mapping given by a smooth function $f(t, \xi)$. Suppose, further, that $D = \{x\colon 0 < x_n < H\}$ and for the parameter ξ it is possible to take any point of the region D. We shall assume that $\xi \in \mathfrak{M}(\xi)$, and for any point $x \in \mathfrak{M}(\xi)$ with $x \neq \xi$ the inequality $0 \leq x_n < \xi_n$ holds; moreover, the image of the boundary of $\Delta(\xi)$ under the mapping realized by the function $f(t, \xi)$ belongs to the hyperplane $x_n = 0$, i.e., to part of the boundary of D. Below we shall specify somewhat more precisely the conditions on the family of manifolds $\mathfrak{M}(\xi)$, but now we give some explanations in connection with the manifolds introduced.

For $m = 1$ each of the manifolds $\mathfrak{M}(\xi)$ is an arc with support plane $x_n = 0$. The point of the arc $\mathfrak{M}(\xi)$ most distant from this plane coincides with the point ξ. Thus, to any point $\xi \in D$ there corresponds a curve $\mathfrak{M}(\xi)$ having the point ξ as apex and joining some pair of points of the plane $x_n = 0$.

For $m = n - 1$, $n \geq 3$, the manifold $\mathfrak{M}(\xi)$ is a hypersurface of "cap" type in n-dimensional space lying in $\overline{D} = \{x\colon 0 \leq x_n \leq H\}$. The boundary of this surface lies on the plane $x_n = 0$, and the point $\xi \in D$ is most distant from this plane. In the general case of arbitrary m the manifold $\mathfrak{M}(\xi)$ is an m-dimensional surface with support plane $x_n = 0$.

The problem we shall study consists in the following. There are given the integrals over the family of manifolds $\mathfrak{M}(\xi)$ of some function $u(x)$ smooth in the region $\overline{D}$:

$$\int_{\mathfrak{M}(\xi)} \rho(t, \xi) u(x)\, dt = v(\xi), \quad \xi \in D, \tag{6.11}$$

where $\rho(t, \xi)$ is a given smooth function. It is required to find the function $u(x)$ from the function $v(\xi)$.

Thus, our problem consists in solving the special integral equation (6.11). It can also be written in another form:

$$\int_{\Delta(\xi)} \rho(t, \xi) u(f(t, \xi))\, dt = v(\xi), \quad \xi \in D. \tag{6.11$'$}$$

2. In studying questions of conditional well-posedness of the above problem we shall make some additional assumptions regarding the structure of the manifolds $\mathfrak{M}(\xi)$ and the weight function $\rho(t, \xi)$. These assumptions will be different in the present subsection and the following one, but they have something in common which we now discuss. We shall assume without special mention that the mapping function $f(t, \xi)$ has sufficiently high smoothness.

We consider a neighborhood of the point ξ on the manifold $\mathfrak{M}(\xi)$. Under the assumption that ξ is a point of elliptic type, we study the character of the analytic representation of the manifold $\mathfrak{M}(\xi)$ in a neighborhood of this point. We shall explain, first of all, what we mean by the term "point of elliptic type." Suppose to the point ξ there corresponds, to be specific, the point $t = 0$—an interior point of the set $\Delta(\xi)$. Let $f_n(t, \xi)$ be the nth component of the function $f(t, \xi)$. We call ξ a point of elliptic type of the manifold $\mathfrak{M}(\xi)$ if there exist positive constants $q(\xi)$ and $Q(\xi)$ such that for any unit vector $\nu = (\nu_1, \ldots, \nu_m)$ the following inequality holds:

$$q(\xi) \leq -\sum_{i,j=1}^{m} \frac{\partial^2 f_n}{\partial t_i \partial t_j}\bigg|_{t=0} \nu_i \nu_j \leq Q(\xi). \tag{6.12}$$

We shall assume in what follows that for the entire family $\mathfrak{M}(\xi)$ there exist positive numbers q_0 and Q_0 such that $q(\xi) \geq q_0$ and $Q(\xi) \leq Q_0$, i.e., we assume a certain uniform ellipticity of the family of manifolds $\mathfrak{M}(\xi)$ in a neighborhood of points ξ. Since by assumption

$$\nabla_t f_n(t, \xi)|_{t=0} = 0,$$

in a neighborhood of the point $t = 0$ we have

$$x_n = \xi_n + \sum_{i,j=1}^{m} \int_0^1 (1-s) \left[\frac{\partial^2}{\partial t_i \partial t_j} f_n(ts, \xi) \right] t_i t_j \, ds. \tag{6.13}$$

Introducing the unit vector ν such that $t = r\nu$, $r = |t|$, we rewrite (6.13) in the form

$$r = (\xi_n - x_n)^{1/2} \left[-\sum_{i,j=1}^{m} \int_0^1 (1-s)\nu_i\nu_j \frac{\partial^2}{\partial t_i \partial t_j} f_n(ts, \xi)\, ds \right]^{-1/2}. \tag{6.14}$$

The denominator of this expression does not vanish at $t = 0$ because of assumption (6.12); therefore, considering (6.14) as an equation for the implicitly defined function $r = r(\sqrt{\xi_n - x_n}, \nu, \xi)$ and applying the implicit function theorem, we find that in a neighborhood of $t = 0$ the variable r is a single-valued function,

$$r = \varphi\left(\sqrt{\xi_n - x_n}, \nu, \xi\right). \tag{6.15}$$

The smoothness of this function depends on the smoothness of $f(t, \xi)$. It follows from (6.14) that

$$\begin{aligned} &\varphi(p, \nu, \xi)|_{p=0} = 0, \\ &\varphi_p(p, \nu, \xi)|_{p=0} = \left(-\frac{1}{2} \sum_{i,j=1}^{m} \frac{\partial^2 f_n}{\partial t_i \partial t_j}\bigg|_{t=0} \nu_i \nu_j \right)^{-1/2}. \end{aligned} \tag{6.16}$$

Using (6.12), we find that

$$0 < \sqrt{2/Q_0} \le \sqrt{2/Q(\xi)} \le \varphi_p(p, \nu, \xi)|_{p=0} \le \sqrt{2/q(\xi)} \le \sqrt{2/q_0}. \tag{6.17}$$

Thus, with respect to the argument $p = \sqrt{\xi_n - x_n}$ the function $\varphi(p, \nu, \xi)$ has a zero of first order at the point $p = 0$.

We set $x = (\overline{x}, x_n)$, $\overline{x} = (x_1, \dots, x_{n-1})$, and similarly $\xi = (\overline{\xi}, \xi_n)$ and $f = (\overline{f}, f_n)$. In this notation the manifold $\mathfrak{M}(\xi)$ in a neighborhood of the point ξ can be represented in the form

$$\overline{x} = g\left(\sqrt{\xi_n - x_n}, \nu, \xi\right), \tag{6.18}$$

where

$$g(0, \nu, \xi) = \overline{\xi}. \tag{6.19}$$

Here $g(p, \nu, \xi) = \tilde{f}[\nu\varphi(p, \nu, \xi), \xi]$.

We have thus shown that in the case of a sufficiently smooth mapping function $f(t, \xi)$ the assumptions made regarding the class of manifolds $\mathfrak{M}(\xi)$ and the character of the points ξ lead to the situation that the analytic structure of the manifold $\mathfrak{M}(\xi)$ in a neighborhood of a point ξ can be described in the form (6.18), (6.19), where $g(p, \nu, \xi)$ is a smooth function of its arguments. Here ν is an arbitrary point of the m-dimensional unit sphere S_m with center at the origin. The analytic structure of the mapping in a neighborhood of ξ is described by the pair of functions g, φ.

Below we shall assume that formulas (6.15)–(6.19) describe the manifolds $\Delta(\xi)$ and $\mathfrak{M}(\xi)$ not only in a neighborhood of a point ξ but also globally. The argument p here varies for each fixed ξ on the segment $[0, \sqrt{\xi_n}]$, and to the value $p = \sqrt{\xi_n}$ there corresponds the boundary of the manifolds $\Delta(\xi)$ and $\mathfrak{M}(\xi)$. This assumption can be weakened to a considerable extent, but it is convenient for us to adopt it for our initial considerations.

We shall present some further clarifications connected with the case $m = 1$. This case can be considered special in a certain sense, since the analogue of the unit sphere S_m here is the set consisting only of the two points $\nu = 1$ and $\nu = -1$. Moreover, to positive values of t there correspond the value $\nu = 1$ and the part of the arc $\mathfrak{M}(\xi)$ from the point ξ to one of the endpoints of the arc (for $x_n = 0$, i.e., $p = \sqrt{\xi_n}$), while to negative values of t there correspond the value $\nu = -1$ and the part of the arc $\mathfrak{M}(\xi)$ from ξ to the other endpoint of the arc. We thus obtain two single-valued functions g: one corresponds to $\nu = 1$, and the other to $\nu = -1$; they completely describe $\mathfrak{M}(\xi)$; we also obtain two functions φ describing the correspondence between points of the segment $0 \leq x_n \leq \xi_n$ and points of the segment $\overline{\Delta}(\xi) = \{t\colon -\varphi(\sqrt{\xi_n}, -1, \xi) \leq t \leq \varphi(\sqrt{\xi_n}, 1, \xi)\}$.

3. We now proceed to study the problem of integral geometry in the formulation related to equation (6.11). In this subsection we consider the special case where the functions ρ and φ do not depend on $\overline{\xi}$, while the function g depends on $\overline{\xi}$ in an additive manner, i.e.,

$$\rho = \rho(t, \xi_n), \qquad \varphi = \varphi(p, \nu, \xi_n), \qquad g = \overline{\xi} + \overline{g}(p, \nu, \xi_n), \quad \overline{g}|_{p=0} = 0. \tag{6.20}$$

This corresponds to the situation that the family of manifolds $\mathfrak{M}(\xi)$ and the weight function ρ are invariant under the group of transformations of the space x connected with parallel translation of the coordinate system to any point of the plane $x_n = 0$.

We have the following uniqueness theorem.

THEOREM 1. *Suppose the function $u(x)$ is continuous in the region $\overline{D}$ and satisfies the inequality*

$$|u(\overline{x}, x_n)| \leq C\psi(\overline{x}), \tag{6.21}$$

where the function $\psi(\overline{x}) \in L_1(R^{n-1})$, while the functions

$$\phi(p, \nu, \xi_n), \qquad \overline{g}(p, \nu, \xi_n), \qquad \rho_0(p, \nu, \xi_n) \equiv \rho(\nu\varphi(p, \nu, \xi_n), \xi_n)$$

are continuous in the region

$$G = \{(p, \nu, \xi_n) \colon p \in [0, \sqrt{H}],\ \nu \in S_m,\ \xi_n \in [0, H]\}$$

together with their derivatives

$$\frac{\partial^s}{\partial p^i \partial \xi_n^j} \rho_0(p, \nu, \xi_n), \qquad \frac{\partial^{s+1}}{\partial p^{i+1} \partial \xi_n^j} \varphi(p, \nu, \xi_n), \qquad \frac{\partial^s}{\partial p^i \partial \xi_n^j} \overline{g}(p, \nu, \xi_n),$$

$$i + j = s, \qquad s = \left[\frac{m+1}{2}\right],$$

and the following inequality holds:

$$\rho(0, \xi_n) \neq 0, \quad \xi_n \in [0, H]. \tag{6.22}$$

Then the function $u(x)$ is uniquely determined in the region $\overline{D}$ by its integrals (6.11).

We write equation (6.11′) in the form

$$\int_0^{\xi_n} dx_n \int_{S_m} u\left[\overline{\xi} + \overline{g}\left(\sqrt{\xi_n - x_n}, \nu, \xi_n\right), x_n\right]$$
$$\times K\left(\sqrt{\xi_n - x_n}, \nu, \xi_n\right) d\omega_\nu = v(\xi), \tag{6.23}$$

where $d\omega_\nu$ is the element of surface area of the unit sphere S_m, and $K(p, \nu, \xi_n)$ is found from the formula

$$K(p, \nu, \xi_n) = -\frac{1}{2} \rho_0(p, \nu, \xi_n)[\varphi(p, \nu, \xi_n)]^{m-1} \frac{1}{p} \varphi_p(p, \nu, \xi_n). \tag{6.24}$$

To obtain (6.23) and (6.24) it suffices, in (6.11), to pass under the integral sign first to a polar coordinate system r, ν and then pass from the variable r to the variable x_n, using (6.15). For $m = 1$ the role of the integral over the sphere S_m is played by the sum on ν ($\nu = -1$, $\nu = 1$). We apply to (6.23) the Fourier transformation in the variable $\overline{\xi}$. We set

$$\tilde{v}(\lambda, \xi_n) = \int_{R^{n-1}} v(\tilde{\xi}, \xi_n) \exp(i\lambda \cdot \tilde{\xi})\, d\overline{\xi}. \tag{6.25}$$

Changing the order of integration on the left side of (6.23), we obtain

$$\int_0^{\xi_n} \tilde{u}(\lambda, x_n) R\left(\sqrt{\xi_n - x_n}, \xi_n, \lambda\right) dx_n = \tilde{v}(\lambda, \xi_n). \tag{6.26}$$

Here

$$R(p, \xi_n, \lambda) = \int_{S_m} K(p, \nu, \xi_n) \exp[-i\lambda \cdot \overline{g}(p, \nu, \xi_n)]\, d\omega_\nu. \tag{6.27}$$

Thus, (6.23) reduces to a family of Volterra integral equations of the first kind. The parameter of this family is the parameter $\lambda = (\lambda_1, \dots, \lambda_{n-1})$ of the Fourier transformation. We shall show that for each fixed λ (6.26) has a unique solution. From the properties of the functions ρ, φ, and $\bar{g}$ noted in the theorem it follows that the kernel $R(p, \xi_n, \lambda)$ can be represented in the form

$$R(p, \xi_n, \lambda) = p^{m-2} R_0(p, \xi_n, \lambda), \tag{6.28}$$

where R_0 has continuous derivatives up to order s with respect to the arguments p, ξ_n in the closed region $G_0 = \{(p, \xi_n)\colon p \in [0, \sqrt{H}],\ \xi_n \in [0, H]\}$, and

$$R_0(0, \xi_n, \lambda) = -\frac{1}{2}\rho(0, \xi_n) \int_{S_m} [\varphi_p(0, \nu, \xi_n)]^m \, d\omega_\nu \neq 0, \quad \xi_n \in [0, H]. \tag{6.29}$$

From the theory of integral equations it is known (see, for example, [192] or [246]) that these properties of the kernel guarantee a unique solution of (6.26). This, in turn, means uniqueness of the solution of (6.11), and the theorem is proved.

It is also possible to obtain a method of solving (6.26). For this it is convenient to go over from (6.26) to a Volterra equation of the second kind. To this end it suffices to apply to (6.26) differentiation with respect to ξ_n of order $m/2$. For m odd, and hence $s = (m+1)/2$, this means that

$$\tilde{v}^{(m/2)}(\lambda, \xi_n) = \frac{1}{\Gamma\left(\frac{1}{2}\right)} \frac{\partial^s}{\partial \xi_n^s} \int_0^{\xi_n} \frac{\tilde{v}(\lambda, z)}{\sqrt{\xi_n - z}} \, dz,$$

where Γ is the Euler gamma function. Differentiating and dividing the left and right sides of the equality obtained by

$$\frac{\Gamma_{(m/2)}}{m/2} R_0(0, \xi_n, \lambda),$$

we find the equation

$$\tilde{u}(\lambda, \xi_n) + \int_0^{\xi_n} \tilde{u}(\lambda, x_n) T\left(\sqrt{\xi_n - x_n}, \xi_n, \lambda\right) dx_n = w(\lambda, x_n), \tag{6.30}$$

in which

$$w(\lambda, x_n) = \frac{m/2}{\Gamma(m/2)} \frac{1}{R_0(0, \xi_n, \lambda)} \tilde{v}^{(m/2)}(\lambda, \xi_n), \tag{6.31}$$

while the kernel of the equation for $m = 2s$ is computed from the formula

$$T\left(\sqrt{\xi_n - x_n}, \xi_n, \lambda\right) = \frac{s}{\Gamma(s)} \frac{1}{R_0(0, \xi_n, \lambda)} \frac{\partial^s}{\partial \xi_n^s} R\left(\sqrt{\xi_n - x_n}, \xi_n, \lambda\right), \tag{6.32}$$

and for $m = 2s - 1$ from

$$T\left(\sqrt{\xi_n - x_n}, \xi_n, \lambda\right) = \frac{s - \frac{1}{2}}{\Gamma(s - \frac{1}{2})} \frac{1}{R_0(0, \xi_n, \lambda)} \frac{\partial^s}{\partial \xi_n^s} \int_{x_n}^{\xi_n} \frac{R\left(\sqrt{z - x_n}, z, \lambda\right)}{\sqrt{\xi_n - z}} \, dz. \tag{6.33}$$

It follows from (6.32), (6.33), and (6.28) that the kernel of equation (6.30) can be represented in the form

$$T\left(\sqrt{\xi_n - x_n}, \xi_n, \lambda\right) = \frac{1}{\sqrt{\xi_n - x_n}} T_0\left(\sqrt{\xi_n - x_n}, \xi_n, \lambda\right),$$

where $T_0(p, \xi_n, \lambda)$ is a continuous function of the arguments p and ξ_n in the region G_0. Hence, if we iterate the kernel in (6.30) once, we obtain the new equation

$$\tilde{u}(\lambda, \xi_n) + \int_0^{\xi_n} \tilde{u}(\lambda, x_n) T_1\left(\sqrt{\xi_n - x_n}, \xi_n, \lambda\right) dx_n = w_1(\lambda, \xi_n), \tag{6.34}$$

in which the kernel $T_1(\sqrt{\xi_n - x_n}, \xi_n, \lambda)$ computed by the formula

$$T_1\left(\sqrt{\xi_n - x_n}, \xi_n, \lambda\right) = -\int_{x_n}^{\xi_n} T\left(\sqrt{\xi_n - z}, \xi_n, \lambda\right) T\left(\sqrt{z - x_n}, z, \lambda\right) dz, \tag{6.35}$$

is continuous in G_0. Therefore, to obtain a solution of this equation it is possible to use the ordinary method of successive approximations. The function $w_1(\lambda, \xi_n)$ is computed here by the formula

$$w_1(\lambda, \xi_n) = w(\lambda, \xi_n) - \int_0^{\xi_n} w(\lambda, x_n) T\left(\sqrt{\xi_n - x_n}, \xi_n, \lambda\right) dx_n. \tag{6.36}$$

4. We observe that the norm of the kernels of equation (6.34) computed in the space $C(G_0)$ grows like $|\lambda|^{2s}$ with increasing $|\lambda|$. Indeed, (6.32) and (6.33) contain s-fold differentiation of the function R which by (6.27) leads to the appearance of a factor estimated like $|\lambda|^s$. The iterated kernel T_1 already has order $|\lambda|^{2s}$. Such growth of the norm of the kernels for large values of $|\lambda|$ leads to the situation that convergence of the method of successive approximations becomes worse with increasing $|\lambda|$. The estimate of stability of the solution of (6.34) also becomes worse. From this equation we can obtain no estimate of stability uniform in λ and hence no estimate of stability of the problem of solving (6.11). Using (6.34), however, it is not hard to obtain an estimate of the conditional stability of this problem of integral geometry. This was first done by A. L. Bukhgeĭm for a problem of integral geometry on a family of curves. A general scheme of obtaining estimates is presented in [59].

THEOREM 2. *Suppose that the function $u(x)$ has continuous derivatives with respect to the variables $x_1, \dots, x_{n-1}$ through order $k > (n-1)$ and*

$$\int_{R^{n-1}} |D^\alpha u(\overline{x}, x_n)|\, d\overline{x} \le M, \quad |\alpha| = k, \tag{6.37}$$

where M is a given constant, and D^α is the operator of differentiation with respect to the variables $x_1, \ldots, x_{n-1}$:

$$D^\alpha = \frac{\partial^{|\alpha|}}{\partial x_1^{\alpha_1} \cdots \partial x_{n-1}^{\alpha_{n-1}}}, \qquad |\alpha| = \alpha_1 + \cdots + \alpha_{n-1}.$$

Suppose, further, that the function $(\partial(m/2)/\partial\xi_n^{(m/2)})v(\xi)$ is continuous in $\overline{D}$ and

$$\int_{R^{n-1}} |\nu_{\xi_n}^{(m/2)}(\overline{\xi}, \xi_n)|\, d\overline{\xi} \le \varepsilon. \tag{6.38}$$

Then the following qualitative estimate of stability of the problem of integral geometry holds:

$$|u(x)| \le CM |\ln(M/\varepsilon)|^{(n-k-1)/2s}. \tag{6.39}$$

Indeed, choosing a sufficiently large positive N and using the representation of $u(x)$ in terms of its Fourier transform with respect to the variables $x_1, \ldots, x_{n-1}$, we have

$$|u(x)| \le \frac{1}{(2\pi)^{n-1}} \int_{|\lambda| \le N} |\tilde{u}(\lambda, x_n)|\, d\lambda + \frac{1}{(2\pi)^{n-1}} \int_{|\lambda| \ge N} |\tilde{u}(\lambda, x_n)|\, d\lambda. \tag{6.40}$$

We shall estimate the first term in this formula using the integral equation (6.34) and the second using a priori information about the class of functions $u(x)$. For sufficiently large N the right side of (6.24) can, by (6.36) and (6.38), be estimated in the form

$$|w_1(\lambda, \xi_n)| \le C_1 N^s \varepsilon, \quad |\lambda| \le N.$$

Since the norm of the kernel T_1 increases with increasing $|\lambda|$ no faster than $|\lambda|^{2s}$, from (6.34) for the solution $\tilde{u}(\lambda, x_n)$ we obtain an estimate of the form

$$|\tilde{u}(\lambda, x_n)| \le C_1 N^s \varepsilon \exp(C_2 N^{2s}), \quad |\lambda| \le N.$$

The a priori assumption about the class of functions $u(x)$ leads to the following estimate:

$$|\tilde{u}(\lambda, x_n)| \le \frac{M}{|\lambda|^k}, \quad |\lambda| > 0.$$

Using these two estimates, from (6.40) we find that

$$|u(x)| \le C_3 N^{s+n-1} \varepsilon \exp(C_2 N^{2s}) + C_4 M N^{-k+n-1}.$$

We now choose $N = N(M/\varepsilon)$ so that

$$C_3 N^{s+k} \exp(C_2 N^{2s}) = (M/\varepsilon) C_4. \tag{6.41}$$

Then

$$|u(x)| \le [N(M/\varepsilon)]^{n-k-1} 2 C_4 M.$$

It is clear that $N(M/\varepsilon) \to \infty$ as $\varepsilon \to 0$. It follows from (6.41) that the leading part of the infinitely large $N(M/\varepsilon)$ has order $(\ln(M/\varepsilon))^{s/2}$. From this we obtain (6.39).

5. Above we considered the scalar case of the formulation of the problem of integral geometry. The case in which a vector-valued function $u(x)$ with components $u_1, \dots, u_r$ is to be determined can be formulated in an altogether analogous way. The analogue of the scalar equation (6.11) is the vector equation

$$\int_{\mathfrak{M}(\xi)} R(t, \xi) u(x)\, dt = v(\xi), \tag{6.42}$$

in which $R(t, \xi)$ is a square matrix of dimension $r \times r$ and v is a vector-valued function with components $v_1, \dots, v_r$. The next theorem is a total analogue of Theorem 1.

THEOREM 3. *Regarding the functions* u, φ, $\overline{g}$, *and* R, *suppose that the assumptions of Theorem* 1 *are satisfied with the natural replacement of condition* (6.22) *by the analogous condition*

$$\det R(0, \xi_n) \neq 0, \quad \xi_n \in [0, H]. \tag{6.43}$$

Then the function $u(x)$ *is uniquely determined from equality* (6.42).

The proof of the theorem is sufficiently obvious, and we omit it. The estimate of stability has the same form.

6. In the case $m = n-1$ it is possible to consider a somewhat more general problem than (6.42); namely, suppose it is required to find the function $u(x)$ from the equation

$$\int_{\mathfrak{M}(\xi)} R(t, \xi) u(x)\, dt + \int_0^{\xi_n} dz \int_{\mathfrak{M}(\overline{\xi}, z)} R_1(t, \overline{\xi}, z) u(x)\, dt = v(\xi) \tag{6.44}$$

with known functions R and R_1. This problem arises in the theory of inverse problems for equations of hyperbolic type (see the next chapter). Under the condition that R_1, just as R, does not depend on $\overline{\xi}$ and possesses smoothness analogous to the smoothness of R, for equation (6.44) there is a uniqueness theorem analogous to Theorem 3. Thus, the part of the operator corresponding to the first term in (6.44) is like the principal part of the entire integral operator, which is responsible for its unique invertibility.

§3. Integral geometry in special classes of functions

In the preceding section the study of the problem of integral geometry was based on special properties of the family of manifolds and the weight function. It is of major interest for applications to study the problem of integral geometry in the formulation of the preceding section while giving up these properties and restricting attention only to the assumptions regarding the family of manifolds presented in subsections 1 and 2. Unfortunately, there are so far no definitive results here. There are investigations of the problem in this rather general formulation only in special classes of functions. Exposition of the results of these investigations is the purpose of the present section.

1. In a somewhat more general formulation than that of §2 the problem of integral geometry was investigated by Yu. E. Anikonov in a class of functions analytic in the variable x_n [15], [22]. Namely, let M be the set of functions $u(x)$ representable in the region $\overline{D} = \{x\colon 0 \le x_n \le H\}$ in the form

$$u(x) = \sum_{s=0}^{\infty} \frac{x_n^s}{s!} \frac{\partial^s}{\partial x_n^s} u(\overline{x}, 0), \tag{6.45}$$

where the $\partial^s u(\overline{x}, 0)/\partial x_n^s$ are continuous for all values $\overline{x} \in E^{n-1}$ and the series (6.45) converges uniformly in $\overline{D}$. There is the following uniqueness theorem.

THEOREM 1. *Suppose the weight function $\rho(t, \xi)$ is continuous in its arguments for all $t \in \Delta(\xi)$ and $\xi \in \overline{D}$, and $\rho(0, \xi) \neq 0$ for $\xi = (\overline{\xi}, 0)$; suppose the mapping function $f(t, \xi)$ is continuous for $t \in \overline{\Delta(\xi)}$, $\xi \in \overline{D}$, and that the region $\Delta(\xi)$ is measurable and contracts uniformly to the point $t = 0$ as $\xi_n \to 0$. Then the solution of* (6.11) *on the set M is unique.*

For the proof we observe that the set M is linear, and thus it suffices to show that for $v(\xi) \equiv 0$, equation (6.11) has on M only the zero solution. We shall consider an arbitrary point $(\overline{x}_0, 0)$ and show that for any function $\overline{u} \in M$ which is a solution of the homogeneous equation the equalities $\partial^s u(\overline{x}_0, 0)/\partial x_n^s = 0$, $s = 0, 1, 2, \ldots,$ hold. It is convenient to prove this by contradiction. Suppose that the derivatives $\partial^s u(\overline{x}_0, 0)/\partial x_n^s = 0$ for $0 \le s \le k-1$, but $\partial^k u(\overline{x}_0, 0)/\partial x_n^k \neq 0$. Then there exists an open half-neighborhood $D_\varepsilon = \{x\colon |\overline{x} - \overline{x}_0|^2 + x_n^2 < \varepsilon^2,\ x_n > 0\}$ of the point $(\overline{x}_0, 0)$ in which the function $u(x)$ preserves the sign of the derivative $\partial^k u(\overline{x}_0, 0)/\partial x_n^k$. We now consider a set $\mathfrak{M}(\xi)$ entirely contained in D_ε. It is obvious that for any $\varepsilon > 0$ such a set $\mathfrak{M}(\xi)$ can be found, since as $\xi \to (\overline{x}_0, 0)$ the set $\Delta(\xi)$ contracts uniformly to the point $t = 0$ which is the preimage of the point ξ under the continuous mapping $f(t, \xi)$. From the properties of the function $\rho(t, \xi)$, it follows that by possibly reducing ε we can always arrange that $\rho(t, \xi) \neq 0$, $t \in \Delta(\xi)$. But then the product $\rho \cdot u$ preserves a definite sign on $\mathfrak{M}(\xi)$, and therefore the left side of (6.11) cannot be equal to zero. Thus, the assumption made is false. It remains to assume only that the derivatives with respect to x_n of all orders (beginning with the zeroth derivative) at the point $(\overline{x}_0, 0)$

vanish. From the representation (6.45) it now follows that $u(\overline{x}_0, x_0) = 0$, and the theorem is proved.

The uniqueness theorem can be extended rather simply to the more general class of functions piecewise analytic in the variable x_n. We do not dwell on this.

2. The study of the problem of integral geometry in other classes of functions is based on obtaining some integro-differential equation. To obtain it we must use more stringent conditions on the manifolds $\mathfrak{M}(\xi)$ and the weight function than in subsection 1. Namely, we shall assume that the analytic structure of the surface and of the mapping function is given by (6.15)–(6.19).

LEMMA 1. *Suppose the function $u(x)$ is continuous and bounded in $\overline{D}$ together with the derivatives*

$$D^\alpha u(x) = \frac{\partial^{|\alpha|}}{\partial x_1^{\alpha_1} \cdots \partial x_{n-1}^{\alpha_{n-1}}} u(x), \qquad |\alpha| \leq s, \qquad s = \left[\frac{m+2}{2}\right],$$

while the functions

$$\varphi(p, \nu, \xi), \qquad g(p, \nu, \xi), \qquad \rho_0(p, \nu, \xi) \equiv \rho(\nu\varphi(p, \nu, \xi), \xi)$$

are continuous and bounded in the region

$$G = \left\{(p, \nu, \xi) \colon p \in \left[0, \sqrt{H}\right], \ \nu \in S_m, \ \xi \in \overline{D}\right\}$$

together with the derivatives

$$\frac{\partial^{s+1}}{\partial p^{i+1} \partial \xi_n^j} \varphi, \qquad \frac{\partial^s}{\partial p^i \partial \xi_n^j} g, \qquad \frac{\partial^s}{\partial p^i \partial \xi_n^j} \rho_0, \qquad i + j = s;$$

$$\rho(0, \xi) \geq \alpha > 0, \qquad \xi \in \overline{D}. \tag{6.46}$$

If a function $u(x)$ is a solution of equation (6.11), *then it satisfies for $m = 2s$ the equation*

$$u(\xi) + \sum_{|\alpha|=0}^{s} \int_0^{\xi_n} \frac{dx_n}{\sqrt{\xi_n - x_n}} \int_{S_m} T_\alpha\left(\sqrt{\xi_n - x_n}, \nu, \xi\right) \times D^\alpha u\left[g\left(\sqrt{\xi_n - x_n}, \nu, \xi\right), x_n\right] d\omega_\nu = w(\xi), \tag{6.47}$$

and for $m = 2s - 1$ the equation

$$u(\xi) + \sum_{|\alpha|=0}^{s} \int_0^{\xi_n} \frac{dx_n}{\sqrt{\xi_n - x_n}} \int_{S_m} d\omega_\nu \int_0^\pi T_\alpha\left(\sqrt{z - x_n}, \nu, \overline{\xi}, z\right) \times D^\alpha u\left[g\left(\sqrt{z - x_n}, \nu, \overline{\xi}, z\right), x_n\right] \big|_{z = \frac{1}{2}[\xi_n + x_n + (\xi_n - x_n)\cos\theta]} \, d\theta = w(\xi). \tag{6.48}$$

In these equations the T_α are bounded continuous functions in the entire domain of variation of their arguments, and

$$w(\xi) = -\frac{m}{\Gamma(m/2)} \frac{\frac{\partial^{(m/2)}}{\partial \xi_n^{(m/2)}} v(\xi)}{\rho(0,\xi) \int_{S_m} [\varphi_p(0,\nu,\xi)]^m \, d\omega_\nu}. \tag{6.49}$$

To prove the lemma it suffices to apply to equation (6.11′), which it is convenient to write in the form

$$\int_0^{\xi_n} dx_n \int_{S_m} u\left[g\left(\sqrt{\xi_n - x_n}, \nu, \xi\right), x_n\right] K\left(\sqrt{\xi_n - x_n}, \nu, \xi\right) d\omega_\nu = v(\xi),$$
$$K(p,\nu,\xi) = -\frac{1}{2}\rho_0(p,\nu,\xi)[\varphi(p,\nu,\xi)]^{m-2}\frac{1}{p}\varphi_p(p,\nu,\xi), \tag{6.50}$$

the operator of differentiation with respect to ξ_n of order $m/2$. Since the function $K(p,\nu,\xi)$ can be represented in the form

$$K(p,\nu,\xi) = p^{m-2}K_0(p,\nu,\xi),$$

where $K_0(p,\nu,\xi)$ is continuous in the region G together with its derivatives with respect to p and ξ_n through order s, and

$$K_0(0,\nu,\xi) = -\frac{1}{2}\rho(0,\xi)\,[\varphi_p(0,\nu,\xi)]^m \neq 0,$$

it follows that in this differentiation, by differentiating with respect to the upper limit, we obtain the term outside the integral

$$u(\xi)\frac{\Gamma(m/2)}{m/2}\int_{S_m} K_0(0,\nu,\xi)\, d\omega_\nu. \tag{6.51}$$

Moreover, for $m = 2s$, calculation of the derivative of order s with respect to the variable ξ_n of the product

$$u\left(g\left(\sqrt{\xi_n - x_n}, \nu, \xi\right) x_n\right) K_0\left(\sqrt{\xi_n - x_n}, \nu, \xi\right)(\xi_n - x_n)^{s-1}$$

leads to the appearance of derivatives to order s of the function $u(x)$ with respect to the variables $x_1, \ldots, x_{n-1}$. The functions occurring as factors in front of these derivatives can be represented in the form of a product of functions $(\xi_n - x_n)^{-1/2}$ with a function continuous in the region G. For $m = 2s - 1$, calculation of the derivative of fractional order $s - \frac{1}{2}$ leads us to the necessity of calculating the following derivative:

$$\frac{\partial^s}{\partial \xi_n^s}\int_{x_n}^{\xi_n} \frac{(z - x_n)^{s-1}}{\sqrt{(\xi_n - z)(z - x_n)}} u\left[g\left(\sqrt{z - x_n}, \nu, \overline{\xi}, z\right), x_n\right]$$
$$\times K_0\left(\sqrt{z - x_n}, \nu, \overline{\xi}, z\right) dz.$$

It is easily computed if the following change of variable is made in the integral:

$$z = \frac{\xi_n + x_n}{2} + \frac{\xi_n - x_n}{2} \cos\theta.$$

This explains the appearance in (6.48) of the inner integral with respect to θ from 0 to π. In differentiating the integrand, all the derivatives $D^\alpha u$, $|\alpha| \le s$, occur in a product with some factors having the same type of smoothness as in the case of even m. The difference here is that the factors, generally speaking, will depend also on θ.

In order to bring the equation obtained after $m/2$-fold differentiation to the normal form (6.47) or (6.48), it suffices to divide both sides of the resulting equality by the factor standing in front of $u(\xi)$ in the term outside the integral.

LEMMA 2. *Suppose the assumptions of Lemma* 1 *are satisfied and* $u(x)$ *is a solution of equation* (6.11) *satisfying the inequalities*

$$\sup_{\overline{x}\in R^{n-1}} |D^\alpha u(x)| \le K \sup_{\overline{x}\in R^{n-1}} |u(x)|, \qquad |\alpha| \le s, \qquad |\alpha| = \alpha_1 + \cdots + \alpha_{n-1}, \tag{6.52}$$

with a constant K *not depending on* x_n. *In* $\overline{D}$ *there is then the following estimate for* $u(x)$:

$$|u(x)| \le C_1(1+K)\exp(C_2K^2x_n) \left\| \frac{\partial^{m/2}}{\partial \xi_n^{m/2}} v(\xi) \right\|_{C(\overline{D})}. \tag{6.53}$$

Proceeding to the proof of the lemma, we set

$$U(x_n) = \sup_{\overline{x}\in R^{n-1}} |u(\overline{x}, x_n)|.$$

Using the boundedness of the functions T_α appearing in (6.47) and (6.48) and the fact that the functions $\rho(0,\xi)$ and $\varphi_p(0,\nu,\xi)$ appearing in (6.49) are bounded below, we find that for a solution of (6.11) satisfying (6.52) we have

$$U(\xi_n) \le C_3 \left\| \frac{\partial^{m/2}}{\partial \xi_n^{m/2}} v(\xi) \right\|_{C(\overline{D})} + C_4K \int_0^{\xi_n} \frac{U(x_n)\,dx_n}{\sqrt{\xi_n - x_n}}.$$

Iterating once in this inequality, we find that

$$U(\xi_n) \le C_1(1+K) \left\| \frac{\partial^{m/2}}{\partial \xi_n^{m/2}} v(\xi) \right\|_{C(\overline{D})} + C_2K^2 \int_0^{\xi_n} U(x_n)\,dx_n,$$

$$C_1 = \max\left(C_3, C_4 \cdot 2\sqrt{H}\right), \qquad C_2 = \pi C_4. \tag{6.54}$$

Applying now Gronwall's Lemma (see [200], §19, Problem 3), we obtain (6.53).

THEOREM 2. *Let M be the linear set of functions $u(x)$ continuous and bounded together with their partial derivatives with respect to $x_1, \dots, x_{n-1}$ to order $s = [(m+1)/2]$ which satisfy inequality* (6.52) *with a constant K depending on the function. Then on the set M the problem* (6.11) *of integral geometry cannot have two distinct solutions under the assumptions of Lemma* 1 *regarding the class of problems in question.*

Indeed, if $u_1(x)$ and $u_2(x)$ are two solutions of the problem belonging to the set M, their difference $\tilde{u} = u_1 - u_2$ also belongs to M because of the linearity of the set M. Because of the linearity of the problem of integral geometry, $\tilde{u}(x)$ satisfies the homogeneous equation; hence it follows from (6.53) that $\tilde{u}(x) \equiv 0$, i.e., $u_1(x) \equiv u_2(x)$.

We note that on linear sets with a universal constant K for the entire set, Lemma 2 gives an estimate of stability of a solution of the problem of integral geometry.

Proceeding from Theorem 2, it is easy to obtain assertions concerning uniqueness of a solution of the problem of integral geometry on concrete sets M.

LEMMA 3. *Let M be the set of functions $u(x)$ which can be represented in $\overline{D}$ in the form*

$$u(x) = \sum_{k=1}^{N} a_k(\overline{x}) b_k(x_n), \tag{6.55}$$

where $a_k(\overline{x}) \in C^s(R^{n-1})$ and $b_k(x_n) \in C[0, H]$. Then for any function $u(x) \in M$ inequality (6.52) *holds.*

Indeed, in the representation (6.55) we may assume that the functions $a_k(\overline{x})$, $k = 1, \dots, N$, are linearly independent in R^{n-1}. Therefore,

$$\sup_{\overline{x} \in R^{n-1}} \left| \sum_{k=1}^{N} \mu_k a_k(\overline{x}) \right| \geq \alpha_0 > 0 \tag{6.56}$$

on the set $\{(\mu_1, \dots, \mu_N) \colon \sum_{k=1}^{N} |\mu_k| = 1\}$. Then for those x_n for which $\sum_{k=1}^{N} |b_k(x_n)| \neq 0$ we have by (6.56)

$$\sup_{\overline{x} \in R^{n-1}} |u(x)| = \left(\sum_{k=1}^{N} |b_k(x_n)| \right) \sup_{\overline{x} \in R^{n-1}} \left| \sum_{k=1}^{N} a_k(\overline{x}) \frac{b_k(x_n)}{\sum_{j=1}^{N} |b_j(x_n)|} \right| \geq \alpha_0 \sum_{k=1}^{N} |b_k(x_n)|.$$

This inequality is obviously satisfied also in the case where $\sum_{k=1}^{N}|b_k(x_n)| = 0$. At the same time

$$\begin{aligned}|D^\alpha u| &\leq \max_{1\leq k\leq N}\sup_{\overline{x}\in R^{n-1}}|D^\alpha a_k(\overline{x})|\sum_{k=1}^{N}|b_k(x_n)| \\ &\leq \frac{1}{\alpha_0}\max_{1\leq k\leq N}\sup_{\overline{x}\in R^{n-1}}|D^\alpha a_k(\overline{x})|\sup_{\overline{x}\in R^{n-1}}|u(x)|,\end{aligned}$$

and the lemma is proved.

COROLLARY. *On the set of functions representable in the form* (6.55) *where* $a_k(\overline{x}) \in C^s(R^{n-1})$ *and* $b_k(x_n) \in C[0, H]$, *a solution of the problem of integral geometry is unique.*

In Lemma 2 we considered a set of functions satisfying for each fixed x_n the inequality

$$\|D^\alpha u\|_{C(R^{n-1})} \leq K\|u\|_{C(R^{n-1})}$$

with a constant K not depending on x_n. It is possible to consider analogous sets connected with the norm in the space $L_2(R^{n-1})$ for each fixed x_n. The following lemma (analogous to Lemma 2) holds.

LEMMA 4. *Let* $u(x)$ *be a solution of equation* (6.11), *and suppose* $u(x)$ *satisfies the inequality*

$$\|D^\alpha u\|_{L_2(R^{n-1})} \leq K\|u\|_{L_2(R^{n-1})}, \quad |\alpha| \leq s, \tag{6.57}$$

with a constant K *not depending on* x_n. *Then in* $\overline{D}$ *the following estimate holds for* $u(x)$:

$$\|u\|_{L_2(R^{n-1})} \leq C_1(1+K)\exp(C_2K^2x_n)\max_{\xi_n\in[0,H]}\left\|\frac{\partial^{(m/2)}}{\partial\xi_n^{(m/2)}}, v(\xi)\right\|_{L_2(R^{n-1})}. \tag{6.58}$$

The proof is altogether analogous to the proof of Lemma 2, and we omit it.

We shall present examples of linear sets for which (6.57) holds.

LEMMA 5. *Let* M *be the set of functions* $u(x)$ *representable in the form* (6.55) *where* $a_k(\overline{x}) \in W_2^s(R^{n-1})$ *and* $b_k(x_n) \in C[0, H]$. *Then the estimate* (6.57) *holds for* $u \in M$.

We obviously may always assume that the system of functions $a_k(\overline{x})$, $k = 1, \dots, N$, is orthonormal in $L_2(R^{n-1})$. Therefore,

$$\|D^\alpha u\|_{L_2(R^{n-1})} \le \left[\sum_{k=1}^{N} \|D^\alpha a_k(\overline{x})\|^2_{L_2(R^{n-1})}\right]^{1/2} \left[\sum_{k=1}^{N} b_k^2(x_n)\right]^{1/2}$$
$$= \left[\sum_{k=1}^{N} \|D^\alpha a_k(\overline{x})\|^2_{L_2(R^{n-1})}\right]^{1/2} \|u\|_{L_2(R^{n-1})}.$$

LEMMA 6. *Let M be the set of functions $u(x)$ having a Fourier transform in the variable $\overline{x}$ which is continuous with respect to x_n and has compact support uniformly bounded with respect to x_n. Then for any function $u \in M$ inequality* (6.57) *holds.*

Suppose the support of the function

$$\tilde{u}(\lambda, x_n) = \int_{R^{n-1}} u(\overline{x}, x_n) e^{i\lambda\overline{x}}\, d\overline{x}$$

is contained in the region $|\lambda| \le N$. From Parseval's equality we then find that

$$\|D^\alpha u\|^2_{L_2(R^{n-1})} = \frac{1}{(2\pi)^{n-1}} \int_{|\lambda| \le N} \lambda^{2|\alpha|} \tilde{u}^2(\lambda, x_n)\, d\lambda$$
$$\le \frac{N^{2|\alpha|}}{(2\pi)^{n-1}} \int_{|\lambda| \le N} \tilde{u}^2(\lambda, x_n)\, d\lambda = N^{2|\alpha|} \|u\|^2_{L_2(R^{n-1})}$$
$$\le N^{2s} \|u\|^2_{L_2(R^{n-1})}.$$

Thus, the set of functions $u(x)$ with Fourier transform in $\overline{x}$ having compact support satisfies condition (6.57). Therefore, uniqueness of a solution of the problem of integral geometry on this set follows from (6.58).

It can be shown that a uniqueness theorem in the problem of integral geometry holds also for sets which can be sufficiently well approximated by sets of the structure considered above (see [230] and [231]).

Theorems analogous to Theorem 2 hold also with respect to equations (6.42) and (6.44) if condition (6.43) is satisfied.

§4. Integral geometry "in the small"

The exposition of this section is based on results obtained by M. M. Lavrent'ev and A. L. Bukhgeĭm [65], [153], [154]. We here consider a problem of integral geometry on surfaces which differs in formulation from the problem studied in §§2 and 3. As will become clear below, this difference in the formulations predetermines a different degree of stability of the problems. In general, it turns out to be possible to classify problems of integral geometry into two types according to the character of their stability. For some problems there are estimates of stability using finitely many derivatives, while for other problems there are no such estimates.

1. We thus consider the following problem. Suppose in the space $x = (x_1, \dots, x_n)$ there is given a bounded region D and Ω is an open region belonging to D. For the case of even-dimensional space we shall assume that the boundaries of D and Ω are uniformly separated from one another by some positive distance h, and thus Ω lies strictly inside D. In the case of odd-dimensional space, D and Ω may coincide. Suppose, further, that for any point $x \in D$ and any unit vector $\nu = (\nu_1, \dots, \nu_n)$ there exists a unique smooth hypersurface $S(x, \nu)$ which passes through the point x and has normal ν at this point.

We denote by U the class of functions $u(x) \in L_2(\Omega)$ whose support is contained in Ω. For a function $u(x) \in U$ we consider the problem of finding it from the equation

$$\int_{S(x,\nu)} \rho(\xi, x, \nu) u(\xi)\, dS = v(x, \nu), \quad x \in D, \quad |\nu| = 1, \tag{6.59}$$

in which $\rho(\xi, x, \nu)$ is a given smooth function of its arguments and dS is the area element of the surface $S(x, \nu)$.

Generally speaking, this problem is overdetermined, since the function $u(x)$ depends on n variables, while the function v depends on $2n - 1$ variables. We note, however, an important case where the number of essential variables of the functions u and v coincide. We consider the surface $S(x, \nu)$ and an arbitrary point $x^0 \in S(x, \nu)$. Let ν^0 be the normal to $S(x, \nu)$ at x^0. Then $S(x^0, \nu^0) = S(x, \nu)$. If $\rho(\xi, x^0, \nu^0) \equiv \rho(\xi, x, \nu)$ for $\xi \in S(x, \nu)$, then $v(x^0, \nu^0) = v(x, \nu)$. Since each point x^0 of the hypersurface $S(x, \nu)$ is characterized by $n - 1$ parameters, this equality shows that there are a total of n essential parameters on which the function $v(x, \nu)$ depends in this case. Thus, if the weight function depends only on the point $\xi \in D$ and the surface $S(x, \nu)$, then the numbers of essential variables of the functions u and v coincide. This suggests that there is apparently a better form for the formulation of the problem of integral geometry with the same family of surfaces that is related to another parametrization of this family, for which the problem is not overdetermined.

In this section the problem of solving equation (6.59) will be considered under the assumption that the diameter of the region D is sufficiently small. This explains the title of the section. Below we shall assume that the equation of the surface $S(x, \nu)$ can be characterized by means of a sufficiently smooth function $\varphi(\xi, x, \nu)$:

$$S(x, \nu) = \{\xi \colon \varphi(\xi, x, \nu) = 0\}, \tag{6.60}$$

where at points of $S(x,\nu)$ we have $|\nabla_\xi\varphi| \neq 0$.

The assumptions regarding the method of parametrization lead to the following necessary conditions on the function φ:

$$\varphi(x,x,\nu)=0, \qquad \nabla_\xi\varphi|_{\xi=x}=\nu|\nabla_\xi\varphi|_{\xi=x}. \tag{6.61}$$

Obviously, it may always be assumed that

$$|\nabla_\xi\varphi|_{\xi=x}=1, \tag{6.62}$$

by dividing φ by $|\nabla_\xi\varphi|_{\xi=x}$ if necessary; therefore, below we shall assume that (6.62) holds. For the function φ we then have the representation

$$\varphi(\xi,x,\nu)=\nu(\xi-x)+\sum_{i,j=1}^{h} a_{ij}(\xi,x,\nu)(\xi_i-x_i)(\xi_j-x_j), \tag{6.63}$$

where

$$a_{ij}(\xi,x,\nu)=\int_0^1 \varphi_{\xi_i\xi_j}(x+t(\xi-x),x,\nu)(1-t)\,dt.$$

We shall use this representation below.

In preceding sections it could be observed that the exposition of results for problems of integral geometry differs somewhat for even- and odd-dimensional spaces. We therefore first formulate the corresponding result and prove it for the case of odd-dimensional space; only afterwards do we prove it for the case of even-dimensional space.

THEOREM 1. *Let n be odd, let $s=(n-1)/2\geq 1$, and suppose that the function $\varphi(\xi,x,\nu)$ is continuous in the region $G=\{(\xi,x,\nu)\colon \xi\in\overline{D},\ x\in\overline{D},\ |\nu|=1\}$ together with its partial derivatives to order $n+2$, while the weight function $\rho(\xi,x,\nu)$ is continuous in G together with its partial derivatives to order n and satisfies in G the condition*

$$\rho(x,x,\nu)\geq\rho_0>0, \quad \rho_0=\text{const}. \tag{6.64}$$

Then there exists a number $d^*>0$, $d^*=d^*(\varphi,\rho)$, such that under the condition $\operatorname{diam} D<d^*$ *a solution of equation* (6.59) *in the class of functions U is unique, and there is the estimate of stability*

$$\|u\|_{L_2(\Omega)}\leq C\|\Delta^s v_1\|_{L_2(\Omega)}, \qquad v_1(x)=\int_{|\nu|=1}\frac{v(x,\nu)}{\rho(x,x,\nu)}\,d\omega_\nu. \tag{6.65}$$

Here $d\omega_\nu$ is the area element of the unit sphere, and Δ^s is the sth power of the Laplace operator.

Using the delta function and the fact that $u(x)$ has compact support, we write (6.59) in the form

$$\int_\Omega \rho(\xi,x,\nu)u(\xi)|\nabla_\xi\varphi(\xi,x,\nu)|\delta(\varphi(\xi,x,\nu))\,d\xi=v(x,\nu) \tag{6.66}$$

and, dividing it by $\rho(x, x, \nu)$, we average equality (6.66) for fixed $x \in D$ over all ν. Using the notation (6.65), we write (6.66) as follows:

$$\int_{\Omega} K(x, \xi) u(\xi)\, d\xi = v_1(x). \tag{6.67}$$

Here

$$\begin{aligned} K(x, \xi) &= \int_{|\nu|=1} \tilde{\rho}(\xi, x, \nu) \delta(\varphi(\xi, x, \nu))\, d\omega_\nu, \\ \tilde{\rho}(\xi, x, \nu) &= \frac{\rho(\xi, x, \nu)}{\rho(x, x, \nu)} |\nabla_\xi \varphi(\xi, x, \nu)|. \end{aligned} \tag{6.68}$$

The kernel of equation (6.67) can be studied rather simply for points x and ξ close to one another. This is also sufficient, since in the proof of the theorem it may be assumed that the diameter of the region D is small. We make use of the invariance of the measure $d\omega_\nu$ under orthogonal transformations of the space $\nu_1, \dots, \nu_n$ and carry out a transformation with an orthogonal matrix Q. The formula for calculating the kernel here assumes the form

$$K(x, \xi) = \int_{|\nu|=1} \tilde{\rho}(\xi, x, Q\nu) \delta(\varphi(\xi, x, Q\nu))\, d\omega_\nu. \tag{6.69}$$

We introduce the unit vector $\nu^0 = (\xi - x)/|\xi - x|$, $\nu^0 = (\nu_1^0, \dots, \nu_n^0)$. In evaluating the integral (6.69), we use two distinct orthogonal transformations depending on the position of the vector ν^0 on the unit sphere. Namely, suppose that $e^1 = (1, 0, \dots, 0)$. For $\nu^0 \cdot e^1 \geq 0$ we then take the transformation defined by

$$Q\nu = \nu + 2(\nu \cdot e^1)\nu^0 - \frac{\nu \cdot \nu^0 + \nu \cdot e^1}{1 + (\nu^0 \cdot e^1)} (\nu^0 + e^1). \tag{6.70}$$

It is easy to verify directly that $(Q\nu \cdot Q\nu) = 1$ for any unit vector ν. Moreover,

$$Qe^1 = \nu^0. \tag{6.71}$$

For $\nu^0 \cdot e^1 < 0$, to evaluate the integral we use the transformation (6.70) in which e^1 is replaced by $-e^1$, and hence $Qe^1 = -\nu^0$.

It obviously suffices to investigate the case of a position of the points x and ξ for which $\nu^0 \cdot e^1 \geq 0$. Below we restrict our attention to just this case. The equation $\varphi(\xi, x, Q\nu) = 0$ for fixed x and ξ defines a surface of dimension $n-2$ on the surface of the sphere $|\nu| = 1$. Indeed, dividing by $|x - \xi|$, this equality by (6.63) can be written in the form

$$\nu^0 \cdot Q\nu + |x - \xi| \sum_{i,j=1}^{n} a_{ij}(x + |x - \xi|\nu^0, x, Q\nu)\nu_i^0 \nu_j^0 = 0. \tag{6.72}$$

We represent the vector ν as follows:

$$\nu = qe^1 + \sqrt{1 - q^2}\,\overline{\nu}, \tag{6.73}$$

where $\overline{\nu}$ is a unit vector orthogonal to e^1. Then by (6.71)

$$\nu^0 \cdot Q\nu = Q^*\nu^0 \cdot \nu = e^1 \cdot \nu = q,$$

and (6.72) assumes the form

$$q + |x-\xi| \sum_{i,j=1}^{n} a_{ij}\left(x + |x-\xi|\nu^0, x, q\nu^0 + \sqrt{1-q^2}Q\overline{\nu}\right)\nu_i^0\nu_j^0 = 0. \qquad (6.74)$$

The next lemma follows easily from the contraction mapping principle.

LEMMA 1. *If the function φ satisfies the conditions of the theorem, then for any $q_0 \in (0,1)$ there exists $d_0 = d_0(\varphi, q_0)$ such that for regions Ω with* $\operatorname{diam}\Omega < d_0$ *the equation $\varphi(x, \xi, Q\nu) = 0$, for $x \in \Omega$ and $\xi \in \Omega$, defines q as a single-valued continuous bounded function of the arguments x, $|x-\xi|$, ν^0, and $\overline{\nu}$ having with respect to these arguments continuous and bounded derivatives to order n and satisfying the inequality*

$$|q(x, |x-\xi|, \nu^0, \overline{\nu})| \leq q_0 < 1. \qquad (6.75)$$

Indeed, since φ satisfies the conditions of the theorem, the $a_{ij}(\xi, x, \nu)$ are uniformly bounded for ξ, $x \in D$ and $|\nu| = 1$ by some constant M. Therefore, writing (6.74) in the form $q = A(q)$, we see that under the condition $Md_0 \leq q_0$ the operator A takes the set of continuous functions satisfying (6.75) into itself. The condition of uniform boundedness of the derivatives of a_{ij} with respect to $\nu_1, \dots, \nu_n$ leads to the situation that for $|x-\xi|$ small the operator A is contractive on this set. Thus, equation (6.74) defines q as a continuous function of the arguments x, $|x-\xi|$, ν^0, and $\overline{\nu}$ satisfying (6.75). The existence and boundedness of the derivatives of q with respect to its arguments to order n follow easily from (6.74) itself, since the functions a_{ij} and the matrix $Q = Q(\nu^0)$ defined by (6.70) have the corresponding derivatives.

It also follows from (6.74) that $q \to 0$ as $|x-\xi| \to 0$.

We now choose a fixed $q_0 \in (0,1)$ and assume below that $\operatorname{diam}\Omega < d_0(\varphi, q_0)$. Formula (6.73), in which q is the solution of (6.74) for an arbitrary unit vector $\overline{\nu}$, $\overline{\nu} \cdot e^1 = 0$, defines (for fixed x and ξ) ν as a smooth, single-valued function of $\overline{\nu}$ and thus defines a surface of dimension $n-2$ on the sphere $|\nu| = 1$. We denote this surface by $\Sigma(x, \xi)$. If $x \to \xi$ so that $(\xi - x)/|\xi - x| \to \nu^0$, then the surface $\Sigma(x, \xi)$ in the limit coincides with the section of the sphere $|\nu| = 1$ by the plane passing through the center of the sphere and orthogonal to the vector e^1.

Let n be the unit vector of the normal to $\Sigma(x, \xi)$ lying in the plane tangent at ν to the sphere $|\nu| = 1$. Taking account of what has been said above, we can write formula (6.69) for the kernel $K(x,\xi)$ of equation (6.67) in the form

$$K(x,\xi)=\int_{\Sigma(x,\xi)}\tilde{\rho}(\xi,x,Q\nu)\frac{d\sigma}{\left|\frac{\partial\varphi}{\partial n}\right|}, \tag{6.76}$$

where $d\sigma$ is the area element of the surface $\Sigma(x,\xi)$. Since on $\Sigma(x,\xi)$ the vector n is parallel to the vector

$$\nabla_\nu\varphi(\xi,x,Q\nu)-\nu(\nu\cdot\nabla_\nu\varphi(\xi,x,Q\nu)),$$

it follows that

$$\left|\frac{\partial\varphi}{\partial n}\right|=\left[|\nabla_\nu\varphi(\xi,x,Q\nu)|^2-|\nu\cdot\nabla_\nu\varphi(\xi,x,Q\nu)|^2\right]^{1/2}. \tag{6.77}$$

From (6.63) we find that

$$\frac{1}{|x-\xi|}\nabla_\nu\varphi(\xi,x,Q\nu)=e^1+|x-\xi|\sum_{i,j=1}^{n}\nabla_\nu a_{ij}(\xi,x,Q\nu)\nu_i^0\nu_j^0.$$

Using (6.77) and the circumstance that $\nu=\nu(x,|\xi-x|,\nu^0,\overline{\nu})$ on $\Sigma(x,\xi)$, we obtain the representation

$$\left|\frac{\partial\varphi}{\partial n}\right|^{-1}=\frac{1}{|x-\xi|}+b(x,|\xi-x|,\nu^0,\overline{\nu}) \tag{6.78}$$

on $\Sigma(x,\xi)$ in which b is a continuous bounded function of its arguments together with its partial derivatives to order $n-1$. Similarly, we have the representation

$$\tilde{\rho}(\xi,x,Q\nu)=1+|x-\xi|c(x,|x-\xi|,\nu^0,\overline{\nu}) \tag{6.79}$$

on $\Sigma(x,\xi)$, where the function c possesses the same properties as b.

The vector $\overline{\nu}$ is orthogonal to e^1 and has unit length; it is therefore completely characterized by the angular spherical coordinates $\psi_1,\dots,\psi_{n-2}$ in the coordinate plane orthogonal to e^1. The surface element $d\sigma$ is computed by the formula

$$d\sigma=\left[\Gamma\left(\frac{\partial\nu}{\partial\psi_1},\frac{\partial\nu}{\partial\psi_2},\dots,\frac{\partial\nu}{\partial\psi_{n-2}}\right)\right]^{1/2}d\psi, \tag{6.80}$$

in which $d\psi=d\psi_1\cdots d\psi_{n-2}$ and Γ denotes the Gram determinant constructed for the vectors $\partial\nu/\partial\psi_1,\dots,\partial\nu/\partial\psi_{n-2}$:

$$\Gamma\left(\frac{\partial\nu}{\partial\psi_1},\dots,\frac{\partial\nu}{\partial\psi_{n-2}}\right)=\begin{vmatrix}\frac{\partial\nu}{\partial\psi_1}\frac{\partial\nu}{\partial\psi_1} & \frac{\partial\nu}{\partial\psi_1}\frac{\partial\nu}{\partial\psi_2} & \cdots & \frac{\partial\nu}{\partial\psi_1}\frac{\partial\nu}{\partial\psi_{n-2}}\\ \frac{\partial\nu}{\partial\psi_2}\frac{\partial\nu}{\partial\psi_1} & \frac{\partial\nu}{\partial\psi_2}\frac{\partial\nu}{\partial\psi_2} & \cdots & \frac{\partial\nu}{\partial\psi_2}\frac{\partial\nu}{\partial\psi_{n-2}}\\ \cdots & \cdots & \cdots & \cdots\\ \frac{\partial\nu}{\partial\psi_{n-2}}\frac{\partial\nu}{\partial\psi_1} & \frac{\partial\nu}{\partial\psi_{n-2}}\frac{\partial\nu}{\partial\psi_2} & \cdots & \frac{\partial\nu}{\partial\psi_{n-2}}\frac{\partial\nu}{\partial\psi_{n-2}}\end{vmatrix}.$$

From (6.76) and (6.78)–(6.80) it follows that the kernel $K(x, \xi)$ can be represented in the form

$$K(x,\xi) = \frac{\omega_{n-1}}{|x-\xi|} + K_0(x, |x-\xi|, \nu^0), \tag{6.81}$$

where K_0 is a bounded continuous function of its arguments together with its derivatives to order $n-1$. Applying then the operator Δ^s, $s = (n-1)/2$, in the variable x to equation (6.67) and using the fact that ([120], p. 43)

$$\Delta^s \frac{1}{|x-\xi|} = (-1)^s (4\pi)^s (s-1)! \delta(x-\xi), \qquad \omega_{n-1} = \frac{2\pi^s}{\Gamma(s)},$$

for $u(x)$ we obtain the equation

$$u(x) + \int_\Omega \tilde{K}(x,\xi) u(\xi)\, d\xi = \frac{(-1)^s}{2^n \pi^{2s}} \Delta^s v_1(x). \tag{6.82}$$

Here

$$\tilde{K}(x,\xi) = \frac{(-1)^s}{2^n \pi^{2s}} \Delta_x^s K_0(x, |x-\xi|, \nu^0).$$

The kernel $\tilde{K}(x, \xi)$ of equation (6.82) has an integrable polar singularity. Indeed, computation of the operator Δ^s leads to differentiation of K_0 to order $2s$ with respect to the variables x, $|x-\xi|$, and ν^0 and to differentiation to order $2s$ of the arguments $|x-\xi|$ and ν^0 with respect to x. But $2s = n-1$, and hence the derivatives with respect to the arguments of the function K_0 are continuous and bounded; at the same time, it is obvious that for the derivatives of $|x-\xi|$ and ν^0 with respect to x we have the estimates

$$\left|\frac{\partial^k}{\partial x_i^k}(|x-\xi|)\right| \le \frac{c}{|x-\xi|^{k-1}}, \qquad \left|\frac{\partial^k}{\partial x_i^k}\nu^0\right| \le \frac{c}{|x-\xi|^k}, \quad 1 \le k \le 2s.$$

From these arguments it follows that the kernel $\tilde{K}(x, \xi)$ is continuous everywhere except for the set of points x and ξ on which $x = \xi$; in a neighborhood of this set we have the following estimate for $\tilde{K}(x, \xi)$:

$$\tilde{K}(x,\xi)| \le \frac{c}{|x-\xi|^{n-1}}.$$

It is known that equation (6.82) with a kernel of this type has a unique solution in $L_2(\Omega)$ if the diameter of the region Ω is sufficiently small. The estimate (6.65) for the solution of (6.82) is obvious.

2. We now consider the case of even-dimensional space. As mentioned above, in this case we assume that Ω lies strictly inside D, so that the distance between Ω and the boundary of D is larger than some $h > 0$. We additionally assume that $h > \operatorname{diam} \Omega$. We denote by Ω_h the open set obtained by taking the union of all open balls of radius h with center at points $x \in \Omega$.

THEOREM 2. *Let $n = 2s$, $s \geq 1$, and suppose the functions φ and ρ satisfy all the conditions of Theorem 1 with n replaced by $n+1$. Then there exists a number $d^* > 0$, $d^* = d^*(\varphi, \rho, h)$, such that under the condition* $\operatorname{diam} \Omega < d^*$ *a solution of equation* (6.59) *in the class of functions U is unique and there is the stability estimate*

$$\|u(x)\|_{L_2(\Omega)} \leq C\|\Delta^s v_2\|_{L_2(\Omega)}, \tag{6.83}$$

in which

$$v_2(x) = \int_{|x-y|\leq h} \frac{v_1(y)\,dy}{|x-y|^{n-1}}, \tag{6.84}$$

and $v_1(x)$ is computed in terms of $v(x,\nu)$ by (6.65).

We note that the estimate (6.83) differs qualitatively from (6.65). Only values of the function $v(x,\nu)$ for $x \in \Omega$ enter in the estimate (6.65), while in (6.83) values of $v(x,\nu)$ from the larger region Ω_h of the variable x are taken into account.

This theorem is proved according to the following scheme. First, just as in the case of n odd, we obtain (6.67). For the kernel $K(x,\xi)$ of this equation there is the representation (6.81), where the smoothness of the function K_0, because of the hypotheses of the theorem, is one greater, namely, K_0 with respect to the arguments x, $|x-\xi|$, and $\nu^0 = (\xi - x)/|\xi - x|$ has bounded continuous derivatives to order n. We now apply the operator of averaging over a ball of radius h defined by (6.84) to both sides of the equality. We then obtain

$$\int_\Omega T(x,\xi)u(\xi)\,d\xi = v_2(x), \quad x \in \Omega, \tag{6.85}$$

in which

$$T(x,\xi) = \int_{|x-y|\leq h} \frac{K(y,\xi)}{|x-y|^{n-1}}\,dy. \tag{6.86}$$

We further show that for $T(x,\xi)$, $x \in \Omega$, $\xi \in \Omega$, there is the representation

$$T(x,\xi) = -\omega_{n-1}\omega_n \ln|x-\xi| + T_0\left(\xi, |\xi - x|, \frac{x-\xi}{|x-\xi|}\right), \tag{6.87}$$

in which $T_0(\xi,\rho,\nu^0)$ is a continuous and bounded function together with its derivatives to order n everywhere except the set $\rho = 0$, in a neighborhood of which $T_0(\xi,\rho,\nu^0)$ satisfies the inequalities ($|\alpha| + k \leq n$)

$$\left|\frac{\partial^k}{\partial \rho^k} D^\alpha_{\xi,\nu^0} T_0(\xi,\rho,\nu^0)\right| \leq C \begin{cases} \ln\rho, & k = 1, \\ \rho^{1-k}, & k \geq 2. \end{cases} \tag{6.88}$$

Applying to (6.85) the operator Δ^s and using the fact that (see, for example, [120], p. 44)

$$\Delta^s_x \ln|x-\xi| = (-1)^{s-1} 2^{n-1} \pi^s (s-1)!\,\delta(x-\xi),$$

we obtain an equation of Fredholm type analogous to (6.82) with a singularity of the form $|x-\xi|^{n-1}\ln|x-\xi|$. From this follows the validity of the theorem.

It thus remains for us to verify that the kernel $T(x,\xi)$ actually possesses the properties mentioned. For this we evaluate the integral (6.86) separately for each of the terms in (6.81) for the function $K(x,\xi)$. Evaluation of the integral of the first term is equivalent to evaluating the integral

$$\begin{aligned}J_1(\rho) &= \int_{|x-y|\le h} \frac{1}{|x-y|^{n-1}}\frac{1}{|y-\xi|}\,dy\\ &= \int_{|\nu|=1} d\omega_\nu \int_0^h \frac{dr}{\sqrt{r^2+\rho^2+2r\rho(\nu^0\cdot\nu)}}\\ &= \int_{|\nu|=1} [\ln|h+\rho(\nu^0\cdot\nu)+\sqrt{h^2+\rho^2+2h\rho(\nu\cdot\nu^0)}|\\ &\quad - \ln|1+\nu\cdot\nu^0|]\,d\omega_\nu - \omega_n\ln\rho.\end{aligned}$$

Here we have introduced spherical coordinates $y=x+r\nu$ and denoted by ρ, $-\nu^0$ the spherical coordinates of the point ξ in this system, i.e., $\rho=|\xi-x|$, $\nu^0=(x-\xi)/|x-\xi|$. As is evident from the formula obtained, evaluation of the integral (6.86) of the first term of the function $K(x,\xi)$ leads to the appearance of the principal part of the kernel $T(x,\xi)$, namely, $\ln\rho$ and some smooth function (even analytic for $\rho<h$) depending only on ρ. Here essential use was also made of the fact that $\rho\le\operatorname{diam}\Omega<h$ for $x\in\Omega$ and $\xi\in\Omega$. We shall now show that evaluation of the integral (6.86) of the second term of the function $K(x,\xi)$ leads to a function possessing the properties of T_0. To evaluate the integral it is here convenient to introduce a spherical system with center at the point ξ ($y=\xi+r\nu$, $\nu=(\nu_1,\dots,\nu_n)$, $x=\xi+\rho\nu^0$):

$$\begin{aligned}J_2(\xi,\rho,\nu^0) &= \int_{|x-y|\le h} \frac{K_0\left(y,|y-\xi|,\frac{\xi-y}{|\xi-y|}\right)}{|x-y|^{n-1}}\,dy\\ &= \int_{|\nu|=1} d\omega_\nu \int_0^{r(\rho,\nu\cdot\nu^0)} \frac{r^{n-1}K_0(\xi+r\nu,r,-\nu)}{[r^2+\rho^2-2r\rho(\nu\cdot\nu^0)]^{(n-1)/2}}\,dr,\end{aligned}$$

$$r(\rho,\nu\cdot\nu^0)=\rho(\nu\cdot\nu^0)+\sqrt{h^2-\rho^2[1-(\nu\cdot\nu^0)^2]}.$$

As we have already done once, we introduce an orthogonal transformation of the coordinate system with matrix Q, so that $Q^*\nu^0=e^1=(1,0,\dots,0)$ (under the condition $\nu^0\cdot e^1\ge 0$). Setting $K_0(\xi+rQ\nu,r,-Q\nu)=\hat K_0(\xi,r,\nu,\nu^0)$, we then find

$$J_2(\xi,\rho,\nu^0)=\int_{|\nu|=1} d\omega_\nu \int_0^{r(\rho,\nu_1)} \frac{r^{n-1}\hat K_0(\xi,r,\nu,\nu^0)}{(r^2+\rho^2-2r\rho\nu_1)^{(n-1)/2}}\,dr.$$

From this it is immediately evident that the derivatives $D^\alpha_{\xi,\nu^0} J_2$, $|\alpha| \le n$, are continuous and bounded for $\xi \in \Omega$ and $x \in \Omega$. Indeed,

$$D^\alpha_{\xi,\nu^0} J_2 = \int_{|\nu|=1} d\omega_\nu \int_0^{r(\rho,\nu_1)} \frac{r^{n-1}}{(r^2+\rho^2-2r\rho\nu_1)^{(n-1)/2}} D^\alpha_{\xi,\nu^0} \hat{K}_0(\xi, r, \nu, \nu^0)\, dr. \tag{6.89}$$

It remains for us to deal with derivatives containing differentiation with respect to ρ. The change of variable $r = \rho r_1$ in the inner integral immediately shows that for $\rho > 0$ the integral $J_2(\xi, \rho, \nu^0)$ depends continuously on ξ, ρ, and ν^0 together with its derivatives to order n. However, as $\rho \to 0$ the derivatives which contain differentiation with respect to ρ are unbounded. Indeed, for sufficiently small ρ we can always choose $\delta > 0$ so that

$$\rho(1+\delta) < r(\rho, \nu_1), \quad \nu_1 \in [-1, 1].$$

Breaking the inner integral into two—one over $[0, \rho(1+\delta)]$ and the second over $[\rho(1+\delta), r(\rho,\nu_1)]$—and making the change of variable $r = \rho r_1$ in the first, we then obtain

$$\begin{aligned} J_2(\xi, \rho, \nu^0) &= \rho \int_{|\nu|=1} d\omega_\nu \int_0^{1+\delta} \frac{r_1^{n-1}}{[r_1^2+1-2r_1\nu_1]^{(n-1)/2}} \hat{K}_0(\xi, \rho r_1, \nu, \nu^0)\, dr_1 \\ &+ \int_{|\nu|=1} d\omega_\nu \int_{\rho(1+\delta)}^{r(\rho,\nu_1)} \frac{r^{n-1}}{(r^2+\rho^2-2r\rho\nu_1)^{(n-1)/2}} \hat{K}_0(\xi, r, \nu, \nu^0)\, dr. \end{aligned} \tag{6.90}$$

The first of the integrals in this formula is a bounded continuous function together with its derivatives to order n. All singularities in the derivatives are generated by the second integral. We now note that in the second integral the factor standing in front of $\hat{K}_0$ can be represented in the form

$$\frac{r^{n-1}}{(r^2+\rho^2-2r\rho\nu_1)^{(n-1)/2}} = \Phi\left(\frac{\rho}{r}, \nu_1\right),$$

where $\Phi(z, \nu_1)$, as a function of the arguments z and ν_1, is bounded together with any finite number of derivatives in the region $|z| \le (1+\delta)^{-1}$, $\nu \in [-1, 1]$. Hence, for its derivatives with respect to ρ we have the estimates

$$\left| \frac{\partial^k}{\partial \rho^k} \Phi\left(\frac{\rho}{r}, \nu_1\right) \right| \le \frac{C}{r^k}, \qquad r \ge \rho(1+\delta), \quad \nu_1 \in [-1, 1]. \tag{6.91}$$

To compute the derivatives $(\partial^k/\partial\rho^k) D^\alpha_{\xi,\nu^0}$ of the second term in (6.90) we can take the symbol D^α_{ξ,ν^0} under the inner integral sign. Computation of the derivative $\partial^k/\partial\rho^k$ of the inner integral leads to (i) the appearance of a number of terms due to computation of the derivatives with respect to the upper and lower limits, and to (ii) an integral due to differentiation of the integrand. The first of these terms are obviously bounded, since the function

$\Phi(\rho/r,\nu_1)$ at the lower limit is bounded and does not depend on ρ, while at the upper limit it coincides with the analytic function $h^{1-n}[r(\rho,\nu_1)]^{n-1}$. The integral arising from differentiation of the integrand has the form

$$\int_{\rho(1+\delta)}^{r(\rho,\nu_1)} \frac{\partial^k}{\partial\rho^k}\Phi\left(\frac{\rho}{r},\nu_1\right) D^\alpha_{\xi,\nu^0}\hat{K}(\xi,r,\nu,\nu^0)\,dr$$

and by (6.91) it can be estimated by the integral

$$c_1\int_{\rho(1+\delta)}^{r(\rho,\nu_1)} \frac{dr}{r^k} = c_1\begin{cases} \ln\frac{\nu_1+\sqrt{(h/\rho)^2-(1-\nu_1^2)}}{1+\delta}, & k=1,\\ \frac{1}{1-k}\{[r(\rho,\nu_1)]^{1-k}-[\rho(1+\delta)]^{1-k}\}, & k>1.\end{cases}$$

From this follows (6.88). The proof of Theorem 2 is thus complete.

REMARK 1. We denote by $D(x,\nu)$ the set of points of the region D for which $\varphi(\xi,x,\nu)\geq 0$. The problem of finding $u(x)$ from the equation

$$\int_{S(x,\nu)}\rho(\xi,x,\nu)u(\xi)\,dS+\int_{D(x,\nu)}\rho_1(\xi,x,\nu)u(\xi)\,d\xi = v(x,\nu),\ x\in D,\ |\nu|=1,$$

is more general than the problem of solving (6.59). Theorems 1 and 2 hold for this problem under the condition that the smoothness of $\rho_1(\xi,x,\nu)$ is at most one worse than the smoothness of $\rho(\xi,x,\nu)$.

REMARK 2. If in (6.59) the weight function $\rho(\xi,x,\nu)$ is replaced by a matrix-valued weight function $R(\rho,x,\nu)$ of arbitrary finite dimension $m\times m$ and the function $u(\xi)$ is replaced by a vector-valued function $\mathbf{u}(\xi)$ with components $u_1,\dots,u_m$, then we obtain a vectorial problem of integral geometry. Theorems 1 and 2 hold for it with the natural replacement of condition (6.64) by the condition

$$|\det R(x,x,\nu)|\geq\rho_0>0,\quad \rho_0=\text{const},$$

and replacement of the function $1/\rho$ in (6.65) by $R^{-1}(x,x,\nu)$.

REMARK 3. From Theorems 1 and 2 we obtain as a corollary the estimate

$$\|u\|_{L_2(\Omega)}\leq C\sup_{|\nu|=1}\|v(x,\nu)\|_{W_2^l(\overline{D})},\qquad l=2[n/2], \tag{6.92}$$

which is valid for any $n\geq 2$.

3. Until now we have considered a formulation of the problem of integral geometry for the case where through each point $x\in D$ and for any direction ν there exists a smooth hypersurface $S(x,\nu)$ with normal ν passing through x. A more general formulation of this problem is possible where through x there pass surfaces $S(x,\nu)$ only for ν belonging to a set $\omega(x)$ of points of the unit sphere. The set $\omega(x)$ here, generally speaking, does not coincide with the unit sphere. It turns out that the question of the character of stability of the problem of integral geometry is closely related to the structure of the set

$\omega(x)$ [65]. For simplicity we consider the case where $\omega(x)$ does not depend on x: $\omega(x) = \omega_0$. Let $\omega_{\delta\alpha} = \{\nu\colon |\nu| = 1,\ |\nu\cdot\alpha| \le \delta\}$ be a spherical band. If for some unit vector α and $\delta > 0$ the condition $\omega_{\delta\alpha}\cap\omega_0 \ne \varnothing$ holds, then an estimate of the type (6.92) (with the set $|\nu| = 1$ under the supremum replaced by the set ω_0) for the problem of integral geometry cannot exist for any l. From this it follows, in particular, that estimates of the form

$$\|u\|_{W_2^k(\Omega)} \le C \sup_{\nu\in\omega_0} \|v(x,\nu)\|_{W_2^l(\overline{D})} \tag{6.93}$$

also cannot exist for any finite k and l and constant $C > 0$.

We shall demonstrate this for the particular case $n = 2$, setting $\varphi(\xi, x, \nu) = \nu(\xi - x)$ and $\rho = 1$. The principle of the proof is manifest to full extent in this example. Suppose $\omega_{\delta\alpha}\cap\omega_0 = \varnothing$ for $\delta > 0$, and let $\alpha = (0, 1)$. We introduce a function $u_\lambda \in C_0^\infty(\Omega)$ of the form

$$u_\lambda(x) = \sin\lambda x_1 \psi(x),$$

where $\psi(x)$ is an infinitely differentiable, compactly supported function with support in Ω which is not identically zero and λ is a sufficiently large numerical parameter. Then

$$\|u_\lambda\|^2_{L_2(\Omega)} = \frac{1}{2}\|\psi\|^2_{L_2(\Omega)} - \frac{1}{2}\int_\Omega \cos 2\lambda x_1 \psi^2(x)\,dx.$$

Integrating by parts a sufficient number of times, we obtain

$$\left|\int_\Omega \cos 2\lambda x_1 \psi^2(x)\,dx\right| \le \frac{C_k}{|\lambda|^k}.$$

On the other hand, for $x \in D$ and $\nu \in \omega_0$ we have

$$\begin{aligned} v_\lambda(x,\nu) &= \frac{1}{|\nu_2|}\int_{-\infty}^{\infty} u_\lambda\left[\xi_1, x_2 - \frac{\nu_1}{\nu_2}(\xi_1 - x_1)\right] d\xi_1 \\ &= \frac{1}{|\nu_2|}\int_{-\infty}^{\infty} \sin\lambda\xi_1\psi\left[\xi_1, x_2 - \frac{\nu_1}{\nu_2}(\xi_1 - x_1)\right] d\xi_1. \end{aligned}$$

Since for $\nu \in \omega_0$, we have the inequality $|\nu\cdot\alpha| = |\nu_2| \ge \delta$, it follows that $1/|\nu_2| \le 1/\delta$, and hence

$$|D^\alpha_{x,\nu} v_\lambda(x,\nu)| \le \frac{C_{\alpha k}}{|\lambda|^k}, \quad x \in D,\ \nu \in \omega_0.$$

Letting λ tend to infinity, we find that $\|v_\lambda\|_{W_2^l(D)} \to 0$ uniformly with respect to all $\nu \in \omega_0$ and any finite l, while

$$\|u_\lambda\|_{L_2(\Omega)} \to \frac{1}{\sqrt{2}}\|\psi\|_{L_2(\Omega)} \ne 0.$$

This implies the assertion made above.

We note that for ω_0 coinciding with the half sphere $|\nu| = 1$, $\nu_1 \geq 0$, estimates of the type (6.93) may nevertheless exist, as individual examples show (see [60] and [61]).

§5. The problem of integral geometry on plane curves and energy inequalities

For the case of the problem of integral geometry on plane curves in a formulation analogous to that considered in §4, the strongest results have been obtained by Mukhometov [187], [189]. In particular, he succeeded in proving a uniqueness theorem and in obtaining an estimate of stability without assuming smallness of the region in which the problem is studied. These stability estimates are based on reducing the problem of integral geometry to an equivalent boundary value problem for a partial differential equation of mixed type and using the method of energy inequalities to study it. It is also possible to investigate the problem of integral geometry on the geodesics of a Riemannian metric in n-dimensional space by means of this technique (see [46], [190], and [236]). In this section we partially present Mukhometov's results, following [187].

In the plane $x = (x_1, x_2)$ we consider a bounded, open, simply connected region D with boundary Γ. Suppose that the boundary is given by an equation of the form $x = g(s)$, where s is the arc length measured from a fixed point on Γ in a positive direction consistent with the choice of orientation on Γ_1 and $g(s)$ is a function of class $C^1[0, T]$, $g(0) = g(T)$; T is the length of Γ. Suppose, further, in the region $\overline{D}$ there is given a two-parameter family of curves $\mathfrak{M} = \{L(t),\ t = (t_1, t_2)\}$ possessing the following properties.

1. A unique curve of the family $\mathfrak{M}$ joins each pair of points x^1, x^2 of the region $\overline{D}$; each curve $L(t)$ intersects Γ at points $g(t_1)$ and $g(t_2)$, while other points of $L(t)$ belong to D; the lengths of the curves $L(t)$ are uniformly bounded.

2. The equation of the curve passing through the point $x^0 \in \overline{D}$ in the direction $\nu^0 = (\cos\theta_0, \sin\theta_0)$ is

$$x = f(s, \theta_0, x^0) \equiv x^0 + s\nu^0 + s^2 \tilde{f}(s, \nu^0, x^0), \tag{6.94}$$

in which $\tilde{f}(s, \nu^0, x^0)$ and its derivatives are continuously differentiable and bounded functions of the arc length s measured from x^0 and with respect to the parameters $\theta_0 \in [0, 2\pi]$, and $x^0 \in \overline{D}$; $f(s, 0, x^0) = f(s, 2\pi, x^0)$, and, moreover,

$$\frac{1}{s}\frac{\partial f}{\partial(s, \theta_0)} \geq c_0 > 0,\ c_0 = \text{const}, \tag{6.95}$$

in the entire domain of the variables s, θ_0, and x^0.

We note that for s close to zero, inequality (6.95) follows in an obvious manner from (6.94); therefore, (6.95) is essential for finite s. In this case it is equivalent to the positivity of the Jacobian $\partial f/\partial(s, \theta_0)$.

The following lemma holds with regard to the family of curves $\mathfrak{M}$.

LEMMA 1. *Suppose that the family of curves satisfies the conditions noted above and, moreover, the function $\tilde{f}(s, \nu^0, x^0)$ has continuous and bounded derivatives to order $l \geq 1$. Then equality (6.94) defines s and ν^0 as single-valued functions, continuously differentiable l times, of the points x^0 and x for all x^0, $x \in \overline{D}$ with $x \neq x^0$, while in a neighborhood of the set $x = x^0$, $x^0 \in \overline{D}$, the following estimates hold for them:*

$$\begin{aligned} |D^\alpha_{x,x^0} s(x^0, x)| &\leq \frac{c}{|x - x^0|^{|\alpha|-1}}, \\ |D^\alpha_{x,x^0} \nu^0(x^0, x)| &\leq \frac{c}{|x - x^0|^{|\alpha|}}, \qquad |\alpha| \leq l. \end{aligned} \tag{6.96}$$

For the proof we observe that, by (6.94),

$$\begin{aligned} s &= \frac{|x - x^0|}{|\nu^0 + s\tilde{f}(s, \nu^0, x^0)|}, \\ \nu^0 &= \frac{x - x^0}{|x - x^0|}|\nu^0 + s\tilde{f}(s, \nu^0, x^0)| + |x - x^0|\frac{\tilde{f}(s, \nu^0, x^0)}{|\nu^0 + s\tilde{f}(s, \nu^0, x^0)|}. \end{aligned}$$

Hence, by the implicit function theorem we find that in a sufficiently small region $|x - x^0| < \delta$ the functions s and ν have the structure

$$\begin{aligned} s &= |x - x^0|\left[1 + |x - x^0|\varphi\left(x^0, |x - x^0|, \frac{x - x^0}{|x - x^0|}\right)\right], \\ \nu &= \frac{x - x^0}{|x - x^0|}\left[1 + |x - x^0|\varphi\left(x^0, |x - x^0|, \frac{x - x^0}{|x - x^0|}\right)\right] \\ &\quad + |x - x^0|\psi\left(x^0, |x - x^0|, \frac{x - x^0}{|x - x^0|}\right), \end{aligned}$$

where φ and ψ are functions of their arguments continuously differentiable l times. From this, in particular, we obtain (6.96). For $|x - x^0| > \delta$ the inequality $s > \delta$ is obviously satisfied, and hence so are the inequalities $\partial f/\partial(s, \theta_0) \geq c_0\delta > 0$. But in this case the smoothness of the functions s and ν^0 stipulated by the lemma follows from the smoothness of $\tilde{f}(s, \nu^0, x^0)$, while the fact that they exist and are single-valued follows from condition 1) on the family of curves.

We denote by $L(x^0, x)$ the segment of the curve of the family $\mathfrak{M}$ passing through the points x^0 and x and contained between them, and by $\nu = (\cos\theta, \sin\theta)$ the unit vector tangent to $L(x^0, x)$ at x: $\nu = f_s(s, \theta_0, x^0)$. The vector ν can also be considered a function of the points x^0 and x which possesses the same smoothness properties as the vector ν^0. Indeed, if $\nu^0 = h(x^0, x)$, then, because of the equal status of the points x^0 and x, it is obvious that $\nu = -h(x, x^0)$.

LEMMA 2.

$$\frac{\partial}{\partial t_1}\theta(g(t_1), x) \geq 0, \qquad t_1 \in [0, T], \; x \in D. \tag{6.97}$$

For the proof we note that $\theta(x^0, x)$, considered as a function of x^0 for fixed $x \in D$, has as its level lines segments of curves of the family $\mathfrak{M}$ passing through x and contained between x and the boundary Γ; hence, $\nabla_{x^0}\theta(x^0, x)$ is directed along the normal to $L(x^0, x)$ at x^0 in the direction of increasing θ. Since the lines $L(x^0, x)$ for fixed x intersect only at the point x, on any closed curve encircling the point $x \in D$ (and, in particular, on Γ), increasing θ corresponds to the positive direction. Setting $x^0 = g(t_1)$, we find that to larger t_1 there correspond larger values of $\theta(g(t_1), x)$. From this we obtain (6.87).

For functions $u(x) \in C^1(\overline{D})$ and $\rho(x^0, x) \in C^1(\overline{D} \times \overline{D})$ we now consider the function $w(x^0, x)$ defined by

$$w(x^0, x) = \int_{L(x^0, x)} \rho(x^0, x') u(x')\, ds, \qquad x^0 \in D, \; x \in D, \tag{6.98}$$

and we pose the following problem: find $u(x)$ on $\overline{D}$ for given functions $\rho(x^0, x)$ and $v(t)$:

$$v(t) = w(g(t_1), g(t_2)), \qquad t \in Q = [0, T] \times [0, T]. \tag{6.99}$$

THEOREM 1. *If the family of curves satisfies assumptions* 1) *and* 2), *and the weight function* $\rho(x^0, x) \in C^1(\overline{D} \times \overline{D})$ *and*

$$\rho(x^0, x) \geq \rho_0 > 0, \qquad x^0 \in \Gamma, \; x \in \overline{D}, \tag{6.100}$$

$$\left| \frac{\partial}{\partial t_1} \ln \rho(g(t_1), x) \right| \leq q \frac{\partial}{\partial t_1}\theta(g(t_1), x), \qquad 0 \leq q < 1, \tag{6.101}$$

then a solution of the above problem is unique for $u \in C^1(\overline{D})$, *and for it there is the stability estimate*

$$\|u\|_{L_2(D)} \leq \frac{1}{\rho_0\sqrt{1-q^2}} \frac{1}{2\sqrt{\pi}} \| \, |\nabla_t v(t)| \, \|_{L_2(Q)}. \tag{6.102}$$

We shall first prove the estimate (6.102) for the case where the smoothness of u, ρ, and the function $\tilde{f}$ contained in (6.94) is one greater than formulated in the hypotheses of the theorem. In this case the following lemma holds for the function $w(x^0, x)$.

LEMMA 3. *Suppose the functions $\tilde{f}$, ρ, and u are twice continuously differentiable with respect to their arguments and are bounded together with their partial derivatives. Then the function $w(x^0, x)$ is twice continuously differentiable with respect to x^0 and x everywhere except at points $x = x^0$, in a neighborhood of which*

$$|D^{\alpha}_{x^0,x} w(x^0, x)| \le C|x - x^0|^{1-|\alpha|}, \quad |\alpha| \le 2. \tag{6.103}$$

The lemma follows from the representation

$$w(x^0, x) = \int_0^{s(x^0,x)} \rho[x^0, f(s, \theta_0(x^0, x), x^0)]u[f(s, \theta_0(x^0, x), x^0)]\, ds$$

by using Lemma 1.

We now obtain a differential equation for $w(x^0, x)$. For this we note that

$$w(x^0, f(s, \theta_0, x^0)) = \int_0^s \rho[x^0, f(s', \theta_0, x^0)]u[f(s', \theta_0, x^0)]\, ds'.$$

Differentiating with respect to s and using the fact that $f(x, \theta_0, x^0) = x$, $f_s(s, \theta_0, x^0) = \nu(x^0, x)$, we find that

$$\nabla_x w(x^0, x) \cdot \nu(x^0, x) = \rho(x^0, x)u(x). \tag{6.104}$$

Dividing both sides of this equality by $\rho(x^0, x)$, setting $x^0 = g(t_1)$, and differentiating the resulting equality with respect to t_1, we obtain an equation for the function $w(g(t_1), x) = \tilde{w}(t_1, x)$:

$$\frac{\partial}{\partial t_1}\left\{\frac{1}{\tilde{\rho}(t_1, x)}[\nabla_x \tilde{w}(t_1, x) \cdot \tilde{\nu}(t_1, x)]\right\} = 0, \quad t_1 \in [0, T], \ x \in \overline{D}, \tag{6.105}$$

in which

$$\tilde{\rho}(t_1, x) = \rho(g(t_1), x), \qquad \nu(t_1, x) = \nu(g(t_1), x).$$

(6.105) is an equation of mixed type, namely, of hyperbolic-parabolic type. It is satisfied in the cylindrical region $[0, T] \times \overline{D}$. On the boundary of this region we have the periodicity condition

$$\tilde{w}(0, x) = \tilde{w}(T, x), \quad x \in \overline{D}, \tag{6.106}$$

and the condition following from (6.99)

$$\tilde{w}(t_1, g(t_2)) = v(t). \tag{6.107}$$

The function $\tilde{w}(t, x)$ is thus a solution of the nonclassical problem (6.105)–(6.107). We note that it follows from (6.98) and (6.99) that $v(t_1, t_1) = 0$, $t_1 \in [0, T]$. If this condition is satisfied, problem (6.105)–(6.107) and the problem of integral geometry posed above are equivalent. It therefore suffices to find $\tilde{w}(t_1, x)$ in order to then find $u(x)$ from (6.104). To investigate the questions of uniqueness and stability of problem (6.105)–(6.107) we apply the method of energy estimates.

For $x \neq g(t_1)$, $x \in \overline{D}$ and $\beta = (-\sin\tilde{\theta}, \cos\tilde{\theta})$, $\tilde{\theta} = \tilde{\theta}(t_1, x) \equiv \theta(g(t_1), x)$, we have the two obvious identities

$$\tilde{\rho}(\nabla_x \tilde{w} \cdot \beta)\frac{\partial}{\partial t_1}\left[\frac{1}{\tilde{\rho}}\nabla_x \tilde{w} \cdot \tilde{\nu}\right] \equiv (\nabla_x \tilde{w} \cdot \beta)(\nabla_x \tilde{w} \cdot \tilde{\nu}_{t_1})$$
$$+(\nabla_x \tilde{w} \cdot \beta)(\nabla_x \tilde{w}_{t_1} \cdot \nu) - (\nabla_x \tilde{w} \cdot \beta)(\nabla_x \tilde{w} \cdot \tilde{\nu})\frac{\partial}{\partial t_1}\ln\tilde{\rho},$$
$$\tilde{\rho}(\nabla_x \tilde{w} \cdot \beta)\frac{\partial}{\partial t_1}\left[\frac{1}{\tilde{\rho}}\nabla_x \tilde{w} \cdot \tilde{\nu}\right] \equiv \frac{\partial}{\partial t_1}[(\nabla_x \tilde{w} \cdot \beta)(\nabla_x \tilde{w} \cdot \tilde{\nu})] - (\nabla_x \tilde{w} \cdot \tilde{\nu})(\nabla_x w \cdot \beta_{t_1})$$
$$-(\nabla_x \tilde{w} \cdot \tilde{\nu})(\nabla_x \tilde{w}_{t_1} \cdot \beta) - (\nabla_x \tilde{w} \cdot \beta)(\nabla_x \tilde{w} \cdot \tilde{\nu})\frac{\partial}{\partial t_1}\ln\tilde{\rho}.$$

Adding these identities and using the fact that

$$\tilde{\nu}_{t_1} = \beta\frac{\partial\tilde{\theta}}{\partial t_1}, \quad \beta_{t_1} = -\tilde{\nu}\frac{\partial\tilde{\theta}}{\partial t_1},$$
$$(\nabla_x \tilde{w} \cdot \beta)(\nabla_x \tilde{w}_{t_1} \cdot \tilde{\nu}) - (\nabla_x \tilde{w} \cdot \tilde{\nu})(\nabla_x \tilde{w}_{t_1} \cdot \beta) \equiv \frac{\partial}{\partial x_1}(\tilde{w}_{t_1}\tilde{w}_{x_2}) - \frac{\partial}{\partial x_2}(\tilde{w}_{t_1}\tilde{w}_{x_1}),$$

we obtain

$$2\tilde{\rho}(\nabla_x \tilde{w} \cdot \beta)\frac{\partial}{\partial t_1}\left(\frac{1}{\tilde{\rho}}\nabla_x \tilde{w} \cdot \tilde{\nu}\right)$$
$$\equiv \left\{\left[(\nabla_x \tilde{w} \cdot \tilde{\nu})^2 + (\nabla_x \tilde{w} \cdot \beta)^2\right]\frac{\partial\tilde{\theta}}{\partial t_1} - 2(\nabla_x \tilde{w} \cdot \tilde{\nu})(\nabla_x \tilde{w} \cdot \beta)\frac{\partial}{\partial t_1}\ln\tilde{\rho}\right\}$$
$$+\frac{\partial}{\partial_{t_1}}[(\nabla_x \tilde{w} \cdot \tilde{\nu})(\nabla_x \tilde{w} \cdot \beta)] + \frac{\partial}{\partial x_1}(\tilde{w}_{t_1}\tilde{w}_{x_2}) - \frac{\partial}{\partial x_2}(\tilde{w}_{t_1}\tilde{w}_{x_1}). \quad (6.108)$$

The left side of this identity vanishes on solutions of (6.105). Under the assumption that $\tilde{w}(t_1, x)$ is a solution of (6.105), we integrate (6.108) over the region G_ε obtained from $G = [0, T] \times \overline{D}$ by discarding the set $\{(t_1, x)\colon t_1 \in [0, T],\ x \in \overline{D},\ |x - g(t_1)| \le \varepsilon\}$ for sufficiently small positive ε. Using the Gauss-Ostrogradskiĭ formula, we then obtain

$$0 = \int_{G_\varepsilon}\left\{[(\nabla_x \tilde{w} \cdot \tilde{\nu})^2 + (\nabla_x \tilde{w} \cdot \beta)^2]\frac{\partial\tilde{\theta}}{\partial t_1} - 2(\nabla_x \tilde{w} \cdot \tilde{\nu})(\nabla_x \tilde{w} \cdot \beta)\right.$$
$$\left.\times\frac{\partial}{\partial t_1}\ln\tilde{\rho}\right\}dx\,dt_1 + \int_{S_\varepsilon}\{[(\nabla_x \tilde{w} \cdot \tilde{\nu})(\nabla_x \tilde{w} \cdot \beta)]\cos(n, t_1)$$
$$+\tilde{w}_{t_1}[\tilde{w}_{x_2}\cos(n, x_1) - \tilde{w}_{x_1}\cos(n, x_2)]\}\,dS.$$

Here S_ε is the boundary of the set D_ε, n is the outer normal to S_ε and dS is the area element.

In this equality we pass to the limit as ε goes to zero. Because of the estimates of Lemma 3, the integrals over G_ε and S_ε converge as improper integrals to integrals over G and S respectively (S is the boundary of G). The integral over the surface S can be broken into two parts: an integral over the upper and lower base of the cylinder G and an integral over the lateral surface $[0,T]\times\Gamma$. On the lower and upper base of the cylinder $\cos(n,x_1)=\cos(n,x_2)=0$, while $\cos(n,t_1)$ has opposite signs at the corresponding points. The periodicity condition (6.106) leads to the situation that the sum of the integrals over the upper and lower base of the cylinder vanishes. On the lateral surface $\cos(n,t_1)=0$ and for $x=g(t_2)$,

$$\tilde{w}_{t_1}=v_{t_1}(t),\qquad \tilde{w}_{x_2}\cos(n,x_1)-\tilde{w}_{x_1}\cos(n,x_2)=\frac{\partial}{\partial t_2}\tilde{w}(t_1,g(t_2))=v_{t_2}(t),$$

while $dS=dt_1\,dt_2$. Thus, we obtain finally

$$\int_G\left\{[(\nabla_x\tilde{w}\cdot\tilde{\nu})^2+(\nabla_x\tilde{w}\cdot\beta)^2]\frac{\partial\tilde{\theta}}{\partial t_1}-2(\nabla_x\tilde{w}\cdot\tilde{\nu})(\nabla_x\tilde{w}\cdot\beta)\frac{\partial}{\partial t_1}\ln\tilde{\rho}\right\}dx\,dt_1$$
$$=-\int_0^T dt_1\int_0^T v_{t_1}v_{t_2}\,dt_2.\quad(6.109)$$

By Lemma 2, $\partial\tilde{\theta}/\partial t_1\ge 0$. At all points where $\partial\tilde{\theta}/\partial t_1=0$ the expression in braces vanishes by (6.101). For points at which $\partial\tilde{\theta}/\partial t_1>0$, we have

$$[(\nabla_x\tilde{w}\cdot\tilde{\nu})^2+(\nabla_x\tilde{w}\cdot\beta)^2]\frac{\partial\tilde{\theta}}{\partial t_1}-2(\nabla_x\tilde{w}\cdot\tilde{\nu})(\nabla_x\tilde{w}\cdot\beta)\frac{\partial}{\partial t_1}\ln\tilde{\rho}$$
$$=(\nabla_x\tilde{w}\cdot\tilde{\nu})^2\left[\frac{\partial\tilde{\theta}}{\partial t_1}-\left(\frac{\partial}{\partial t_1}\ln\tilde{\rho}\right)^2\frac{1}{\partial\tilde{\theta}/\partial t_1}\right]$$
$$+\left[(\nabla_x\tilde{w}\cdot\tilde{\nu})\frac{\frac{\partial}{\partial t_1}\ln\tilde{\rho}}{\sqrt{\frac{\partial}{\partial t_1}\tilde{\theta}}}-(\nabla_x\tilde{w}\cdot\beta)\sqrt{\frac{\partial\tilde{\theta}}{\partial t_1}}\right]^2$$
$$\ge(\nabla_x\tilde{w}\cdot\tilde{\nu})^2(1-q^2)\frac{\partial\tilde{\theta}}{\partial t_1}\ge\rho_0^2(1-q^2)u^2(x)\frac{\partial\tilde{\theta}}{\partial t_1}.$$

Therefore, strengthening (6.109), we obtain

$$\rho_2^0(1-q^2)\int_D u^2(x)\,dx\int_0^{2\pi}\frac{\partial\tilde{\theta}}{\partial t_1}\,dt_1\le\frac{1}{2}\int_0^T dt_1\int_0^T|\nabla_t v|^2\,dt_2.$$

From this we obtain (6.102).

To complete the proof of the theorem we note that the expression on the right side of (6.102) is meaningful for $u\in C^1(\overline{D})$, $\rho(x^0,x)\in C^1(\Gamma\times\overline{D})$, and a family of curves $\mathfrak{M}$ satisfying conditions 1) and 2). Therefore, approximating the functions u, ρ, and $\tilde{f}$ satisfying the hypotheses of the theorem by functions u_n, ρ_n, and $\tilde{f}_n$ for which (6.102) has already been established and passing to the limit in it as $n \to \infty$, we obtain the estimate (6.102) under the hypotheses of the theorem. From this estimate, in particular, we obtain uniqueness of a solution of the problem of integral geometry in the space $C^1(\overline{D})$.

CHAPTER VII

Multidimensional Inverse Problems for Linear Differential Equations

With respect to differential equations, problems of the following type are called inverse problems: there is given a class of differential equations characterized by a finite collection of functions; there is given some information regarding solutions of one of the differential equations of the class in question; it is required to find this equation. The problem thus consists in finding a specific collection of functions defining a differential equation. With regard to linear differential equations, the inverse problem may consist in finding part or even all of the coefficients of this equation. According to the presently adopted terminology, an inverse problem is called n-dimensional if at least one of the functions to be determined depends on n variables.

The definition of the term "inverse problems for differential equations" presented above is not the most general. This definition does not include, for example, such problems as the problem of determining boundary or initial conditions. Similar new "inverse" problems arise naturally if in place of a class of differential equations we consider the classes of differential operators connected with them (see, for example, §6 of this chapter).

The creation of a theory of inverse problems for differential equations started comparatively recently. As we have already noted in Chapter I, one of the first inverse problems considered was the one-dimensional kinematic inverse problem of seismology. The study of another inverse problem, the inverse problem of potential theory, began in the forties of our century. The first result connected with this problem was obtained by P. S. Novikov in 1938 [198]. It was subsequently studied by A. N. Tikhonov, M. M. Lavrent'ev, V. K. Ivanov, L. N. Sretenskiĭ, A. I. Prilepko, V. N. Strakhov, and others. Mathematically, this problem consists in determining a compactly supported right side of a linear second-order equation of elliptic type.

A problem that has become widely known is the one-dimensional inverse Sturm-Liouville problem connected with the ordinary differential operator L_q:

$$L_q y \equiv -y'' + q(x)y, \qquad x \in [a, b],$$
$$y'(a) - hy(a) = 0, \qquad y'(b) + Hy(b) = 0.$$

In this problem it is required to find L_q on the basis of its spectral function (the formulation of V. A. Marchenko). Investigation of this problem was begun by V. A. Ambartsumyan and G. Borg in a somewhat different formulation (find L_q from its spectrum) and was then continued by V. A. Marchenko, M. G. Kreĭn, I. M. Gel′fand, B. M. Levitan, and others [165], [177]. There is also another inverse problem connected with a Sturm-Liouville differential operator—the inverse scattering problem where it is required to find L_q from the scattering data at infinity. It was investigated mainly in the fifties [275]. Recently, interest in this problem has been renewed in connection with the discovery that by solving this problem one can obtain exact solutions of a number of important nonlinear equations of physics such as the Korteweg-deVries equation, the sine-Gordon equation, and others.

The first result connected with the study of a multidimensional inverse problem distinct from the inverse problem of potential theory was obtained by Yu. M. Berezanskiĭ [42]–[44] at the beginning of the fifties. The systematic development of the theory of multidimensional inverse problems connected with finding the coefficients of linear differential equations began in the mid-sixties. In connection with this we mention the papers of M. M. Lavrent′ev [147]–[150], V. G. Romanov [215]–[241], Yu. E. Anikonov [9]–[22], A. D. Iskenderov [116]–[119], N. Ya. Beznoshchenko [34]–[37], S. Elubaev [99], V. G. Yakhno [244], [293]–[296], M. V. Klibanov [125]–[131], L. P. Nizhnik [193]–[197], and A. S. Zapreev [101], [102].

One-dimensional inverse problems for equations in multidimensional space occupy an intermediate position between inverse problems for ordinary differential equations and multidimensional inverse problems. Such problems often reduce to a Sturm-Liouville inverse problem. In this direction we mention the works of A. S. Alekseev [1], [2], [5]–[7], A. S. Blagoveshchenskiĭ [47]–[53], B. M. Budak and A. D. Iskenderov [55]–[57], A. G. Megrabov [169]–[176], K. G. Reznitskaya [206]–[208], and V. G. Romanov [211], [214], [216], [220].

The chapter here submitted to the reader cannot, of course, claim completeness of the exposition of material on this question. Its main purpose is to give an idea of the circle of problems being solved and the methods of solving them.

§1. Examples of formulations of multidimensional inverse problems. Mathematical problems connected with investigating them

In this section we present several examples of multidimensional inverse problems for the simplest equations of mathematical physics. For simplicity, here and below we assume that the point x is a point of three-dimensional space x_1, x_2, x_3. We remark that a number of results for two-dimensional space x_1, x_2 follow from the considerations presented in three-dimensional space by means of the method of descent in inverse problems presented in §5.

1. We consider the differential equation

$$u_{tt} = \Delta u + q(x)u + 4\pi\delta(x - x^0, t) \tag{7.1}$$

and the Cauchy problem for it in the whole space with zero initial data

$$u(x, t, x^0)|_{t=0} = u_t(x, t, x^0)|_{t=0} = 0. \tag{7.2}$$

Here Δ is the Laplace operator in the variables x_1, x_2, x_3, and $\delta(x - x^0, t) = \delta(x - x^0)\delta(t)$ is the Dirac delta function concentrated at the point x^0 at time $t = 0$. For a known function $q(x)$, problem (7.1), (7.2) has a unique solution depending continuously on $q(x)$. The solution $u(x, t, x^0)$ depends on the point x^0 as a parameter of the problem. We now suppose that the function $q(x)$ is unknown; it is required to find it if a solution of problem (7.1), (7.2) is known at points of some plane S in x-space at all times $t > 0$:

$$u(x, t, x^0) = f(x, t, x^0), \quad x \in S, \ t > 0. \tag{7.3}$$

Regarding the parameter x^0, we assume that $x^0 \in S$. We assume the function $q(x)$ to be continuous in the whole space.

We shall study the structure of the solution of problem (7.1), (7.2). For this we use Kirchhoff's formula which for the inhomogeneous wave equation

$$u_{tt} = \Delta u + \varphi(x, t)$$

provides the solution of the Cauchy problem with data (7.2) in the form of a retarded potential

$$u(x, t) = \frac{1}{4\pi} \int_{|x-\xi| \le t} \frac{\varphi(\xi, t - |x - \xi|)}{|x - \xi|} \, d\xi. \tag{7.4}$$

Here $\xi = (\xi_1, \xi_2, \xi_3)$ and $d\xi = d\xi_1 \, d\xi_2 \, d\xi_3$. Applying Kirchhoff's formula to (7.1), we reduce problem (7.1), (7.2) to the equivalent integral equation

$$u(x, t, x^0) = \frac{1}{4\pi} \int_{|x-\xi| \le t} \frac{q(\xi)u(\xi, t - |x - \xi|, x^0) + 4\pi\delta(\xi - x^0, t - |x - \xi|)}{|x - \xi|} \, d\xi.$$

Using properties of the delta function, we reduce this equation to the form

$$u(x,t,x^0) = \frac{\delta(t-|x-x^0|)}{|x-x^0|} + \frac{1}{4\pi}\int_{|x-\xi|\le t} \frac{q(\xi)u(\xi,t-|x-\xi|,x^0)}{|x-\xi|}\,d\xi. \quad (7.5)$$

This is an equation of Volterra type. To solve it we use the method of successive approximations. For this we represent $u(x,t,x^0)$ in the form of the series

$$u(x,t,x^0) = \sum_{n=0}^{\infty} u_n(x,t,x^0), \quad (7.6)$$

where the terms $u_n(x,t,x^0)$ of the series are found from the formulas

$$u_0(x,t,x^0) = \frac{\delta(t-|x-x^0|)}{|x-x^0|},$$

$$u_n(x,t,x^0) = \frac{1}{4\pi}\int_{|x-\xi|\le t} \frac{q(\xi)u_{n-1}(\xi,t-|x-\xi|,x^0)}{|x-\xi|}\,d\xi, \quad n=1,2,\dots. \quad (7.7)$$

We shall now show that all $u_n(x,t,x^0) \equiv 0$, $n = 0,1,2,\dots$, in the region $t < |x-x^0|$. It is most convenient to do this by induction. For $n = 0$ the assertion is obvious. We now assume that it is true for all $u_k(x,t,x^0)$ with $k = 1,\dots,n-1$, and show that for $k = n$ it is also satisfied. By the induction hypothesis the integrand

$$u_{n-1}(\xi,t-|x-\xi|,x^0) \equiv 0, \qquad \forall\xi: t-|x-\xi| < |\xi-x^0|.$$

But in the region $t < |x-x^0|$ for arbitrary ξ we have

$$t < |x-x^0| \le |x-\xi| + |\xi-x^0|.$$

Hence, for $|x-x^0| > t$ the integrand in (7.7) is identically zero. Therefore, $u_n(x,t,x^0) \equiv 0$ for $t < |x-x^0|$. From (7.6) we obtain

$$u(x,t,x^0) \equiv 0, \qquad t < |x-x^0|.$$

From a physical point of view this result is altogether obvious. Indeed, perturbations in x-space propagate with finite speed equal to one; therefore, if the distance from the point x to the point x^0 is greater than t, then a perturbation created at x^0 has not had time to reach x. If we tentatively picture the four-dimensional space ξ, τ as shown in Figure 8, then the set of points satisfying the inequality $\tau < |\xi - x^0|$ is the exterior of a cone having vertex at the point $(x^0,0)$ and directed upward. Outside this cone $u \equiv 0$. The region of integration in (7.7) is the projection onto the plane $\tau = 0$ of the interior of the cone with vertex at the point (x,t) directed downward.

From this it is clear that the actual region of integration in (7.7) (the set of points ξ at which the integrand is not identically zero) is the projection onto

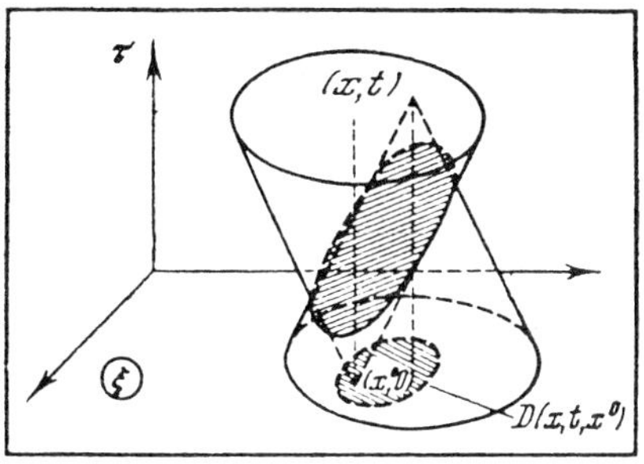

Figure 8

the plane $\tau = 0$ of the interior of the intersection of these two cones and has the form of the interior of an ellipsoid of rotation with foci at the points x^0 and x:

$$D(x,t,x^0) = \{\xi: |x-\xi| + |\xi - x^0| \le t\}.$$

Thus,

$$u_n(x,t,x^0) = \frac{1}{4\pi}\int_{D(x,t,x^0)} \frac{q(\xi)u_{n-1}(\xi, t-|x-\xi|, x^0)}{|x-\xi|}\,d\xi, \quad n = 1, 2, \dots. \tag{7.7'}$$

We now proceed to the analysis of how the $u_n(x,t,x^0)$ depend on q. From the formula for u_0 it is evident that u_0 does not depend on q. From (7.7′) for $n = 1$ it follows that u_1 depends on q in linear fashion:

$$u_1(x,t,x^0) = \frac{1}{4\pi}\int_{D(x,t,x^0)} \frac{q(\xi)\delta(t-|x-\xi|-|\xi-x^0|)}{|x-\xi|\,|\xi-x^0|}\,d\xi, \tag{7.8}$$

all the remaining u_n for $n \ge 2$ depend on q in nonlinear fashion, and this dependence on q for $q = \text{const}$ has the form q^n. Each of the u_n in (7.6) has the physical interpretation of a term corresponding to n-fold scattering by inhomogeneities of the medium (the inhomogeneities are related to the function $q(x)$). It can be shown (see, for example, [217]) that the u_n, $n \ge 1$, are continuous in the region $|x - x^0| \le t$ and $\sum_1^\infty u_n$ converges uniformly in each finite region of the variables x, t and $\|u_n\|_C$ can be estimated in terms of $(\|q\|_C)^n$ and the dimensions of this region. We make use of this circumstance and, under the assumption that $\|q\|_C$ is small, we replace the original inverse problem by a linearized problem. The basis for the linearization is the approximate representation of the solution of problem (7.1), (7.2) in the form

$$u(x,t,x^0) = u_0(x,t,x^0) + u_1(x,t,x^0), \tag{7.9}$$

which is the more precise the smaller $\|q\|_C$. This representation, from the point of view of functional analysis, is equivalent to replacing the nonlinear operator $U(q)$ assigning to each function q the solution of problem (7.1), (7.2) by its first differential computed at the element $q \equiv 0$. The linearized inverse problem is to find $q(x)$ if in place of (7.3) the function u_1 is known on S:

$$u_1(x, t, x^0) = \varphi(x, t, x^0), \quad x \in S, \ t > 0. \tag{7.10}$$

Thus, in (7.8) for $x \in S$ the left side is a known function φ; it is required to find $q(x)$ from this equation. We note that the integrand in (7.8) is concentrated on the boundary of the region $D(x, t, x^0)$, i.e., on the set of points belonging to the ellipsoid

$$S(x, t, x^0) = \{\xi : |x - \xi| + |\xi - x^0| = t\}.$$

Therefore, the integral in (7.8) of a generalized function can be transformed into an integral over the surface

$$u_1(x, t, x^0) = \frac{1}{4\pi} \int_{S(x,t,x^0)} \frac{q(\xi)}{|x - \xi|\,|\xi - x^0|} \frac{dS}{|\nabla_\xi(|x - \xi| + |\xi - x^0|)|}. \tag{7.11}$$

Here dS is the element of area of the surface $S(x, t, x^0)$. The weight factor occurring here can easily be computed in terms of the cosine of the angle α between the vectors $\xi - x^0$ and $\xi - x$, and the latter, by means of the law of cosines, can be expressed in terms of t, $|x - x^0|$, $|\xi - x^0|$, and $|\xi - x|$:

$$\begin{aligned}
|\nabla_\xi(|x - \xi| + |x^0 - \xi|)| &= \left| \frac{\xi - x^0}{|\xi - x^0|} + \frac{\xi - x}{|\xi - x|} \right| \\
&= \sqrt{2(1 + \cos a)} \\
&= \sqrt{2\left[1 + \frac{|x^0 - \xi|^2 + (t - |x^0 - \xi|)^2 - |x - x^0|^2}{2|x^0 - \xi|(t - |x^0 - \xi|)}\right]} \\
&= \frac{\sqrt{t^2 - |x - x^0|^2}}{\sqrt{|x^0 - \xi|\,|\xi - x|}}.
\end{aligned}$$

Hence, (7.11) can be written in the form

$$u_1(x, t, x^0) = \frac{1}{4\pi\sqrt{t^2 - |x^0 - x|^2}} \int_{S(x,t,x^0)} \frac{q(\xi)}{\sqrt{|x - \xi|\,|x^0 - \xi|}}\, dS. \tag{7.12}$$

The formula for $u_1(x, t, x^0)$ has the simplest form if the area element dS is expressed in terms of the element of solid angle with center at x^0. To this end we introduce a spherical coordinate system r, θ, φ, $r = |\xi - x^0|$, directing the polar axis through the point x. Using the fact that $S(x, t, x^0)$ is a surface of revolution about the polar axis, we represent dS as the product of two linear elements, one of which is the element of arc length of the section of the surface

$S(x,t,x^0)$ by the plane $\varphi = \text{const}$ and the other the element of arc length of the perpendicular section:

$$dS = \frac{r\,d\theta}{\cos(n,\nu)} r \sin\theta\, d\varphi = \frac{r^2}{\cos(n,\nu)}\, d\omega.$$

Here (n,ν) denotes the angle between the normal n to the surface $S(x,t,x^0)$ and the vector $\nu = (\xi - x^0)/|\xi - x^0|$, and $d\omega = \sin\theta\, d\theta\, d\varphi$. But

$$n = \frac{\nabla_\xi(|\xi - x^0| + |\xi - x|)}{|\nabla_\xi(|\xi - x^0| + |\xi - x|)|} = \frac{(\xi - x)/|\xi - x| + (\xi - x^0)/|\xi - x^0|}{\sqrt{2}\sqrt{1 + \frac{\xi - x}{|\xi - x|}\frac{\xi - x^0}{|\xi - x^0|}}},$$

and hence

$$\cos(n,\nu) = \frac{1}{\sqrt{2}}\sqrt{1 + \frac{\xi - x}{|\xi - x|}\frac{\xi - x^0}{|\xi - x^0|}} = \frac{1}{\sqrt{2}}\sqrt{1 + \cos\alpha} = \frac{1}{2}\frac{\sqrt{t^2 - |x - x^0|^2}}{\sqrt{|x - \xi|\,|x^0 - \xi|}}.$$

For dS we finally obtain

$$dS = \frac{2\sqrt{|\xi - x^0|^5|\xi - x|}}{\sqrt{t^2 - |x - x^0|^2}}\, d\omega.$$

The formula for $u_1(x,t,x^0)$ then takes the form

$$u_1(x,t,x^0) = \frac{1}{2\pi(t^2 - |x - x^0|^2)} \int_{S(x,t,x^0)} |\xi - x^0|^2 q(\xi)\, d\omega. \tag{7.13}$$

We return now to considering the inverse problem in the linearized formulation. Taking $x \in S$ in (7.13), we find

$$\begin{gathered} \int_{S(x,t,x^0)} |\xi - x^0|^2 q(\xi)\, d\omega = \psi(x,t,x^0), \quad x \in S, \\ \psi(x,t,x^0) = 2\pi(t^2 - |x - x^0|^2)\varphi(x,t,x^0). \end{gathered} \tag{7.14}$$

Under the assumption that $x^0 \in S$ is a fixed point, the problem of determining q reduces, as is evident from (7.14), to the following problem of integral geometry: for the function $q(\xi)$, all integrals with weight function $|\xi - x^0|^2$ are known over all possible ellipsoids of rotation $S(x,t,x^0)$ having one focus fixed while the other runs through the set of points of the plane S; on the basis of these integrals it is required to find $q(\xi)$. This problem was studied by V. G. Romanov [209] (see also [217]). It was shown that any continuous function q which is even relative to the plane S is uniquely determined by the function ψ. It can be shown that evenness of q relative to the plane S can be given up; to uniquely determine q it then suffices to give, together with the function $u_1(x,t,x^0)$ on S, its normal derivative $\partial u_1(x,t,x^0)/\partial n$, $x \in S$; n is the normal to S.

Other formulations of the linearized inverse problem which refine the original formulation of the problem are possible as well. Namely, we suppose that the point x^0, which is a parameter of the problem, itself varies, running through the plane S. Then ψ is a function of a substantially greater number of variables than $q(x)$. In the space x, t, x^0, $x \in S$, $x^0 \in S$, it is possible to consider manifolds of the same dimension as the number of independent variables of the function $q(x)$ (in the present case this number is equal to three) and to study the question of unique determination of $q(x)$ from the given information. We consider, for example, the manifold $\mathfrak{M}_0 = \{(x,t,x^0): x = x^0 \in S,$ $t \geq 0\}$. The problem of finding the function $q(x)$ on the basis of the function $\psi(x,t,x^0)$ given on $\mathfrak{M}_0$ is then equivalent in linearized formulation to the following problem: given the integrals over spheres $S(x^0,t,x^0)$ with centers at points $x^0 \in S$ of radius $t/2$,

$$(t/2)^2 \int_{S(x^0,t,x^0)} q(\xi)\, d\omega = \psi(x^0,t,x^0), \quad x^0 \in S,\ t \geq 0,$$

it is required to find $q(\xi)$. We considered this problem in §1 of the preceding chapter. In particular, a uniqueness theorem for the inverse problem in the linearized formulation in the class of continuous functions $q(x)$ which are even relative to the plane S follows from the results obtained there. It is also possible to consider still more general ways of defining the three-dimensional manifold $\mathfrak{M}$, for example, the following: suppose β is a fixed vector lying in the plane S; we set $\mathfrak{M}_\beta = \{(x,t,x^0): x \in S,\ x^0 \in S,\ x - x^0 = \beta,\ t \geq 0\}$. The manifold $\mathfrak{M}_0$ we introduced earlier is obtained from this for $\beta = 0$. To the manifold $\mathfrak{M}_\beta$ there corresponds a family of ellipsoids of rotation invariant under translation of the coordinate system as a rigid whole along S. In this case we can invoke the results of Chapter VI, §2, to investigate the problem of integral geometry obtained here. From this, in particular, we obtain the following result: if $S = \{x: x_3 = 0\}$, $q(x)$ is known in the region $x_3 \leq \varepsilon$, $\varepsilon > 0$, and if in the region $x_3 > \varepsilon$, it has for each fixed x_3 a Fourier transform with respect to x_1 and x_2 continuously depending on x_3, then in the region $x_3 > \varepsilon$ the function $q(x)$ is uniquely determined by giving the function ψ on the set $\mathfrak{M}_\beta$.

We have thus considered a linearized version of the formulation of the inverse problem for equation (7.1), and it turned out that its investigation is closely related to the problems of integral geometry in Chapter VI. It can, of course, be thought that this is a special feature of just the linear part of the problem, i.e., it is a result of linearization. However, this is not the case. We shall see later that investigation of the inverse problem (7.1)–(7.3) as a whole (and not only the linear part) also reduces to a problem of integral geometry which is somewhat more general than problem (7.14). To corroborate what has been said above we shall consider for equation (7.1) an example of a formulation of the inverse problem where in pure form it reduces to a problem of integral geometry.

2. Suppose that, regarding a solution of problem (7.1), (7.2), the following is known: for points x, x^0, running successively through the set of points of some closed, convex surface S, at all points $x, x^0 \in S$ the solution $u(x,t,x^0)$ is known for times t contained in the interval

$$|x - x^0| - \varepsilon \le t \le |x - x^0| + \varepsilon,$$

where ε is any arbitrarily small positive number. On the basis of the given function

$$\varphi(x,t,x^0) = u(x,t,x^0), \quad x \in S,\ x^0 \in S,\ t \in [|x - x^0| - \varepsilon, |x - x^0| + \varepsilon],$$

it is required to find $q(x)$ inside the region bounded by the surface S.

To investigate this problem we use the representation for $u(x,t,x^0)$ in the form of the series (7.6). In this series the first term u_0 is known and does not depend on q. We transformed the expression for u_1 to the form (7.13). For our purpose a more convenient representation is the one obtained from (7.13) if in the integral over the variables of integration θ, φ we go over to the variables of integration r, φ ($r = |\xi - x^0|$). For this we use the equation of the ellipsoid $S(x,t,x^0)$ in a polar coordinate system:

$$r = \frac{t^2 - |x - x^0|^2}{2(t - |x - x^0| \cos\theta)}.$$

Hence,

$$dr = -\frac{t^2 - |x - x^0|^2}{2(t - |x - x^0| \cos\theta)^2} |x - x^0| \sin\theta \, d\theta = -\frac{2r^2 |x - x^0| \sin\theta \, d\theta}{t^2 - |x - x^0|^2},$$

and (7.13) takes the form

$$\begin{aligned} u_1(x,t,x^0) &= \frac{1}{4\pi|x - x^0|} \int_{S(x,t,x^0)} q(\xi) \, dr \, d\varphi \\ &= \frac{1}{4\pi|x - x^0|} \int_0^{2\pi} d\varphi \int_{\frac{1}{2}(t - |x - x^0|)}^{\frac{1}{2}(t + |x - x^0|)} q(\xi) \, dr. \end{aligned} \tag{7.15}$$

Continuity of $u_1(x,t,x^0)$ in the region $t \ge |x - x^0|$ and its boundedness in any finite region of the space x, t, x^0 follow from this formula. If $|q(x)| \le q_0$, then $u_1(x,t,x^0) \le \frac{1}{2} q_0$.

By means of analogous transformations the expressions for the u_n $(n \geq 2)$ can be represented in the form

$$u_n(x,t,x^0) = \frac{1}{4\pi|x-x^0|} \int_{|x-x^0|}^{t} d\tau \int_{S(x,\tau,x^0)} rq(\xi)u_{n-1}$$

$$\times(\xi, t-|x-\xi|, x^0)\,dr\,d\varphi = \frac{1}{4\pi|x-x^0|} \int_{|x-x^0|}^{t} d\tau \int_0^{2\pi} d\varphi$$

$$\times \int_{\frac{1}{2}(\tau-|x-x^0|)}^{\frac{1}{2}(\tau+|x-x^0|)} q(\xi)u_{n-1}(\xi, t-|x-\xi|, x^0) r\,dr, \qquad t \geq |x-x^0|.$$

From this we find that the u_n, $n \geq 2$, are continuous in the region $t \geq |x-x^0|$ and obtain the estimates

$$|u_n(x,t,x^0)| \leq \frac{1}{8} q_0(t^2 - |x-x^0|^2) \max_{(\xi,\tau)\in D(x,t,x^0)} |u_{n-1}(\xi,\tau,x^0)|.$$

It is also possible to obtain considerably finer estimates which show that the series (7.6) converges uniformly in any finite region of the space x, t, x^0 (see, for example, [217]).

We now consider the limit of all the $u_n(x,t,x^0)$, $n \geq 1$, as $t \to |x-x^0|$. The surface of the ellipsoid of revolution $S(x,t,x^0)$ here contracts to the line segment $L(x^0,x)$ joining the points x^0 and x. As is evident from (7.15), here

$$u_1(x,|x-x^0|,x^0) = \frac{1}{2|x-x^0|} \int_{L(x^0,x)} q(\xi)\,ds,$$

where ds is the element of length of $L(x^0,x)$. From the estimates for u_n, $n \geq 2$, it follows that all the u_n, $n \geq 2$, tend to zero as $t \to |x-x^0|$. We thus obtain

$$\frac{1}{2|x-x^0|} \int_{L(x^0,x)} q(\xi)\,ds = \lim_{t\to|x-x^0|+0} [\varphi(x,t,x^0) - u_0(x,t,x^0)],$$

and the integrals of the function $q(\xi)$ turn out to be known along lines joining an arbitrary pair of points $x, x^0 \in S$. From the results of Chapter VI, §5, it follows that $q(\xi)$ is uniquely determined inside S by these integrals.

3. We shall consider an inverse problem for the simplest equation of elliptic type. For this we use the wave equation. Suppose that at a point x^0 of three-dimensional space there is a source of periodic oscillations with angular frequency ω. The wave process is then described by the equation

$$q(x)u_{tt} = \Delta u + 4\pi\delta(x-x^0)e^{i\omega t}, \quad q(x) \geq q_0 > 0.$$

During the course of time, a periodic regime of oscillations with frequency ω is established in space, i.e.,

$$u(x,t,x^0) = v(x,x^0,\omega)e^{i\omega t}.$$

The function $v(x, x^0, \omega)$ satisfies the equation of elliptic form

$$\Delta v + \omega^2 q(x) v = -4\pi\delta(x - x^0) \tag{7.16}$$

and a radiation condition at infinity. The inverse problem for equation (7.16) can here be formulated as the problem of finding $q(x)$ from the function $v(x, x^0, \omega)$ given on some manifold. We shall give a more precise formulation of the problem a little later. We now suppose that the function $q(x)$ can be represented in the form

$$q(x) = q_0 + \tilde{q}(x), \quad q_0 > 0,$$

where q_0 is a given constant, and $\tilde{q}(x)$ satisfies the inequality

$$|\tilde{q}(x)| \leq \frac{C}{(1 + |x|)^{1+\alpha}}, \quad C > 0, \ \alpha > 0.$$

The problem of finding the function $v(x, x^0, \omega)$ from (7.16) can be reduced to solving an integral equation by inverting the Helmholtz operator with wave number $k = \omega\sqrt{q_0}$:

$$v(x, x^0, \omega) = \frac{\exp\left(i\omega\sqrt{q_0}|x - x^0|\right)}{|x - x^0|} + \omega^2 \int_{R^3} \frac{\exp\left(i\omega\sqrt{q_0}|x - \xi|\right)}{|x - \xi|} \tilde{q}(\xi) v(\xi, x^0, \omega)\, d\xi. \tag{7.17}$$

Hence, obviously,

$$\lim_{\omega \to 0} \frac{1}{\omega^2} \left[v(x, x^0, \omega) - \frac{\exp\left(i\omega\sqrt{q_0}|x - x^0|\right)}{|x - x^0|} \right] = \int_{R^3} \frac{\tilde{q}(\xi)\, d\xi}{|x - \xi|\, |x^0 - \xi|}. \tag{7.18}$$

We shall now refine the formulation of the inverse problem. Suppose the function $v(x, x^0, \omega)$ is known for x and x^0 belonging to some fixed set m in the space x, x^0 for values of ω arbitrarily close to zero; it is required to find $\tilde{q}(x)$. In this case we can compute the left side of (7.18) on the set m and after this use (7.18) to find $\tilde{q}(x)$. Thus, here the problem of constructing $q(x)$ reduces to solving a Fredholm integral equation of the first kind.

In particular, if m is a product of regions D_1 and D_2 of the space x, where $x \in D_1$, $x^0 \in D_2$, and $\operatorname{supp} q(x) \subset D_3$, where D_3 is a region having no common points with D_1 and D_2, then the continuous function $\tilde{q}(x)$ is uniquely determined from (7.18) [147].

It turns out that the problem of determining the coefficients of a linear parabolic equation depending only on spatial variables also reduces to problems of investigating Fredholm integral equations of the first kind. At the same time, the structure of the kernels of these equations is such that in certain cases they can be reduced to problems of integral geometry (see the next section). If the coefficients to be determined depend on the time coordinate t, then the equations arising are operator Volterra equations of the first kind.

§2. A general approach to investigating questions of uniqueness and stability of inverse problems

In this section we present a general method for reducing questions of uniqueness and stability in inverse problems to questions of determining the coefficients of linear differential equations in a certain auxiliary linear inverse problem. The method is not new; it is usually used in investigating classical problems for nonlinear differential equations.

1. We consider the equation

$$u_{tt} - L_q u = f(x,t), \tag{7.19}$$

in which L_q is a uniformly elliptic operator of the form

$$L_q u = \sum_{i,j=1}^{3} a_{ij}(x) u_{x_i x_j} + \sum_{i=1}^{3} b_i(x) u_{x_i} + c(x) u, \qquad a_{ij} = a_{ji},$$

$$0 < \mu_0 \leq \sum_{i,j=1}^{3} a_{ij}(x) \alpha_i \alpha_j \leq M_0 < \infty, \qquad \sum_{i=1}^{3} \alpha_i^2 = 1. \tag{7.20}$$

Here q denotes the ordered collection of coefficients of the differential operator:

$$q = (a_{11}, a_{12}, a_{13}, a_{22}, a_{23}, a_{33}, b_1, b_2, b_3, c).$$

Thus, giving L_q is equivalent to giving the vector q. We consider a set Q consisting of elements q, and pose the problem of distinguishing in the set of operators L_q, $q \in Q$, that operator which corresponds to the given information regarding solutions of (7.19). Rather than going into a very general situation, we consider a special case of giving such information. Suppose, to be specific, that all components of the vector q are unknown; from general considerations it is then clear that to determine all components of q (in the present case there are ten) it is necessary to provide information of no less dimension than the dimension of q. For example, it is possible to consider as many solutions of (7.19) as there are components of q and prescribe them on some three-dimensional manifold, or to consider a family of solutions of (7.19) depending on one or several parameters $\lambda = (\lambda_1, \dots, \lambda_m)$ and prescribe them on some three-dimensional manifold in the space x, t, λ. Concrete formulations of inverse problems are still more varied. To be specific, we shall consider the situation where as many solutions of (7.19) are given as there are components of q. Suppose these solutions for equation (7.19) are constructed as solutions of the Cauchy problem corresponding to different initial data and, generally speaking, to different right sides. In this case it is convenient to introduce the vector-valued function $U = (u_1, \dots, u_{10})$ whose components consist of solutions of (7.19) corresponding to different initial data. For the function U we have the decomposing system of equations

$$U_{tt} - L_q U = F(x,t) \tag{7.21}$$

with Cauchy data

$$U|_{t=0} = U_0(x), \quad U_t|_{t=0} = U_1(x). \tag{7.22}$$

Suppose the solution of problem (7.21), (7.22) is known on some manifold $\mathfrak{M}$ of the space x, t:

$$U|_{\mathfrak{M}} = \varphi(x,t); \tag{7.23}$$

it is required to find $q \in Q$ on the basis of the function $\varphi(x,t)$. The functions U_0, U_1, and F are here assumed known.

LEMMA 1. *Investigation of questions of uniqueness and stability of the inverse problem* (7.21)–(7.23) *on the basis of the data* (7.23) *reduces to investigation of analogous questions for the problem of determining the function* $\tilde{q}(x)$ *from the relations*

$$\tilde{U}_{tt} - L_q \tilde{U} = \tilde{q}(x) R_{\hat{q}}(x,t), \tag{7.24}$$

$$\tilde{U}|_{t=0} = 0, \ \tilde{U}_t|_{t=0} = 0, \tag{7.25}$$

$$\tilde{U}|_{\mathfrak{M}} = \tilde{\varphi}(x,t), \tag{7.26}$$

in which L_q *is the given operator* ($q \in Q$); $\tilde{\varphi}(x,t)$ *and* $R_{\hat{q}}(x,t)$ *are given functions* ($R_{\hat{q}}$ *is a matrix*).

Indeed, if (q, U) and $(\hat{q}, \hat{U})$ are two solutions of problem (7.21)–(7.23) corresponding to the same functions F, U_0, and U_1 but different data (7.23): φ and $\hat{\varphi}$, then, setting $\tilde{q} = q - \hat{q}$, $\tilde{\varphi} = \varphi - \hat{\varphi}$, and $\tilde{U} = U - \hat{U}$, we obtain (7.24)–(7.26) in which

$$L_{\tilde{q}} \hat{U} \equiv \tilde{q}(x) R_{\hat{q}}(x,t). \tag{7.27}$$

The matrix $R_{\hat{q}}$ depends on $\hat{U}$, i.e., on the final analysis of $\hat{q} \in Q$ and the functions U_0, U_1, and F.

If for problem (7.24)–(7.26) an estimate of stability of $\tilde{q}$ in terms of $\tilde{\varphi}$ is known, then this stability estimate is thus also an estimate of stability on the basis of data (7.23) of problem (7.21)–(7.23).

Thus, in the sense of investigating questions of uniqueness and stability, the problem of determining the differential operator L_q reduces to the problem of determining a special right side of the differential equation. This is also an inverse problem, but it is now linear. Questions of its stability are to a considerable extent connected with the function $R_{\hat{q}}(x,t)$. An altogether natural and minimal condition here is the requirement $\det R_{\hat{q}}(x,t) \neq 0$. Actually, even in questions of whether a solution is unique we must impose more stringent conditions on $R_{\hat{q}}(x,t)$. An altogether natural condition, for example, is that $\det R_{\hat{q}}(x,t)$ be nonzero at all points of some smooth surface of the form $t = t(x)$. Under particular conditions in place of $R_{\hat{q}}(x,t)$, its partial derivative with respect to t of some order may satisfy an analogous condition.

Problem (7.24)–(7.26) is a problem of integral geometry. Indeed, we invert the operator $(\tilde{U}_{tt} - L_q\tilde{U})$, using the fundamental solution $G_q(x,t;\xi,\tau)$ of the Cauchy problem for equation (7.24). We then find that the problem of finding $\tilde{q}(x)$ from conditions (7.24)–(7.26) is equivalent to solving the equation

$$\int_{D(x,t)} \tilde{q}(\xi)R_{\hat{q}}(\xi,\tau)G_q(x,t;\xi,\tau)\,d\xi\,d\tau = \tilde{\varphi}(x,t), \quad (x,t) \in \mathfrak{M}. \tag{7.28}$$

Here $D(x,t)$ is the closed region bounded by a piece of the plane $\tau = 0$ and the characteristic conoid with vertex at the point (x,t) and having its lateral surface directed toward the plane $\tau = 0$. In the present case the set of points of the region $D(x,t)$ can be described by the inequality $0 \le \tau \le t - \tau_q(x,\xi)$, where $\tau_q(x,\xi)$ is the distance between the points x and ξ in the Riemannian metric in which the element $d\tau$ of arc length is computed by the formula

$$d\tau = \sqrt{\sum_{i,j=1}^{3} b_{ij}(x)\,dx_i\,dx_j},$$

where the $b_{ij}(x)$ give rise to the matrix B inverse to the matrix of coefficients a_{ij}, $1 \le i,\ j \le 3$, of the operator L_q. From another point of view $\tau_q(x,\xi)$ is the solution of the first-order characteristic equation

$$\sum_{i,j=1}^{3} a_{ij}(x)\tau_{x_i}\tau_{x_j} = 1,$$

with Cauchy data corresponding to the degenerate case of a surface contracted to the point ξ:

$$\tau = O(|x-\xi|), \quad x \to \xi.$$

Using the fact that the fundamental solution $G_q(x,t;\xi,\tau)$ in three-dimensional space has the structure (see, for example, [217] and [223])

$$G_q(x,t;\xi,\tau) = [\sigma_q(x,\xi)\delta(t-\tau-\tau_q(x,\xi)) + G_q^0(x,\xi,t-\tau)\varepsilon(t-\tau-\tau_q(x,\xi))],$$

where $\varepsilon(t)$ is the Heaviside function,

$$\varepsilon(t) = \begin{cases} 1, & t \ge 0, \\ 0, & t < 0, \end{cases}$$

equation (7.28) can be brought to the form

$$\int_{\tau_q(x,\xi)\le t} \tilde{q}(\xi) R_{q\hat{q}}(\xi, x, t)\, d\xi = \tilde{\varphi}(x,t), \quad (x,t) \in \mathfrak{M}, \tag{7.29}$$

in which the weight function $R_{q\hat{q}}(\xi, x, t)$ is computed from the formula

$$R_{q\hat{q}}(\xi, x, t) = R_{\hat{q}}(\xi, t - \tau_q(x,\xi))\sigma_q(x,\xi) + \int_{\tau_q(x,\xi)}^{t} G_q^0(x,\xi,\tau) R_{\hat{q}}(\xi, t-\tau)\, d\tau. \tag{7.30}$$

Questions of uniqueness and stability of the solution of (7.29) depend strongly on the set $\mathfrak{M}$ and properties of the weight function $R_{q\hat{q}}(\xi, x, t)$; the latter are determined by properties of the functions U_0, U_1, F and the set Q to which q and $\hat{q}$ belong.

We shall consider several cases among those which may occur in the study of inverse problems. We shall assume here that the set $\mathfrak{M}$ is the direct product of a surface S in x-space with a segment $[0, T]$ of the t-axis. This assumption enables us to compute the derivative with respect to t of both sides of (7.29).

We additionally assume that the functions U_0, U_1 , F and the set Q are such that the solution of problem (7.21), (7.22) is sufficiently smooth.

1) $U_0(x)$ satisfies the condition

$$\det R_{\hat{q}}(x, 0) \neq 0. \tag{7.31}$$

The fact that (7.31) is a restriction only on $U_0(x)$ follows from (7.27) for $t = 0$. If (7.31) is satisfied, by differentiating (7.29) with respect to t, we reduce it to the form

$$\int_{\tau_q(x,\xi)=t} \tilde{q}(\xi) R^0_{q\hat{q}}(\xi, x)\, dS + \int_{\tau_q(x,\xi)\le t} \tilde{q}(\xi) R^1_{q\hat{q}}(\xi, x, t)\, d\xi = \tilde{\varphi}_t(x,t), \quad (x,t) \in \mathfrak{M}, \tag{7.32}$$

in which

$$R^0_{q\hat{q}}(\xi, x) = R_{\hat{q}}(x, 0)\sigma_q(x,\xi), \tag{7.33}$$

$$R^1_{q\hat{q}}(\xi, x, t) = \sigma_q(x,\xi)\frac{\partial}{\partial t} R_{\hat{q}}[\xi, t - \tau_q(x,\xi)] + \int_{\tau_q(x,\xi)}^{t} \left[\frac{\partial}{\partial t} R_{\hat{q}}(\xi, t-\tau)\right] G_q^0(x,\xi,\tau)\, d\tau. \tag{7.34}$$

Since $\sigma_q(x,\xi) \neq 0$ (see [217]), from (7.33) it follows that $\det R^0_{q\hat{q}}(x,\xi) \neq 0$. Equation (7.32) is the normal form of writing the problem of integral geometry when there is given a linear combination of integrals of the function $\tilde{q}$ over surfaces and the regions they bound. The condition $\det R^0_{q\hat{q}}(\xi, x) \neq 0$ on the weight function, as was found in Chapter VI, is a sufficient condition for the investigation of questions of uniqueness and stability of the solution of problem (7.32). Because of this, equation (7.32) can be investigated for each specific $\mathfrak{M}$ by the methods of Chapter VI.

2) $U_0(x) \equiv 0$, and $U_1(x)$ satisfies the condition

$$\det\left(\frac{\partial}{\partial t}R_{\hat{q}}(x,t)\bigg|_{t=0}\right) = 0. \tag{7.35}$$

In this case $R_{\hat{q}}(x,0) \equiv 0$, and hence $R^0_{q\hat{q}}(\xi,x) \equiv 0$. Equation (7.32) has precisely the same form as (7.29), only in place of $R_{q\hat{q}}(\xi,x,t)$ it contains

$$R^1_{q\hat{q}}(\xi,x,t) = \frac{\partial}{\partial t}R_{q\hat{q}}(\xi,x,t).$$

Differentiating (7.32) with respect to t, we obtain an equation of the same type in which

$$R^0_{q\hat{q}}(\xi,x) = \sigma_q(x,\xi)\frac{\partial}{\partial t}R_{\hat{q}}(x,t)|_{t=0}, \tag{7.36}$$

$$\begin{aligned} R^1_{q\hat{q}}(\xi,x,t) &= \sigma_q(x,\xi)\frac{\partial^2}{\partial t^2}R_{\hat{q}}(\xi,t-\tau_q(x,\xi)) \\ &\quad + \int_{\tau_q(x,\xi)}^{t}\left[\frac{\partial^2}{\partial t^2}R_{\hat{q}}(\xi,t-\tau)\right]G^0_q(x,\xi,\tau)\,d\tau, \end{aligned} \tag{7.37}$$

while the right side contains $\tilde{\varphi}_{tt}$ in place of $\tilde{\varphi}_t$. Condition (7.35) leads to the inequality $\det R^0_{q\hat{q}}(x,\xi) \neq 0$.

3) $U_0 \equiv 0$ and $U_1 \equiv 0$; $F(x,0)$ satisfies the condition

$$\det\left(\frac{\partial^2}{\partial t^2}R_{\hat{q}}(x,t)|_{t=0}\right) \neq 0. \tag{7.38}$$

In this case $R_{\hat{q}}(x,0) = \partial R_{\hat{q}}(x,t)/\partial t|_{t=0} \equiv 0$. Differentiating (7.29) three times with respect to t, we obtain an equation whose left side coincides with (7.32), while $R^0_{q\hat{q}}$ and $R^1_{q\hat{q}}$ are computed from formulas analogous to (7.36) and (7.37) with the replacement of the first derivative of $R_{\hat{q}}$ with respect to t by the second derivative and replacement of the second derivative by the third. Under condition (7.38) we again have $\det R^0_{q\hat{q}} \neq 0$.

The more general situation, $U_0 = U_1 \equiv 0$, $F(x,0) = 0$, $\partial^k F(x,t)/\partial t^k|_{t=0} = 0$, $k = 1,\dots,n-1$, while $\partial^n F(x,t)/\partial t^n|_{t=0}$ satisfies the condition

$$\det\left(\frac{\partial^{n+2}}{\partial t^{n+2}}R_{\hat{q}}(x,t)|_{t=0}\right) \neq 0,$$

is investigated similarly.

The case where the Cauchy data or the right side of (7.21) are generalized functions (corresponding, for example, to a source of oscillations concentrated at a fixed point of space) is of major interest for practical applications. In this case $R_{\hat{q}}(x,t)$ is also a genearlized function. Accordingly, the weight function $R_{q\hat{q}}(\xi,x,t)$ in (7.29) must be considered a generalized function. In connection with this, instead of integral geometry on surfaces $\tau_q(x,\xi)=t$, we may obtain integral geometry on other manifolds, for example, on ellipsoids [217].

2. We now consider the equation of parabolic type

$$u_t - L_q u = f(x,t), \tag{7.39}$$

in which the operator L_q is defined by (7.20). For equation (7.39) we pose the problem of finding L_q, $q \in Q$, analogous in formulation to problem (7.21)–(7.23): the vector-valued function U, which is a solution of the Cauchy problem

$$U_t - L_q U = F(x,t), \tag{7.40}$$

$$U|_{t=0} = U_0(x), \tag{7.41}$$

is given on some manifold $\mathfrak{M}$,

$$U|_{\mathfrak{M}} = \varphi(x,t). \tag{7.42}$$

It is required to find L_q, $q \in Q$, on the basis of φ.

With regard to the inverse problem (7.40)–(7.42) there is a lemma altogether analogous to Lemma 1.

LEMMA 2. *Investigation of questions of uniqueness and stability of the inverse problem* (7.40)–(7.42) *on the basis of the data* (7.42) *reduces to the investigation of analogous questions for the problem of determining the function* $\tilde{q}(x)$ *from the relations*

$$\tilde{U}_t - L_q\tilde{U} = \tilde{q}(x)R_{\hat{q}}(x,t), \tag{7.43}$$

$$\tilde{U}|_{t=0} = 0, \tag{7.44}$$

$$\tilde{U}|_{\mathfrak{M}} = \tilde{\varphi}(x,t), \tag{7.45}$$

in which L_q *is the given operator* ($q \in Q$); $\tilde{\varphi}(x,t)$ *and* $R_{\hat{q}}(x,t)$ *are given functions.*

The proof of this lemma is practically no different than the proof of Lemma 1. The formula for computing $R_{\hat{q}}(x,t)$ coincides with (7.27).

With the help of the fundamental solution $H_q(x,t;\xi,\tau)$ of equation (7.39), the linear problem (7.43)–(7.45) can be reduced to a Fredholm equation of the first kind

$$\int_{R^3} \tilde{q}(\xi)R_{q\hat{q}}(\xi,x,t)\,d\xi = \tilde{\varphi}(x,t), \quad (x,t)\in\mathfrak{M}, \tag{7.46}$$

in which

$$R_{q\hat{q}}(\xi, x, t) = \int_0^t R_{\hat{q}}(\xi, \tau) H_q(x, t; \xi, \tau)\, d\tau. \tag{7.47}$$

Thus, the inverse problem (7.40)–(7.42) reduces to the Fredholm equation (7.46).

We mention a case where the investigation of (7.46) can be reduced to the study of the problem (7.29) of integral geometry. Suppose $\mathfrak{M} = S \times [0, \infty]$, where S is a surface in x-space. We restrict ourselves to considering only a class Q of differential operators $(\partial/\partial t - L_q)$ and functions $U_0(x)$ and $F(x, t)$ to which there correspond solutions growing as $t \to \infty$ no faster than exponentially:

$$|U(x, t)| \le Ce^{\lambda^2 t}, \qquad |U_{x_i x_j}(x, t)| \le Ce^{\lambda^2 t}. \tag{7.48}$$

In this case the Laplace transform in the variable t can be applied to both sides of (7.46). Denoting by $\overline{f}(p)$ the Laplace transform of the function $f(t)$, using the fact that the function $H_q(x, t; \xi, \tau)$ in the case $q = q(x)$ depends only on the difference $t - \tau$,

$$H_q(x, t; \xi, \tau) \equiv H_q(x, t - \tau; \xi, 0),$$

and applying the Laplace transform to (7.46), we obtain

$$\int_{R^3} \tilde{q}(\xi) \overline{R}_{\hat{q}}(\xi, p) \overline{H}_q(x, p; \xi, 0)\, d\xi = \overline{\varphi}(x, p), \quad x \in S. \tag{7.49}$$

We now use the rather obvious relation between the Laplace transforms of the fundamental solutions of the operator $(\partial/\partial t - L_q)$ and the operator $(\partial^2/\partial t^2 - L_q)$:

$$\overline{H}_q(x, p^2; \xi, 0) = \overline{G}_q(x, p; \xi, 0). \tag{7.50}$$

Replacing p by p^2 in (7.49) and using (7.50), we find that

$$\int_{R^3} \tilde{q}(\xi) \overline{R}_q(\xi, p^2) \overline{G}_q(x, p; \xi, 0)\, d\xi = \overline{\varphi}(x, p^2), \quad x \in S. \tag{7.51}$$

We note that the functions $\overline{R}_q(\xi, p^2)$, $\overline{G}_p(x, p; \xi, 0)$, and $\overline{\varphi}(x, p^2)$ are analytic in the region $\operatorname{Re} p^2 \ge \lambda^2$, i.e., in the region of the complex plane $p = \sigma + ir$ lying inside the hyperbola $r = \pm\sqrt{\sigma^2 - \lambda^2}$. For a function $f(t)$ with Laplace transform $\overline{f}(p)$ analytic for $\operatorname{Re} p^2 \ge \lambda^2$, we have

$$f(t) = \lim_{n \to \infty} \left[\frac{(-1)^n}{n!} p^{n+1} \frac{d^n}{dp^n} \overline{f}(p) \right]_{p = n/t}. \tag{7.52}$$

Applying the inverse Laplace transform (7.52) in the variable p to both sides of (7.51), we obtain

$$\int_{R^3} \tilde{q}(\xi) \hat{R}_{q\hat{q}}(\xi, x, t)\, d\xi = \hat{\varphi}(x, t), \quad (x, t) \in \mathfrak{M}, \tag{7.53}$$

in which

$$\hat{R}_{q\hat{q}}(\xi, x, t) = \int_0^t \hat{R}_{\hat{q}}(\xi, t-\tau) G_q(x, \tau; \xi, 0)\, d\tau,$$

$$\hat{R}_{\hat{q}}(\xi, t) = \lim_{n\to\infty} \left[\frac{(-1)^n}{n!} p^{n+1} \frac{d^n}{dp^n} \int_0^\infty e^{-p^2\tau} R_{\hat{q}}(\xi, \tau)\, d\tau \right]_{p=n/t},$$

$$\hat{\varphi}(x, t) = \lim_{n\to\infty} \left[\frac{(-1)^n}{n!} p^{n+1} \frac{d^n}{dp^n} \int_0^\infty e^{-p^2\tau} \tilde{\varphi}(x, \tau)\, d\tau \right]_{p=n/t}. \tag{7.54}$$

Using the structure of the fundamental solution $G_q(x, \xi; t, \tau)$, with the help of the first of formulas (7.54) we find that $\hat{R}_{q\hat{q}}(\xi, x, t) \equiv 0$ for $t < \tau_q(x, \xi)$. Equation (7.53) thus reduces to the form

$$\int_{\tau_q(x,\xi)\le t} \tilde{q}(\xi) \hat{R}_{q\hat{q}}(\xi, x, t)\, d\xi = \hat{\varphi}(x, t), \quad (x, t) \in \mathfrak{M}, \tag{7.55}$$

which differs from (7.29) only in the weight function.

§3. Inverse problems for hyperbolic equations of second order

In this section we consider two formulations of the inverse problem for the equation

$$u_{tt} = L_q u, \tag{7.56}$$

in which the operator L_q is defined by (7.20). Investigation of the problems will be based on using the fundamental solution for equation (7.56). We wrote out the structure of the fundamental solution earlier. Here we shall require finer properties of the function G_q, namely, the connection between the smoothness of the coefficients of the operator L_q and the differential properties of the function G_q. In particular, we shall be interested in the behavior of the derivatives of G_q inside the characteristic cone up to its boundary. In this chapter we restrict ourselves only to the case of a constant principal part of the operator L_q; namely, we shall assume that $a_{ij}(x) = \delta_{ij}$. Under this assumption, analysis of the differential properties of the function G_q presents no special difficulties. The content of the present section is based on [67], [68], [217], [247], and [296].

1. We recall the structure of the functions G_q for the case where the coefficients of the operator L_q do not depend on t:

$$G_q(x, t; \xi, \tau) = [\sigma_q(x, \xi)\delta(t - \tau - \tau_q(x, \xi)) + G_q^0(x, \xi, t-\tau)\varepsilon(t - \tau - \tau_q(x, \xi))]. \tag{7.57}$$

We henceforth omit the index q in investigating the properties of the function G_q. Moreover, since below we consider an operator L_q with fixed principal part and the function $\tau_q(x, \xi)$ is defined precisely by the coefficients of the principal part of the operator L_q, the index q on the function $\tau(x,\xi)$ is not needed at all. In the present case $\tau(x,\xi) = |x-\xi|$, while the function $\sigma(x,\xi)$ is defined by (see, for example, [217], Chapter II, §1)

$$\sigma(x,\xi) = \frac{1}{|x-\xi|}\exp\left[\frac{1}{2}\int_0^1 \sum_{i=1}^3 b_i(\xi + t(x-\xi))(x_i-\xi_i)\,dt\right]. \tag{7.58}$$

As is evident from this formula, the product $\sigma(x,\xi)\tau(x,\xi)$ is a smooth function if the $b_i(x)$ are smooth functions; namely, if $b_i(x) \in C^s(\overline{D})$, then $(\sigma\cdot\tau) \in C^s(\overline{D}\times\overline{D})$. We note that $\sigma(x,\xi)$ also has a similar property in the case of variable coefficients $a_{ij}(x)$ if $a_{ij}(x) \in C^{s+2}(\overline{D})$ and the central field of rays corresponding to the function $\tau(x,\xi)$ is regular. For arbitrary, sufficiently smooth coefficients there is also the equality

$$L^*(\sigma(x,\xi)\tau(x,\xi)) = \tau(x,\xi)L^*\sigma(x,\xi), \tag{7.59}$$

which in the present case can easily be verified directly. Here L^* denotes the operator which is adjoint in the Lagrange sense to L. Equality (7.59) shows that the product $\tau(x,\xi)L^*\sigma(x,\xi)$ is a smooth function of smoothness s if $b_i \in C^{s+2}(\overline{D})$ and $c(x) \in C^s(\overline{D})$.

We now proceed to the analysis of the function $G^0(x,\xi,t-\tau)$. It satisfies the integral equation (see, for example, [217], Chapter II, §1, formula (15))

$$G^0(x,\xi,t) = \frac{1}{4\pi}\int_{D(x,\xi,t)} \sigma(\eta,\xi)L^*_\eta\sigma(\eta,x)\delta[t-\tau(\eta,x)-\tau(\eta,\xi)]\,d\eta$$
$$+\frac{1}{4\pi}\int_{D(x,\xi,t)} G^0[\eta,\xi,t-\tau(\eta,x)]L^*_\eta\sigma(\eta,x)\,d\eta, \tag{7.60}$$

in which $\eta = (\eta_1,\eta_2,\eta_3)$ is the variable point of integration, and $D(x,\xi,t)$ is the region in η-space bounded by the surface $S(x,\xi,t) = \{\eta\colon \tau(\eta,x)+\tau(\eta,\xi) = t\}$.

Singularities in the structure of $G^0(x,\xi,t)$ are connected only with singularities of the first of the integrals in (7.60). We transform it to a more convenient form. We note that we were concerned with the transformation of a similar integral in §1. In the present case, to analyze the integral, it is convenient for us to represent it in a coordinate system connected with the points x and ξ. Together with the coordinate system η we introduce a Cartesian system η' whose origin is placed at the point ξ and such that the η'_3-axis is directed along the line joining the points x and ξ in the direction of x. We denote the unit vector $(x-\xi)/|x-\xi|$ by $\nu^0(x,\xi)$. We use a system in which

η and η' are connected by the transformation equations

$$\eta = \xi + Q\eta',$$

$$Q = \left\| \begin{array}{ccc} \cos\theta_0 \cos\varphi_0 & -\sin\theta_0 \sin\varphi_0 & \sin\theta_0 \cos\varphi_0 \\ \cos\theta_0 \sin\varphi_0 & \sin\theta_0 \cos\varphi_0 & \sin\theta_0 \sin\varphi_0 \\ -\sin\theta_0 & 0 & \cos\theta_0 \end{array} \right\|. \tag{7.61}$$

Here θ_0, φ_0 are the angular coordinates of the vector ν^0 in the spherical coordinate system connected with the Cartesian coordinate system η in the usual manner.

We associate with the coordinate system η' the spherical coordinate system r, θ, φ in the usual way. Comparing the first of the integrals of (7.60) with the integrals defined by (7.8) and (7.15), and using the invariance of the volume element $d\eta$ and the distances $\tau(\eta, x)$ and $\tau(\eta, \xi)$ under the transformation (7.61), we find that

$$J(x, \xi, t) = \frac{1}{4\pi} \int_{D(x,\xi,t)} \sigma(\eta, \xi) L_\eta^* \sigma(\eta, x) \delta[t - \tau(\eta, x) - \tau(\eta, \xi)] \, d\eta$$

$$= \frac{1}{\tau(x, \xi)} \int_0^{2\pi} d\varphi \int_{\frac{t-\tau(x,\xi)}{2}}^{\frac{t+\tau(x,\xi)}{2}} P(x, \xi, \eta) \, dr.$$

Here,

$$P(x, \xi, \eta) = \frac{1}{4\pi} \tau(\eta, \xi) \sigma(\eta, \xi) \tau(\eta, x) L_\eta^* \sigma(\eta, x)$$

is a smooth function of the variables x, ξ, and η belonging to $C^s(\overline{D} \times \overline{D} \times \overline{D})$ if $b_i \in C^{s+2}(\overline{D})$ and $c \in C^3(\overline{D})$. The variable point η is connected with η' by (7.61). In turn

$$\eta' = r(\sin\theta \cos\varphi, \sin\theta \sin\varphi, \cos\theta),$$

where θ and r are connected by the single-valued relation

$$r = \frac{t^2 - \tau^2(x, \xi)}{2(t - \tau(x, \xi) \cos\theta)}.$$

In the last of the integrals we change the variable of integration from r to the variable z:

$$r = (1/2)(t + \tau(x, \xi) z).$$

Then

$$J(x, \xi, t) = \frac{1}{2} \int_0^{2\pi} d\varphi \int_{-1}^{1} P(x, \xi, \eta) \, dz, \tag{7.62}$$

and η' can be expressed in terms of the variables of integration φ, z by

$$\eta_1' = \frac{1}{2} \sqrt{t^2 - \tau^2(x, \xi)} \sqrt{1 - z^2} \cos\varphi,$$

$$\eta_2' = \frac{1}{2} \sqrt{t^2 - \tau^2(x, \xi)} \sqrt{1 - z^2} \sin\varphi, \qquad \eta_3' = \frac{1}{2} [tz + \tau(x, \xi)]. \tag{7.63}$$

Formulas (7.61)–(7.63) enable us to completely characterize the analytic structure of the function $J(x,\xi,t)$ and hence also the function $G^0(x,\xi,t)$. Namely, $G^0(x,\xi,t)$ is a jointly smooth function in arguments x, ξ, t in that region of the variables for which $S(x,\xi,t) \subset \overline{D}$ and $\tau(x,\xi) \neq 0$, $\tau(x,\xi) \neq t$. In this region it has s continuous partial derivatives with respect to all arguments provided only that $b_i(x) \in C^{s+2}(\overline{D})$ and $c(x) \in C^3(\overline{D})$. Here the function $G^0(x,\xi,t)$ itself is continuous for each fixed ξ up to the boundary of the cone $\tau(x,\xi) = t$, while its partial derivatives have singularities determined by the singularities of the corresponding derivatives of the functions $\nu^0(x,\xi)\sqrt{t^2 - \tau^2(x,\xi)}$ and $\nu^0(x,\xi)$, namely,

$$|D^\alpha_{x,\xi,t} G^0| \leq K_1 |D^\alpha_{x,\xi,t} \nu^0 \sqrt{t^2 - \tau^2(x,\xi)}| + K_2 |D^\alpha_{x,\xi} \nu^0(x,\xi)|, \quad |\alpha| \leq s, \tag{7.64}$$

where K_1 and K_2 are constants depending only on the norms of the $b_i(x)$ and $c(x)$ in the space $C^s(\overline{D})$.

2. We now proceed to the formulation and analysis of inverse problems for equation (7.56). As already mentioned earlier, here we shall assume that $a_{ij} = \delta_{ij}$, $i,j = 1,2,3$. In this case the operator L_q is determined by the coefficients $(b_1, b_2, b_3, c) = q$, so that the problem of finding the operator L_q consists in determining one or several of the coefficients b_i, c. We shall here consider two formulations of the problem of determining all the coefficients of the operator L_q. These formulations are essentially a specialization of the formulations of §2.

PROBLEM 1. Find the coefficients $b_i(x)$ and $c(x)$ of the operator L_q if solutions $u_i(x,t)$, $i = 1,2,3,4$, of the Cauchy problems for equation (7.56) with data

$$u|_{t=0} = \varphi_i(x), \qquad u_t|_{t=0} = \psi_i(x), \quad i = 1,2,3,4, \tag{7.65}$$

are known on the plane $x_3 = 0$:

$$u_i|_{x_3=0} = f_i(x_1, x_2, t), \quad t > 0, \ i = 1,2,3,4. \tag{7.66}$$

It was shown in §2 that this problem reduces to the investigation of a problem of integral geometry. In the present case the set $\mathfrak{M}$ is the plane $x_3 = 0$ in the space x, t. Let $\Phi(x)$ be the matrix

$$\Phi(x) = \left\| \begin{matrix} (\varphi_1)_{x_1} & (\varphi_1)_{x_2} & (\varphi_1)_{x_3} & \varphi_1 \\ \multicolumn{4}{c}{\cdots\cdots\cdots\cdots\cdots\cdots} \\ (\varphi_4)_{x_1} & (\varphi_4)_{x_2} & (\varphi_4)_{x_3} & \varphi_4 \end{matrix} \right\|,$$

let $D_H = \{x \colon |x_3| \leq H,\ H > 0\}$, and let $D_{Hh} = \{x \colon h \leq x_3 \leq H,\ 0 < h < H\}$. We say that a function $g(x)$ belongs to the set $M(H,h)$ if in the region D_{Hh}

it can be represented as a finite sum of the form

$$g(x) = \sum_{k=1}^{N} g_k(x_1, x_2) r_k(x_3).$$

THEOREM 1. *Suppose that the coefficients $b_i(x) \in C^3(D_H)$ and $c(x) \in C^1(D_H)$ are known in the region $D_H \backslash D_{Hh}$. If $\varphi_i(x) \in C^3(D_H)$, $\psi_i(x) \in C^2(D_H)$, and*

$$|\det \Phi(x)| \geq \mu > 0, \quad x \in D_{Hh}, \tag{7.67}$$

then in the region D_{Hh}, coefficients b_i and c belonging to the set $M(H,h)$ are uniquely determined by giving the functions $f_i(x_1, x_2, t)$, $i = 1,2,3,4$, for values $t \in [h, H]$.

To prove the theorem we reduce problem (7.56), (7.65), (7.66) according to the scheme of §2 to the investigation of an auxiliary problem of integral geometry. To this end we introduce the two operators L_q and $L_{\hat{q}}$, the solutions in u_i and $\hat{u}_i$ of problems (7.56), (7.65) corresponding to them, and the differences $q - \hat{q} = \tilde{q}$ and $u_i - \hat{u}_i = \tilde{u}_i$. In accordance with the notation of §2 we set

$$R_{\hat{q}}(x,t) = \left\| \begin{matrix} (\hat{u}_1)_{x_1} & (\hat{u}_1)_{x_2} & (\hat{u}_1)_{x_3} & \hat{u}_1 \\ \cdots & \cdots & \cdots & \cdots \\ (\hat{u}_4)_{x_1} & (\hat{u}_4)_{x_2} & (\hat{u}_4)_{x_3} & \hat{u}_4 \end{matrix} \right\|.$$

The auxiliary problem of integral geometry then has the form (see (7.32))

$$\int_{\tau(x,\xi)=t} \tilde{q}(\xi) R^0_{q\hat{q}}(\xi, x)\, d\xi + \int_{\tau(x,\xi)\leq t} \tilde{q}(\xi) R^1_{q\hat{q}}(\xi, x, t)\, d\xi = 0,$$
$$(x,t) \in \mathfrak{M},\ h \leq t \leq H, \tag{7.68}$$

where

$$\begin{aligned} R^0_{q\hat{q}}(\xi, x) &= \Phi(x) \sigma_q(x, \xi), \\ R^1_{q\hat{q}}(\xi, x, t) &= \sigma_q(x,\xi) \frac{\partial}{\partial t} R_{\hat{q}}(\xi, t - \tau(x,\xi)) \\ &\quad + \int_{\tau(x,\xi)}^{t} \frac{\partial}{\partial t} R_{\hat{q}}(\xi, t-\tau) G^0_q(x, \xi, \tau)\, d\tau. \end{aligned} \tag{7.69}$$

We observe that the actual integration (7.68) goes only over that portion of the surface and interior of the sphere $\tau(x,\xi) = t$ which is situated in the region D_{Hh}, since $\tilde{q} \equiv 0$ outside D_{Hh}. Here the point $\xi \in D_{Hh}$ and $x \in \{x \colon x_3 = 0\}$. Hence, $\tau(x,\xi) \geq h > 0$, and the function $\sigma_q(x,\xi)$ is bounded for any x and ξ. Moreover, it follows from (7.58) that $\sigma_q(x,\xi) \geq \beta_q > 0$. Because of this, $|\det R^0_{q\hat{q}}| \geq \alpha \beta_q^3 > 0$. It follows from the hypotheses of the theorem that $R_{\hat{q}}(x,t)$ has continuous second derivatives, while the function $G^0_q(x,\xi,\tau)$ has first derivatives for values $t \neq \tau(x, \xi)$. In a neighborhood of the characteristic cone for G_q^0 there is the estimate

$$|\nabla_{x,\xi,t} G_q^0| \leq \frac{K}{\sqrt{t^2 - \tau^2(x, \xi)}}, \qquad \tau(x, \xi) \geq h.$$

It follows from the results of Chapter VI, §3, that in this case, by differentiation with respect to t, equation (7.68) can be reduced to an equation of the second kind, but the first derivatives of the function $\tilde{q}$ with respect to the variables x_1 and x_2 occur under the integral sign. It is not hard to verify that in this case the hypotheses of the uniqueness theorem for (7.68) are satisfied. Indeed, $|\det R_{q\hat{q}}^0(x, \xi)| \geq \alpha \beta_q^3 > 0$, the first derivatives of $R_{q\hat{q}}^0$ with respect to the variable ξ exist and are continuous for $\xi \in D_{Hh}$ and x belonging to the plane $x_3 = 0$, and the function $R_{q\hat{q}}^1(\xi, x, t)$ has a continuous derivative with respect to the variable t. Due to the fact that the surfaces $\tau_q(x, \xi) = t$ satisfy all the conditions we imposed on them in §3 of Chapter VI, these conditions are sufficient for uniqueness of a solution of (7.68). Thus, $\tilde{q} \equiv 0$ in D_{Hh}, and hence $q \equiv \hat{q}$. The theorem is proved.

REMARK. In the case where in the data (7.65) $\varphi_i = 0$, $i = 1, 2, 3, 4$, by one differentiation with respect to t, problem (7.56), (7.65), (7.66) can be reduced to an analogous problem in which the functions $\psi_i(x)$ play the role of the functions φ_i. Therefore, if $\varphi_i(x) \equiv 0$ and the determinant of the matrix Ψ constructed from the functions ψ_i by the same rule as the matrix Φ is constructed from the φ_i is uniformly bounded away from zero, then a uniqueness theorem entirely analogous to Theorem 1 holds. It must only be assumed that $\psi_i \in C^3(D_H)$.

PROBLEM 2. Find the coefficients $b_i(x)$ and $c(x)$ of the operator L_q if solutions of problems (7.56), (7.65) are known on the set $\mathfrak{M} = S \times [0, T]$, where S is a closed convex surface of class C^5 in x-space

$$u_i = f_i(x, t), \quad x \in S, \ t \in [0, T]. \tag{7.70}$$

We denote by D the region bounded by S.

THEOREM 2. *If* $\operatorname{supp} q(x) \subset \Omega$, *where* Ω *is a region entirely contained in* D *and at a positive distance* $h > 0$ *from* S, *and* $b_i(x) \in C^4(\Omega)$ *and* $c(x) \in C^2(\Omega)$, *then under the condition that* $\varphi_i \in C^3(R^3)$, $\psi_i \in C^2(R^3)$, $|\det \Phi(x)| \geq \mu > 0$, $x \in \overline{\Omega}$, $T \geq \operatorname{diam} D$, *and* $\operatorname{diam} \Omega$ *is small, the operator* L_q *is uniquely determined by the information* (7.70).

In this case the proof of the theorem also reduces to establishing uniqueness of a solution of (7.68) with the sole difference that $x \in S$ and $t \in [0, T]$. By the way, here it suffices to assume that $t \in [h, T]$, since only in this case does the sphere $\tau(x, \xi) = t$, $x \in S$, have nonempty intersection with the region

Ω. We use the results of §4 to investigate equation (7.68). It is easy to see that through any point $x^0 \in D$ and in a direction orthogonal to any vector ν^0 there passes exactly one surface $\tau(x,\xi) = t$ for suitable choice of the point $(x,t) \in \mathfrak{M}$. Indeed, through a point x^0 we pass a line in the direction of the vector $-\nu^0$ to the intersection with the surface S. Then x is the point of intersection, and t is equal to the distance from x^0 to the point of intersection. Here $x = x(x^0, \nu^0)$ and $t = t(x^0, \nu^0)$ are smooth functions of class $C^5(\overline{D}' \times \omega)$, where $\overline{D}'$ is any closed region lying strictly within D and ω is the surface of the unit sphere. As a result the equation of the surface passing through x^0 with direction of the normal ν^0 can be written in the form $\varphi(\xi, x^0, \nu^0) = 0$, where

$$\varphi(\xi, x^0, \nu^0) = \tau(x(x^0, \nu^0), \xi) - t(x^0, \nu^0).$$

With respect to the variables x^0 and ν^0 the function φ has derivatives to fifth order, as the hypotheses of the uniqueness theorem require, and it is analytic in ξ.

The weight function $R^0_{q\hat{q}}(\xi, x)$ (here we have a matrix version of the problem of integral geometry studied in §4 of Chapter VI) has determinant uniformly bounded away from zero. By the conditions imposed on the functions φ_i and ψ_i and the coefficients of the operator L_q, the function $R^0_{q\hat{q}}$ with respect to the arguments ξ and x has continuous derivatives to third order and hence continuous derivatives to third order with respect to the arguments ξ, x^0, and ν^0. The function $\partial R_{\hat{q}}(\xi, t)/\partial t$ is continuous together with its derivatives to second order. The function $G^0_q(s, \xi, t)$ is continuous and with respect to the arguments x, ξ, and t has derivatives to second order which are continuous for $\tau(x,\xi) \neq t$. For its derivatives in a neighborhood of the surface $\tau(x,\xi) = t$ we have the estimates

$$|D^\alpha_{x,\xi,t} G^0| \le K[t^2 - \tau^2(x,\xi)]^{-1/2}, \quad |\alpha| = 1,$$
$$|D^\alpha_{x,\xi,t} G^0| \le K[t^2 - \tau^2(x,\xi)]^{-3/2}, \quad |\alpha| = 2.$$

Using (7.69), from this we conclude that $R^0_{q\hat{q}}(\xi, x, t)$ has continuous derivatives with respect to its arguments to second order for $\tau(x,\xi) \neq t$, and the function itself and its first derivatives are continuous up to the boundary of the cone $\tau(x,\xi) = t$, while for its second derivatives in a neighborhood of the cone there is the estimate

$$|D^\alpha_{\xi,x,t} R^1_{q\hat{q}}| \le K_1[t^2 - \tau^2(x,\xi)]^{-1/2}, \quad |\alpha| = 2. \tag{7.71}$$

Thus, the requirements on the weight functions $R^0_{q\hat{q}}$ and $R^1_{q\hat{q}}$ almost coincide with the requirements on them in the hypotheses of the uniqueness theorem of Chapter VI, §4 (see Remarks 1 and 2 of that section). Only the condition of continuity of the second derivatives is not satisfied. It was shown in [68] that this circumstance is not essential, namely, fulfillment of (7.71) is sufficient for the uniqueness theorem to be valid. We shall not consider the proof of this fact here. Thus, it follows from (7.68) that $\tilde{q} \equiv 0$, i.e., $q(x) \equiv \hat{q}(x)$, $x \in \Omega$.

There are a large number of works in which specific formulations of inverse problems for second-order hyperbolic equations are studied (see [18], [47]–[53], [115], [122], [147], [158], [159], [205], and [223]). The majority of them are connected with the problem of determining one of the coefficients b_i or c.

§4. Inverse problems for first-order hyperbolic systems

Many physical processes are described by systems of partial differential equations of first order, for example, the system of equations of acoustics, or electromagnetic oscillations, or the dynamical equations of elasticity theory. As a rule, equations of second order are derived from them under certain additional assumptions. In connection with this it is desirable to create a theory of inverse problems for first-order systems of equations. Systematic investigations in this direction began comparatively recently. For hyperbolic systems one-dimensional inverse problems have been considered in the case of two independent variables by L. P. Nizhnik [193]–[197] and V. G. Romanov and L. I. Slinyucheva [243]; in the case of n variables by V. G. Romanov and S. P. Belinskiĭ [41], [242]. In this section we present a survey of the results obtained in the theory of multidimensional inverse problems for hyperbolic systems. The exposition of the material is based on the work of Romanov [233]–[235], [238]–[240] and Belinskiĭ [39], [40].

We consider the system of equations

$$u_t + \sum_{i=1}^{n} A_i u_{x_i} + Qu = F, \tag{7.72}$$

in which $u = (u_1, \dots, u_m)$, $F = (F_1, \dots, F_m)$, the A_i and Q are $m \times m$ square matrices, and $x = (x_1, \dots, x_n)$. The coefficients of the system (7.72) and the right side are, generally speaking, functions of x and t; in each concrete formulation of an inverse problem for the system (7.72) the dependence of the coefficients on x and t will be of special nature and will be specifically mentioned. Inverse problems for the system (7.72) consist in finding the matrices A_i and Q or the right side F on the basis of a family of its solutions given on some manifold of the variables x and t.

We shall henceforth assume that the system (7.72) is t-hyperbolic [199] or symmetric [139]. In the case where the results are valid only for a symmetric system this will be noted separately.

In this section we study only questions of uniqueness of a solution for a number of inverse problems. However, for all formulations of the problems considered here estimates of stability can also be derived rather simply from the method of establishing the uniqueness theorems.

PROBLEM 1. Suppose $G = \{(x,t): 0 < x_n < h,\ t > 0\}$, $h > 0$, Γ is the boundary of G, and in the region G the matrices A_i and Q depend only on x_n, while the vector F can be represented in the form

$$F(x,t) = \Phi(x_n,t)f(x), \tag{7.73}$$

where $f = (f_1, \dots, f_m)$ and Φ is an $m \times m$ square matrix. For given matrices A_i, Q, and Φ, find the vector F in the region G if it is known that a solution of system (7.72), continuous in G together with its partial derivatives u_t and u_{x_i}, coincides on Γ with a given function ψ:

$$u|_\Gamma = \psi(x,t). \tag{7.74}$$

THEOREM 1. *Suppose the matrices A_i, $\partial A_n/\partial x_n$, and Q are continuous on $[0,h]$, $\det A_n \neq 0$, and the vector F is continuous in G together with its derivative $\partial F/\partial t$, while $F(x,0)$ has bounded support and $\det \Phi(x_n,0) \neq 0$, $x_n \in [0,h]$. Then there exists $h^* > 0$ such that for all $h < h^*$ the vector F is uniquely determined by giving the function ψ.*

The proof is based on applying the Fourier transform in the variables $x_1, \dots, x_{n-1}$ to the system (7.72) with subsequent investigation of the inverse problem for the system with two independent variables x_n, t by a method close to that of [243]. We set

$$\Gamma_0 = \{(x,t): 0 \le x_n \le h, t = 0\}, \qquad \Gamma_1 = \{(x,t): x_n = 0,\ t > 0\},$$
$$\Gamma_2 = \{(x,t): x_n = h,\ t > 0\}.$$

Then $\Gamma = \Gamma_1 \cup \Gamma_0 \cup \Gamma_2$.

Because of the linearity of problem (7.72)–(7.74), it suffices to show that to the function $\psi \equiv 0$ there corresponds the only possible solution $F \equiv 0$. Because the function $f(x)$ is compactly supported, which follows from the hypotheses of the theorem, it follows that a solution of problem (7.72)–(7.74) for $\psi \equiv 0$ is compactly supported for each fixed t. This is a consequence of the finite domain of dependence for system (7.72) for finite t. Therefore, the Fourier transform in the variables $x_1, \dots, x_{n-1}$ can be applied for any t to a function $u(x,t)$ which is a solution of (7.72). Denoting by $\mu = (\mu_1, \dots, \mu_{n-1})$ the parameter of the Fourier transform and by $\tilde{u}(\mu, x_n, t)$ the result of applying it to the function $u(x,t)$, we transform (7.72) and (7.74) to the form

$$\tilde{u}_t + A_n \tilde{u}_{x_n} + \left[Q - \sum_{j=1}^{n-1} (i\mu_j)A_j\right] \tilde{u} = \tilde{F}, \tag{7.75}$$

$$\tilde{u}|_{\tilde{\Gamma}} = 0. \tag{7.76}$$

Here $\tilde{\Gamma}$ denotes the projection of Γ onto the (x_n, t)-plane. We additionally denote by $\tilde{\Gamma}_k$ the projection of Γ_k onto (x_n, t) and by $\tilde{G}$ the projection of G onto the (x_n, t)-plane. We henceforth consider the system (7.75) in the region $\tilde{G}$, assuming μ to be fixed.

Because of the hyperbolicity of the system (7.72), the matrix A_n is similar to a diagonal matrix $\Lambda = T^{-1}A_nT$. The matrix $T(x_n)$ consists of columns which are the eigenvectors of A_n. In connection with this (7.75) can be reduced to the form

$$v_t + \Lambda v_{x_n} + Dv = T^{-1}\tilde{F}, \tag{7.75$'$}$$

where

$$v = T^{-1}u, \qquad D = T^{-1}\left[\left(Q - \sum_{j=1}^{n-1} i\mu_j A_j\right)T + A_nT_{x_n}\right], \qquad v|_{\tilde{\Gamma}} = 0. \tag{7.76$'$}$$

We note that the smoothness of the matrix A_n assumed in the hypotheses of the theorem suffices for the continuity of the matrix T_{x_n} for $x_n \in [0, h]$. Therefore, the matrix D is continuous for all $x_n \in [0, h]$. The matrix Λ has the eigenvalues of A_n as diagonal elements. We denote them by $\lambda_1, \dots, \lambda_n$. Since $\det A_n \neq 0$, $x_n \in [0, h]$, it follows that $\lambda_k = \lambda_k(x_n) \neq 0$ for $x_n \in [0, h]$. The smoothness of the matrix A_n implies continuity of all the $\lambda_k(x_n)$ for $x_n \in [0, h]$.

To simplify subsequent notation we denote the coordinate x_n by z. We introduce the function $w = v_t$. The function w is a solution (generally speaking, generalized) of the system

$$w_t + \Lambda w_z + Dw = T^{-1}\tilde{F}_t, \tag{7.77}$$

which is continuous in $\tilde{G} \cup \tilde{\Gamma}$. By virtue of (7.75′) and (7.76′), its trace on $\tilde{\Gamma}$ can be found by the formulas

$$w = \begin{cases} T^{-1}\tilde{F}, & (z, t) \in \tilde{\Gamma}_0, \\ 0, & (z, t) \in \tilde{\Gamma}_1 \cup \tilde{\Gamma}_2. \end{cases} \tag{7.78}$$

We shall go over from (7.77) and (7.78) to integral relations for the functions w and $\tilde{F}$. To this end for arbitrary $k = 1, \dots, m$ we consider the characteristic of the system (7.77) corresponding to $\lambda_k(z)$ and passing through an arbitrary point $(z_0, t_0) \in \tilde{G} \cup \tilde{\Gamma}$. Its equation can be represented in the form

$$t = t_0 + \int_{z_0}^{z} \frac{dz}{\lambda_k(z)}. \tag{7.79}$$

Since $\lambda_k(z)$ has constant sign along the characteristic, t is a monotone function of z. Therefore, only the following two versions of the intersection of the characteristic with $\tilde{\Gamma}$ are possible: 1) the characteristic intersects $\tilde{\Gamma}_0$ at one point and $\tilde{\Gamma}_1 \cup \tilde{\Gamma}_2$ at one point; 2) the characteristic has two points of intersection with $\tilde{\Gamma}_1 \cup \tilde{\Gamma}_2$ and none with $\tilde{\Gamma}_0$. In the first case we denote by (z_k, t_k) the point of intersection of the characteristic with $\tilde{\Gamma}_1 \cup \tilde{\Gamma}_2$ (thus, $z_k = 0$ or $z_k = h$), while in the second case (z_k, t_k) shall denote whichever of the two points of intersection with $\tilde{\Gamma}_1 \cup \tilde{\Gamma}_2$ corresponds to the smaller value of t. The oriented segment of the characteristic contained between the points (z_0, t_0) and (z_k, t_k) with positive direction from (z_k, t_k) to (z_0, t_0) we denote by $L_k(z_0, t_0)$.

We introduce the function

$$\varphi(\mu, z) = T^{-1}(z)\tilde{F}(\mu, z, 0).$$

By (7.73) we have

$$\tilde{F}(\mu, z, t) = \Phi(z, t)\Phi^{-1}(z, 0)T(z)\varphi(\mu, z) \equiv R(z, t)\varphi(\mu, z).$$

It suffices for us to show that $\varphi \equiv 0$ in order to demonstrate the validity of the theorem.

For any point $(z_0, t_0) \in \tilde{G}$, integrating the k-component of (7.77) along the curve $L_k(z_0, t_0)$, we obtain

$$w_k(\mu, z_0, t_0) = \int_{L_k(z_0, t_0)} (T^{-1}R_t\varphi - Dw)_k \frac{dz}{\lambda_k(z)}, \quad k = 1, 2, \dots, m. \tag{7.80}$$

On the other hand, performing an analogous integration along $L_k(z_0, 0)$ for any point $(z_0, 0)$, we have, with (7.78) taken into account,

$$\varphi_k(\mu, z_0) = \int_{L_k(z_0, 0)} (T^{-1}R_t\varphi - Dw)_k \frac{dz}{\lambda_k(z)}, \quad k = 1, 2, \dots, m. \tag{7.81}$$

It is not hard to find a closed region contained in $\tilde{G} \cup \tilde{\Gamma}$ such that the system of equations (7.80), (7.81) will be closed in it. Such a region (not the smallest possible!) is, for example, the rectangle $\Pi_h = \{(z, t): 0 \leq z \leq h, 0 \leq t \leq h/\lambda_0\}$, where

$$\lambda_0 = \min_{1 \leq k \leq m} \min_{0 \leq z \leq h} |\lambda_k(z)|.$$

In this region we obtain estimates for w and φ which follow from (7.80) and (7.81). To this end we introduce the norms

$$\|\varphi(\mu)\| = \max_{1 \leq k \leq m} \max_{0 \leq z \leq h} |\varphi_k(\mu, z)|,$$

$$\|w(\mu)\| = \max_{1 \leq k \leq m} \max_{(z,t) \in \Pi_h} |w_k(\mu, z, t)|.$$

Moreover, for any smooth matrix $S = (s_{ij})$, $S = S(\mu, z, t)$, we introduce the matrix norm

$$\|S(\mu)\| = \max_i \left(\sum_j \max_{(z,t)\in\Pi_h} |s_{ij}(\mu, z, t)| \right).$$

From (7.80) and (7.81) we easily find that

$$\|w(\mu)\| \leq \left(\|T^{-1}R_t\|\,\|\varphi(\mu)\| + \|D\|\,\|w(\mu)\|\right)\frac{h}{\lambda_0},$$

$$\|\varphi(\mu)\| \leq \left(\|T^{-1}R_t\|\,\|\varphi(\mu)\| + \|D\|w(\mu)\right)\frac{h}{\lambda_0}.$$

Adding these two inequalities, we obtain

$$\|w(\mu)\| + \|\varphi(\mu)\| \leq \Delta(\mu, h)\left[\|w(\mu)\| + \|\varphi(\mu)\|\right], \tag{7.82}$$

where

$$\Delta(\mu, h) = \frac{2h}{\lambda_0}\left(\|T^{-1}R_t\| + \|D\|\right).$$

Let h^* be chosen from the condition $\Delta(0, h^*) = 1$. From the continuity of Δ in the variable μ it then follows that for any positive $h < h^*$ it is not hard to find positive numbers $\varepsilon(h)$ and $\gamma(h)$ such that for all $|\mu| \leq \varepsilon(h)$ the inequality $\Delta(\mu, h) \leq 1 - \gamma(h)$ holds. From (7.82) it follows then that $\|w(\mu)\| = \|\varphi(\mu)\| = 0$, $|\mu| \leq \varepsilon(h)$. Therefore, $\tilde{F}(\mu, z, t) \equiv 0$ for $|\mu| \leq \varepsilon(h)$. But by the conditions imposed on the function $F(x, t)$, its Fourier transform depends analytically on μ, and hence $\tilde{F}(\mu, z, t) \equiv 0$ for all μ, and thus $F(x, t) \equiv 0$.

THEOREM 2. *Suppose the matrices A_i, Q and the vector F satisfy the hypotheses of Theorem 1 and, moreover, all eigenvalues of the matrix A_n are negative. Then for any h the vector F is uniquely determined by giving the function ψ on $\Gamma_1 \cup \Gamma_2$.*

Under the hypotheses of this theorem all curves $L_k(z_0, t_0)$ intersect $\tilde{\Gamma}_1$. Therefore, the system of equations (7.80), (7.81) can be written as a Volterra system with respect to the variable z_0:

$$w_k(\mu, z_0, t_0) = \int_0^{z_0} (T^{-1}R_t\varphi - Dw)_k|_{t=t_k(z,z_0,t_0)}\frac{dz}{\lambda_k(z)}, \quad k = 1, 2, \dots, m, \tag{7.80'}$$

$$\varphi_k(\mu, z_0) = \int_0^{z_0} (T^{-1}R_t\varphi - Dw)_k|_{t=t_k(z,z_0,0)}\frac{dz}{\lambda_k(z)}, \quad k = 1, 2, \dots, m. \tag{7.81'}$$

In these equalities $t_k(z, z_0, t_0)$ denotes the function defined by the right side of (7.79).

In the triangular region $T_h = \{(z,t): 0 \le z \le h,\ 0 \le t \le z/\lambda_0\}$ the system (7.80), (7.81) is closed. Carrying out estimates for it in this region according to the scheme typical for Volterra equations, we obtain $\varphi_k(\mu, z) \equiv 0$, $z \in [0,h]$. Therefore, $F(x,t) \equiv 0$.

It follows from Theorem 2, in particular, that it is formally possible to set $h = \infty$ and consider the problem in the region $G' = \{(x,t): x_n > 0, t > 0\}$. The vector F is here uniquely determined by giving the solution on the boundary of G'.

A uniqueness theorem valid for any h can also be formulated in the case where the number of linearly independent components of the function $f(x)$ is less than m. The conditions of the theorem can be stated most simply for the case of a diagonal matrix A_n. We remark that any hyperbolic system can easily be reduced to such a form with a diagonal matrix A_n.

THEOREM 3. *Suppose the hypotheses of Theorem 1 are satisfied and, moreover, the first r ($r \ge 1$) elements of the diagonal matrix A_n are negative, while the remainder are positive. If the components $f_{r+1}, \dots, f_m$ of the function $f(x)$ can be linearly expressed in terms of the components $f_1, \dots, f_r$, then the vector $F(x,t)$ is uniquely determined by giving the function ψ on $\Gamma_0 \cup \Gamma_1$.*

A proof of this theorem is given in [234]; we shall not present it here.

PROBLEM 2. For known matrices $A_i(x)$ and vector $F(x,t)$ find the matrix $Q(x)$ in the region $G = \{(x,t): 0 < x_n < h, t > 0\}$ if m distinct solutions u^l, $l = 1, \dots, m$, of the system (7.72) on the boundary Γ of the set G are known:

$$u^l|_\Gamma = \psi^l(x,t), \quad l = 1, 2, \dots, m. \tag{7.83}$$

We denote by $\Psi(x,t)$ the square matrix composed of the ψ^l, $l = 1, \dots, m$. Let $\Gamma = \Gamma_0 \cup \Gamma_1 \cup \Gamma_2$, where the Γ_k have the meaning of the sets introduced earlier in the proof of Theorem 1. We denote by $\mathfrak{M}(K)$ the class of matrices $Q(x)$ such that the norm of $Q(x)$ in C does not exceed K.

THEOREM 4. *Suppose the matrices $A_i = (a^i_{kj})$ are diagonal,*

$$a^i_{kj} = \lambda_{ik}(x)\delta_{kj}. \tag{7.84}$$

Here δ_{kj} is the Kronecker symbol. Suppose the matrices A_i, Q and Ψ, the vector F, and their partial derivatives of first order are continuous and bounded in $G \cup \Gamma$. If, moreover,

$$|\det \Psi(x,0)| \ge \alpha > 0, \quad (x,0) \in \Gamma_0, \tag{7.85}$$

$$\begin{aligned} -\lambda_{ns}(x) &\ge \mu > 0, \quad s = 1, 2, \dots, r, \\ \lambda_{ns}(x) &\ge \mu > 0, \quad s = r+1, \dots, m, \end{aligned} \tag{7.86}$$

then for each class $\mathfrak{M}(K)$ of matrices Q there exists $h^* > 0$ such that for $h < h^*$ any matrix $Q \in \mathfrak{M}(K)$ is uniquely determined by the information (7.83).

We prove the theorem by arguing by contradiction; namely, we suppose that for any $h > 0$ there exist at least two matrices Q and $\hat{Q}$ to which there corresponds the same information (7.83). We denote the solutions u^l corresponding to the system (7.72) with matrix $\hat{Q}$ by $\hat{u}^l$. We set $\tilde{u}^l = u^l - \hat{u}^l$ and $\tilde{Q} = Q - \hat{Q}$, and obtain the relations connecting $\tilde{u}^l$ and $\tilde{Q}$:

$$\tilde{u}_t^l + \sum_{i=1}^n A_i \tilde{u}_{x_i}^l + Q\tilde{u}^l + \tilde{Q}\hat{u}^l = 0, \quad l = 1, 2, \dots, m, \tag{7.87}$$

$$\tilde{u}^l|_\Gamma = 0, \quad l = 1, 2, \dots, m. \tag{7.88}$$

It is convenient to use a differential consequence of the system (7.87). Setting $\tilde{v}^l = \tilde{u}_t^l$ and $\hat{v}^l = \hat{u}_t^l$, we find that

$$\tilde{v}_t^l + \sum_{i=1}^n A_i \tilde{v}_{x_i}^l + Q\tilde{v}^l + \tilde{Q}\hat{v}^l = 0, \quad l = 1, 2, \dots, m, \tag{7.89}$$

$$\tilde{v}^l|_\Gamma = \begin{cases} -\tilde{Q}\psi^l, & (x,t) \in \Gamma_0, \\ 0, & (x,t) \in \Gamma_1 \cup \Gamma_2. \end{cases} \tag{7.90}$$

We now consider an arbitrary point $(x^0, t^0) \in G \cup \Gamma$. Through it there can pass m characteristics connected with equation (7.72). To find the kth characteristic it suffices to solve the Cauchy problem

$$dx_i/dt = \lambda_{ik}(x), \qquad x_i|_{t=t^0} = x_i^0, \quad i = 1, 2, \dots, n. \tag{7.91}$$

The constraints imposed on the smoothness of the functions by the hypotheses of the theorem guarantee the existence of a unique solution of problem (7.91) and its extendability to any t both for $t > t^0$ and for $t < t^0$, not taking the solution beyond the confines of the region $\overline{G} = G \cup \Gamma$ (see [200], §14, Remark 3). It is easy to see also that conditions (7.86) ensure that any characteristic passing through the point (x^0, t^0) hits the boundary Γ in finite time t. Indeed, it follows from (7.86) that on each characteristic between x_n and t there is a one-to-one and strictly monotone correspondence. At the same time it follows from (7.91) that along a characteristic

$$|x_n - x_n^0| \geq \mu |t - t^0|,$$

and hence after a finite interval of change of the variable t not exceeding h/μ, the point $(x(t), t)$ hits either Γ_0 or $\Gamma_1 \cup \Gamma_2$. Because of the monotone correspondence between x_n and t, there are only two possibilities for the intersection of a characteristic with the boundary Γ: 1) a characteristic intersects

Γ_0 at one point and $\Gamma_1 \cup \Gamma_2$ at one point; 2) it intersects Γ_1 at one point and Γ_2 at one point. In the first case we denote by (ξ^k, τ^k) the point of intersection of the kth characteristic with $\Gamma_1 \cup \Gamma_2$, and in the second case (ξ^k, τ^k) shall denote whichever of the points of intersection of the characteristic with $\Gamma_1 \cup \Gamma_2$ corresponds to the smaller value of t. The segment of the characteristic contained between (ξ^k, τ^k) and (x^0, t^0) with positive direction toward (x^0, t^0) we denote by $L_k(x^0, t^0)$.

Further, we set

$$\tilde{Q}(x)\Psi(x,0) = B(x), \qquad B = (b_{kl}).$$

Then

$$\tilde{Q}(x) = B(x)\Psi^{-1}(x,0), \qquad v_{kl}|_{\Gamma_0} = -b_{kl}(x).$$

Integrating the kth equation of (7.89) along the curve $L_k(x^0, t^0)$, we obtain

$$\tilde{v}_{kl}(x^0, t^0) = -\int_{L_k(x^0,t^0)} [Q\tilde{v}^l + B\Psi^{-1}(x,0)\hat{v}^l]_k\, dt, \quad k, l = 1, 2, \ldots, m. \tag{7.92}$$

Using condition (7.90), from this we find that

$$b_{kl}(x^0) = \int_{L_k(x^0,0)} [Q\hat{v}^l + B\Psi^{-1}(x,0)\hat{v}^l]_k\, dt, \quad k, l = 1, 2, \ldots, m. \tag{7.93}$$

The system of equalities (7.92), (7.93) in the region $\Pi_h = \{(x^0, t^0): 0 \le x_n^0 \le h,\ 0 \le t \le h/\mu\}$ is closed relative to the functions $\tilde{v}_{kl}(x^0, t^0)$ and $b_{kl}(x^0)$. At the same time under the conditions of the theorem it is not hard to obtain for $\hat{v}^l$ an estimate in Π_h which is uniform in $C(\Pi_h)$. For this it suffices to construct for $\hat{v}^l$ a system of integral equations by integrating a system of equalities analogous to (7.89) along the characteristics. The system so obtained for each fixed t is a system of integral equations of Volterra type. The kernels of the system are continuous and depend only on the functions λ_{ik} and the matrix $\hat{Q}(x)$. From this it is clear that for $\hat{Q}(x) \in \mathfrak{M}(K)$, the norm in C of the function $\hat{v}^l$ is finite and depends on the constant K. Passing from (7.92) and (7.93) to the estimate, we find that

$$\begin{aligned} \|\tilde{V}\| &\le \frac{h}{\mu}\left(\|Q\|\,\|\tilde{V}\| + \|B\|\,\|\Psi^{-1}(x,0)\|\,\|\hat{V}\|\right), \\ \|B\| &\le \frac{h}{\mu}\left(\|Q\|\,\|\tilde{V}\| + \|B\|\,\|\Psi^{-1}(x,0)\|\,\|\tilde{V}\|\right), \end{aligned} \tag{7.94}$$

$$\tilde{V} = (\tilde{v}_{kl},\ k, l = 1, 2, \ldots, m), \quad \hat{V} = (\hat{v}_{kl},\ k, l = 1, 2, \ldots, m).$$

The matrix norms are here understood as follows:

$$\|\tilde{V}\|^2 = \sum_{k,l=1}^{m} \max_{(x,t)\in\Pi_h} \tilde{v}_{kl}^2(x,t).$$

We set

$$h^* = \mu[K + \|\Psi^{-1}(x,0)\|\beta(K)],$$

where $\beta(K)$ denotes the constant bounding the norm of $\hat{V}(x,t)$ for $\hat{Q} \in \mathfrak{M}(K)$. Then for $h < h^*$ the system of inequalities (7.94) has only the zero solution. Hence, $\|B\| = 0$. Therefore, $\tilde{Q} \equiv 0$, i.e., $Q \equiv \hat{Q}$, and the theorem is proved.

REMARK. Problem 2 can also be considered in a more general region than $G = \{(x,t)\colon 0 < x_n < h,\ t > 0\}$. In particular, for the region G it is possible to take the direct product of a compact region in x-space with the infinite segment $[0,\infty)$ in t-space. A theorem analogous to Theorem 4 is valid also in this case if in place of the conditions (7.86) it is required that the region D be uniformly convex and that

$$\sum_{i=1}^{n} \lambda_{ik}^2(x) \geq \mu > 0, \quad k = 1,2,\ldots,m.$$

The diameter of the region D here plays the role of the small parameter h [240].

PROBLEM 3. Find the matrix $Q = Q(x)$ for given matrices $A_i(x)$ and vector $F(x,t)$ if m solutions u^l, $l = 1,\ldots,m$, of system (7.72) are known on the two planes $t = 0$ and $t = T$ $(T > 0)$:

$$u^l(x,0) = \varphi^l(x), \qquad u^l(x,T) = \Psi^l(x), \quad l = 1,2,\ldots,m. \tag{7.95}$$

The following simple example shows that without indispensable assumptions regarding the matrix $Q(x)$ this problem does not have a unique solution. Let $m = n = 1$, and let $u(x,t)$ be a solution of the homogeneous equation

$$u_t + u_x + q(x)u = 0, \tag{7.96}$$

assuming values identically equal to one for $t = 0$ and $t = T$. Then any periodic function $q(x)$ with period T whose integral over the periodicity interval is equal to zero is a solution of the problem. For example, for $q(x) = \sin(k2\pi x/T)$, $k = 1,2,\ldots$, a solution of (7.96) satisfying the given conditions for $t = 0$ and $t = T$ has the form

$$u(x,t) = \exp\left[\frac{T}{2k\pi}\cos k\frac{2\pi}{T}x - \cos k\frac{2\pi}{T}(x-t)\right].$$

Conditions distinguishing in a set of matrices Q sets of uniqueness for the solution of Problem 3 are connected with restrictions on the support of the matrix Q.

THEOREM 5. *Let $A_i = A_i(x)$ and $F = F(x,t)$ be continuous together with their partial derivatives, and suppose they are bounded in $G = \{(x,t)\colon x \in R^n,$ $t > 0\}$; let $Q(x), \varphi^l(x)$ and $\psi^l(x)$ be continuous and bounded in G, where*

$$|\det \Psi(x)| \geq \alpha > 0,$$

$\Psi(x)$ being the matrix composed of the $\psi^l(x)$, $l = 1, \ldots, m$, and suppose that the elements of the matrices $A_i = (a^i_{kj})$ satisfy conditions (7.84) and (7.86). Then for each class of matrices $Q(x) \in \mathfrak{M}(K)$ there exists an $h^* > 0$ such that any function $Q(x) \in \mathfrak{M}(K)$ whose support is contained in the strip $D_h = \{x: 0 \le x_n \le h\}$, $h < h^*$, is uniquely determined by the information (7.95).

For the proof we use an argument by contradiction and the notation adopted in the proof of Theorem 4. In this notation for the matrix $\tilde{V}(x,t)$ formed from the columns $\tilde{v}^l = \tilde{u}^l_t$, $l = 1, \ldots, m$, we have the equation

$$\tilde{V}_t + \sum_{i=1}^{n} A_i \tilde{V}_{x_i} + Q\tilde{V} + \tilde{Q}\hat{V} = 0 \tag{7.97}$$

and the conditions

$$\tilde{V}|_{t=0} = -\tilde{Q}\Phi(x), \qquad \tilde{V}|_{t=T} = -\tilde{Q}\Psi(x). \tag{7.98}$$

Here $\Phi(x)$ is the matrix composed of the columns $\varphi^l(x)$, $l = 1, \ldots, m$.

We denote by $L_k(x^0, t^0)$ the segment of the kth characteristic contained between the point (x^0, t^0) and the plane $t = 0$ with positive direction toward the point (x^0, t^0). We denote by $(\xi^k, 0)$ the point of intersection of the curve $L_k(x^0, t^0)$ with the plane $t = 0$. Thus, $\xi^k = \xi^k(x^0, t^0)$. We additionally set

$$B(x) = \tilde{Q}(x)\Psi(x), \qquad \tilde{Q}(x) = B(x)\Psi^{-1}(x).$$

By the hypotheses of the theorem, the support of $B(x)$ is contained in the region D_h.

Integrating the kth row of the system (7.97) along $L_k(x^0, t^0)$, we find that

$$\begin{aligned} \tilde{v}_{kl}(x^0, t^0) &= -(B(\xi^k)\Psi^{-1}(\xi^k)\Phi(\xi^k))_{kl} \\ &\quad - \int_{L_k(x^0,t^0)} [Q\tilde{V} + B\Psi^{-1}\hat{V}]_{kl}\, dt, \quad k, l = 1, 2, \ldots, m. \end{aligned} \tag{7.99}$$

Setting $t^0 = T$ and $x^0 \in D_h$ in this equality and using conditions (7.98), we obtain a second group of integral relations:

$$\begin{aligned} b_{kl}(x^0) &= [B(\xi^k(x^0,T))\Psi^{-1}(\xi^k(x^0,T))\Phi(\xi^k(x^0,T))]_{kl} \\ &\quad + \int_{L_k(x^0,T)} [Q\tilde{V} + B\Psi^{-1}\hat{V}]_{kl}\, dt, \ x^0 \in D_h, \quad k, l = 1, 2, \ldots, m. \end{aligned} \tag{7.100}$$

We henceforth assume that $h < \mu T$. Then for $x^0 \in D_h$ the point $\xi^k(x^0, T)$ lies outside D_h, and hence $B(\xi^k(x^0, T)) = 0$. Therefore, equalities (7.100) assume the form

$$b_{kl}(x^0) = \int_{L_k(x^0,T)} [Q\tilde{V} + B\Psi^{-1}\hat{V}]_{kl}\, dt, \qquad x^0 \in D_h; \quad k, l = 1, 2, \ldots, m. \tag{7.100$'$}$$

We note that the system of relations (7.99), (7.100′) is closed relative to the matrices $\tilde{V}(x^0,t)$ and $B(x^0)$ in the region

$$G_h = \{(x^0,t^0)\colon -\mu_1(T-t) \le x_n^0 \le h + \mu_1(T-t),\ 0 \le t^0 \le T\},$$

where

$$\mu_1 = \max_{1\le k\le m}\max_{x\in R^n} |\lambda_{nk}(x)|.$$

Here, since $Q = B = 0$ outside the region D_h, the integration in (7.99) and (7.100) goes only over that part of the characteristic lying in the region $\Pi_n = \{(x^0,t^0)\colon 0 \le x_n^0 \le h,\ 0 \le t \le T\}$. The interval of variation of the variable of integration t here does not exceed h/μ. In view of this circumstance, proceeding from relations (7.99), (7.100′), we obtain estimates for $\tilde{V}(x^0,t^0)$ and $B(x^0)$ in the region G_h:

$$\begin{aligned}
&\|\tilde{V}\| \le \|B\|\,\|\Psi^{-1}\Phi\| + \frac{h}{\mu}\left(\|Q\|\,\|\tilde{V}\| + \|B\|\,\|\Psi^{-1}\|\,\|\hat{V}\|\right),\\
&\|B\| \le \frac{h}{\mu}\left(\|Q\|\,\|\tilde{V}\| + \|B\|\,\|\Psi^{-1}\|\,\|\hat{V}\|\right).
\end{aligned}\tag{7.101}$$

For $Q, \hat{Q} \in \mathfrak{M}(K)$ we have

$$\|Q\| \le K, \qquad \|\hat{V}\| \le \beta(K).$$

Choosing h^* from the condition

$$h^* = \min[\mu T, \mu(\|\Psi^{-1}\|\beta(K) + K(1 + \|\Psi^{-1}\Phi\|))]^{-1},$$

we find that for $h < h^*$ the system (7.101) has only the trivial solution. Hence, $B \equiv 0$; thus $\tilde{Q} \equiv 0$, and so $Q(x) \equiv \hat{Q}(x)$.

REMARK. In the case where all the $\lambda_{nk}(x) \ge \mu > 0$, $k = 1,\dots,m$, under the conditions of Theorem 5 any matrix $Q \in \mathfrak{M}(K)$ whose support is contained in the region $x_n \ge 0$ is uniquely determined by the information (7.95) [233].

PROBLEM 4. Let $G = \{(x,t,\overline{x}^0)\colon 0 < x_n < h,\ t > 0,\ \overline{x}^0 \in R^{n-1}\}$ and $\overline{x}^0 = (x_1^0,\dots,x_{n-1}^0)$, and let Γ be the boundary of G, where $\Gamma = \Gamma_0 \cup \Gamma_1 \cup \Gamma_2$,

$$\begin{aligned}
\Gamma_0 &= \{(x,t,\overline{x}^0)\colon 0 \le x_n \le h,\ t = 0,\ \overline{x}^0 \in R^{n-1}\},\\
\Gamma_1 &= \{(x,t,\overline{x}^0)\colon x_n = 0,\ t > 0,\ \overline{x}^0 \in R^{n-1}\},\\
\Gamma_2 &= \{(x,t,\overline{x}^0)\colon x_n = h,\ t > 0,\ \overline{x}^0 \in R^{n-1}\};
\end{aligned}$$

suppose that the matrices A_i and Q depend only on x and that $F = F(\overline{x} - \overline{x}^0, x_n, t)$ and $\overline{x} = (x_1,\dots,x_{n-1})$. Suppose, further, that the matrix A_n is diagonal, and its diagonal elements $k_1,\dots,k_m$ have the property $k_s > 0$ for $s = 1,\dots,r$ and $k_s < 0$ for $s = r+1,\dots,m$. For $l = 1,\dots,2m$ we consider

$2m$ boundary value problems for the system (7.72) consisting in finding for fixed l a solution of (7.72) in $\overline{G} = G \cup \Gamma$ on the basis of data on Γ:

$$\begin{aligned} u_{sl}|_{\Gamma_0} &= \varphi_{sl}(\overline{x} - \overline{x}^0, x_n), \quad s = 1, 2, \dots, m, \\ u_{sl}|_{\Gamma_1} &= \psi_{sl}(\overline{x} - \overline{x}^0, t), \quad s = 1, 2, \dots, r, \\ u_{sl}|_{\Gamma_2} &= \psi_{sl}(\overline{x} - \overline{x}^0, t), \quad s = r + 1, \dots, m. \end{aligned} \tag{7.102}$$

It is required to find matrices A_i and Q in the region $D_h = \{x : 0 \leq x_n \leq h\}$ if the traces on Γ of the following components of a solution of problem (7.72), (7.102) corresponding to $2m$ distinct data (7.102) are known:

$$\begin{aligned} u_{sl}|_{\Gamma_1} &= \chi_{sl}(x, t, \overline{x}^0), \quad s = r + 1, \dots, m, \\ u_{sl}|_{\Gamma_2} &= \chi_{sl}(x, t, \overline{x}^0), \quad s = 1, 2, \dots, r. \end{aligned} \tag{7.103}$$

We note that the data (7.102) and (7.103) complement one another on Γ, so that the traces of all components of the $2m$ distinct solutions of the system (7.72) are known on Γ.

Following [235], we consider a linearized version of the formulation of this problem which consists in the following. We suppose that the matrices A_i and Q can be represented in the form

$$A_i = A_i^0(x_n) + A_i^1(x), \qquad Q = Q^0(x_n) + Q^1(x), \quad i = 1, 2, \dots, n, \tag{7.104}$$

where the matrices $A_i^0(x_n)$ and $Q^0(x_n)$ are known and such that the system of equations (7.72) corresponding to them is t-hyperbolic, where the structure of the matrix A_n^0 is analogous to the structure of the matrix A_n, i.e., its diagonal elements k_i^0 are such that $k_s^0(x_n) > 0$ for $s \leq r$ and $k_s^0(x_n) < 0$ for $s > r$, while the matrices A_i^1 and Q^1 are small (in the sense of smallness of their elements in the norm of C). Then any solution u^l of problem (7.72), (7.102) can be represented in the form

$$u^l = (u^0)^l + v^l,$$

where $(u^0)^l$ is a solution of the equation

$$u_t + \sum_{i=1}^{n} A_i^0 u_{x_i} + Q^0 u = F \tag{7.105}$$

with data (7.102), while v^l up to second-order smallness satisfies the equation (7.106) and the homogeneous conditions (7.102).

We remark that because of the special form of the right side and the data (7.102) a solution of problem (7.102), (7.105) depends on $\overline{x}$ and $\overline{x}^0$ only in the combination $\overline{x} - \overline{x}^0$:

$$(u^0)^l \equiv (u^0)^l(\overline{x} - \overline{x}^0, x_n, t). \tag{7.107}$$

The linearized formulation of the inverse Problem 4 thus consists in finding the matrices A_i^1 and Q^1 from the traces of the components of the functions v^l on Γ:

$$\begin{aligned} &v_{sl}|_{\Gamma_1} = \chi_{sl}(x,t,\overline{x}^0) - u_{sl}^0(\overline{x}-\overline{x}^0,0,t) \equiv g_{sl}(\overline{x},t,\overline{x}^0),\\ &s = r+1,\dots,m; \quad l = 1,2,\dots,2m,\\ &v_{sl}|_{\Gamma_2} = \chi_{sl}(x,t,\overline{x}^0) - u_{sl}^0(\overline{x}-\overline{x}^0,h,t) \equiv g_{sl}(\overline{x},t,\overline{x}^0),\\ &s = 1,2,\dots,r; \quad l = 1,2,\dots,2m. \end{aligned} \tag{7.108}$$

We denote by $\Phi(x)$ the square matrix of dimension $2m \times 2m$ composed of the system of functions $\varphi_{sl}(x)$ that occur in (7.102), according to the following rule: the lth column of the matrix Φ consists of the elements $\varphi_{1l},\dots,\varphi_{ml}$, $\partial\varphi_{1l}/\partial x_n,\dots,\partial\varphi_{ml}/\partial x_n$.

Let $\tilde{\Phi}(\lambda, x_n)$ denote the Fourier transform with respect to the variable $\overline{x}$ of the matrix $\Phi(x)$, $x = (\overline{x}, x_n)$.

THEOREM 6. *Suppose the functions $F(\overline{x},x_n,t)$, $\varphi_{sl}(x)$ and $\psi_{sl}(\overline{x},t)$ contained on the right side of* (7.72) *and in the data* (7.102) *and the matrices A_i^1 and Q^1 are compactly supported in $\overline{x}$ for each fixed x_n, t and together with A_i^0 and Q^0 possess sufficient smoothness for the existence of a classical solution of problem* (7.72), (7.102). *Suppose, moreover, that the functions $\varphi_{sl}(x)$ in the data* (7.102) *are chosen so that the square matrix $\tilde{\Phi}(\lambda, x_n)$ possesses the property*

$$\det \tilde{\Phi}(0, x_n) \neq 0, \quad x_n \in [0,h]. \tag{7.109}$$

Then there exists $h^* > 0$ such that for $h < h^*$ the matrices A_i^1 and Q^1 are uniquely determined by giving the functions g_{sl}, $s = 1,\dots,m$, $l = 1,\dots,2m$.

We note that the set of functions φ_{sl} satisfying the conditions presented in the theorem is nonempty, as the following simple example shows. Let

$$\begin{aligned} &\varphi_{sl} = \delta_{sl}\omega_R(\overline{x}), \quad s,l = 1,2,\dots,m,\\ &\varphi_{sl} = \delta_{s(l-m)}(1+x_n)\omega_R(\overline{x}), \qquad s = 1,2,\dots,m; \quad l = m+1,\dots,2m, \end{aligned}$$

where δ_{sl} is the Kronecker symbol, and $\omega_R(\overline{x})$ is a function of the type of the Sobolev averaging kernel:

$$\omega_R(\overline{x}) = \begin{cases} \exp \frac{|\overline{x}|^2}{|\overline{x}|^2 - R^2}, & |\overline{x}| < R,\\ 0, & |\overline{x}| \geq R. \end{cases}$$

Then

$$|\det \tilde{\Phi}(\lambda, x_n)| = |\tilde{\omega}_R(\lambda)|^{2m}$$

and obviously $|\tilde{\omega}_R(0)| \neq 0$.

We now proceed to the proof of the theorem. By the linearity of the inverse problem it suffices to prove that to functions $g_{sl} \equiv 0$ there correspond $A_i^1 \equiv Q^1 \equiv 0$. We thus set $g_{sl} = 0$, $s = 1, \dots, m$, $l = 1, \dots, 2m$. Using the fact that the v^l satisfy the homogeneous conditions (7.102), we then obtain

$$v^l|_\Gamma \equiv 0, \quad l = 1, 2, \dots, 2m. \tag{7.110}$$

From the existence for the system (7.72) of a finite domain of dependence and the fact that the right side of (7.72) and the data (7.102) are compactly supported in $\overline{x}$, it follows that solutions u^l of problem (7.72), (7.102) and solutions $(u^0)^l$ of problem (7.102), (7.105) for each fixed $\overline{x}^0$ have compact support in $\overline{x}$. In (7.106) we change the variable $\overline{x}$ to the variable $\xi = \overline{x} - \overline{x}^0$. To the function $v(\overline{x}^0 + \xi, x_n, t, \overline{x}^0) \equiv \hat{v}(\xi, x_n, t, \overline{x}^0)$ it is then obviously possible to apply the Fourier transform with respect to $\overline{x}^0$ because of the compact support of the matrices A_i^1 and Q^1, and the Fourier transform with respect to ξ because of the compact support in ξ of the functions $(u^0)^l(\xi, x_n, t)$. Denoting by μ and λ the parameters of the Fourier transforms of $\hat{v}(\xi, x_n, t, \overline{x}^0)$ in the variables $\overline{x}^0$ and ξ respectively and by $\tilde{v}(\lambda, x_n, t, \mu)$ its Fourier transform, we find that the function $\tilde{v}^l$ satisfies the equation

$$\tilde{v}_t + A_n^0 \tilde{v}_{x_n} + C^0 \tilde{v} + \tilde{A}_n^1(\mu, x_n) \tilde{u}_{x_n}^0(\lambda - \mu, x_n, t) \\ + C^1(\lambda, \mu, x_n)(\tilde{u}^0)^l(\lambda - \mu, x_n, t) = 0. \tag{7.111}$$

Here

$$C^0(\lambda, x_n) = Q^0(x_n) - i \sum_{j=1}^{n-1} \lambda_j A_j^0(x_n),$$

$$C^1(\lambda, \mu, x_n) = \tilde{Q}^1(\mu, x_n) - i \sum_{j=1}^{n-1} (\lambda_j - \mu_j) \tilde{A}_j^1(\mu, x_n),$$

and $\tilde{A}_j^1(\mu, x_n)$, $\tilde{Q}^1(\mu, x_n)$ and $(\tilde{u}^0)^l(\mu, x_n, t)$ denote the Fourier transforms of the corresponding matrices and vector-valued functions.

To condition (7.110) there corresponds the condition

$$\tilde{v}^l(\lambda, x_n, t, \mu)_{\tilde{\Gamma}} \equiv 0, \quad l = 1, 2, \dots, 2m, \tag{7.112}$$

where $\tilde{\Gamma}$ is the projection of Γ onto the (x_n, t)-plane. We further denote by $\tilde{G}$ the projection of G and by $\tilde{\Gamma}_k$, $k = 0, 1, 2$, the projection of Γ_k onto the (x_n, t)-plane. We shall henceforth consider relations (7.111) in the region $\tilde{G} \cup \tilde{\Gamma}$ under conditions (7.112) on $\tilde{\Gamma}$ for fixed values of the parameters λ and μ. Because of condition (7.109) and the continuity of $\tilde{\Phi}(\lambda, x_n)$ there exists $\varepsilon > 0$ such that

$$|\det \tilde{\Phi}(\lambda, x_n)| \neq 0, \quad x_n \in [0, h], \quad |\lambda| \leq \varepsilon. \tag{7.113}$$

Suppose $|\lambda| \leq \varepsilon/2$ and $|\mu| \leq \varepsilon/2$. Then $|\lambda-\mu| \leq \varepsilon$, and by (7.113) there exists $\tilde{\Phi}^{-1}(\lambda-\mu, x_n)$. We introduce the matrix $H = H(\lambda-\mu, x_n, t)$ of dimension $2m \times 2m$ whose columns are formed by the components of the vector $(u^0)^l(\lambda-\mu, x_n, t)$ and of its derivative with respect to x_n:

$$u_{1l}^0, u_{2l}^0, \dots, u_{ml}^0, \frac{\partial}{\partial x_n} u_{1l}^0, \frac{\partial}{\partial x_n} u_{2l}^0, \dots, \frac{\partial}{\partial x_n} u_{ml}^0.$$

We note that, since the equalities (7.102) are satisfied,

$$H(\lambda-\mu, x_n, 0) = \tilde{\Phi}(\lambda-\mu, x_n). \tag{7.114}$$

We denote by $B(\lambda, \mu, x_n)$ the rectangular matrix of m rows and $2m$ columns formed by the matrices C^1 and $\tilde{A}_n^1$,

$$B = (C^1, \tilde{A}_n^1), \tag{7.115}$$

and by $\tilde{V}$ the matrix consisting of the $\tilde{v}^l$, $l = 1, \dots, 2m$. In this notation we have

$$\tilde{V}_t + A_n^0 \tilde{V}_{x_n} + C^0 \tilde{V} + B(\lambda, \mu, x_n) H(\lambda-\mu, x_n, t) = 0, \tag{7.116}$$

$$\tilde{V}|_{\tilde{\Gamma}} = 0. \tag{7.117}$$

We shall show that (7.116) and (7.117) imply $B \equiv 0$ for λ and μ sufficiently small. This is most easily done by writing out the equalities for the function $W = \tilde{V}_t$ corresponding to (7.116) and (7.117). Differentiating (7.116) and using (7.114) and (7.117), we find that

$$W_t + A_n^0 W_{x_n} + C^0 W + PD = 0, \tag{7.118}$$

$$W|_{\tilde{\Gamma}} = \begin{cases} -P(\lambda, \mu, x_n), & (x_n, t) \in \tilde{\Gamma}_0, \\ 0, & (x_n, t) \in \tilde{\Gamma}_1 \cup \tilde{\Gamma}_2. \end{cases} \tag{7.119}$$

Here P and D are defined by

$$\begin{aligned} &P(\lambda, \mu, x_n) = B(\lambda, \mu, x_n) \tilde{\Phi}(\lambda-\mu, x_n), \\ &D(\lambda-\mu, x_n, t) = \Phi^{-1}(\lambda-\mu, x_n) H_t(\lambda-\mu, x_n, t). \end{aligned} \tag{7.120}$$

We shall go over from (7.118) and (7.119) to integral relations for the components of the matrices W and P. For this we consider an arbitrary point $(x_n^0, t^0) \in \tilde{G} \cup \tilde{\Gamma}$ and the characteristic passing through it corresponding to the sth diagonal element of the matrix A_n^0; we denote by $L_s(x_n^0, t)$ the oriented segment of this characteristic in accordance with the rule we adopted in the proof of Theorem 1. Integrating the s-component of (7.118), we find that

$$w_{sl}(\lambda, x_n^0, t^0, \mu) = -\int_{L_s(x_n^0, t^0)} (C^0 W + PD)_{sl}\, dt,$$

$$s = 1, 2, \dots, m; \ l = 1, 2, \dots, 2m. \tag{7.121}$$

Setting here $t^0 = 0$ and using (7.119), we have

$$p_{sl}(\lambda, \mu, x_n^0) = \int_{L_s(x_n^0, t^0)} (C^0 W + PD)_{sl}\, dt, \quad s = 1, 2, \dots, m;\ l = 1, 2, \dots, 2m. \tag{7.122}$$

The system (7.121), (7.122) is closed relative to the functions w_{sl} and p_{sl} in the rectangle

$$\Pi_h = \left\{ (x_n, t) \colon 0 \le x_n \le h,\ 0 \le t \le \frac{1}{k_0} h \right\},$$

where

$$k_0 = \min_{1 \le s \le m} \min_{0 \le x_n \le h} |k_s(x_n)|.$$

Carrying out the estimates in this region, we obtain

$$\|W(\lambda, \mu)\| \le \frac{h}{k_0} \left(\|C^0(\lambda)\| \, \|W(\lambda, \mu)\| + \|P(\lambda, \mu)\| \, \|D(\lambda, \mu)\| \right),$$
$$\|P(\lambda, \mu)\| \le \frac{h}{k_0} \left(\|C^0(\lambda)\| \, \|W(\lambda, \mu)\| + \|P(\lambda, \mu)\| \, \|D(\lambda, \mu)\| \right).$$

Here the norms are computed for fixed λ and μ.

Let

$$h^* = \frac{k_0}{\|D(0,0)\| + \|C^0(0)\|}.$$

Then for any positive $h < h^*$, because of the continuous dependence of the matrices D and G^0 on λ and μ, it is possible to find $\varepsilon_0 = \varepsilon_0(h)$, $0 < \varepsilon_0 < \varepsilon/2$, and $\gamma(h) > 0$ such that

$$\frac{h}{k_0} \left(\|C^0(\lambda)\| + \|D(\lambda, \mu)\| \right) < 1 - \gamma(h), \quad |\lambda| \le \varepsilon_0,\ |\mu| \le \varepsilon_0.$$

From these inequalities it then follows that

$$P(\lambda, \mu, x_n) \equiv 0, \quad x_n \in [0, h],\ |\lambda| \le \varepsilon_0,\ |\mu| \le \varepsilon_0.$$

Using the notation (7.120) and (7.115) and the concrete form of the dependence of the matrix C^1 on λ, we find that

$$\tilde{A}_j^1(\mu, x_n) = \tilde{Q}^1(\mu, x_n) = 0, \quad x_n \in [0, h],\ |\mu| \le \varepsilon_0. \tag{7.123}$$

According to the assumptions of the theorem, the A_j^1 and Q^1 have compact support in $\overline{x}$, and hence their Fourier transforms are analytic in μ. From (7.123) it then follows that $\tilde{A}_j^1(\mu, x_n) \equiv \tilde{Q}^1(\mu, x_n) \equiv 0$ for any μ. Hence,

$$A_j^1(x) \equiv Q(x) = 0, \quad j = 1, 2, \dots, m;\ x \in D_h,$$

and the theorem is proved.

THEOREM 7. *If the conditions of Theorem 6 are satisfied and the number* $r \geq 1$ *while the elements of the last* $m - r$ *rows of the matrices* A_i^1 *and* Q^1 *can be linearly expressed in terms of the elements of the first* r *rows of* A_i^1 *and* Q^1, *then in the class of matrices continuous and compactly supported in* D_h *the matrices* A_i^1 *and* Q^1 *are uniquely determined by giving the functions* g_{sl} *only on* $\Gamma_1 \cup \Gamma_0$ [235].

PROBLEM 5. Suppose $G = \{(x,t): x \in R^n,\ t \geq 0\}$, $x = (\overline{x}, x_n)$, $\overline{x} = (x_1, \dots, x_{n-1})$, the matrices A_i are symmetric, and $A_i = A_i(\overline{x}, t)$, $Q = Q(\overline{x}, t)$ and $F = F(x,t)$. Find one of the matrices A_i, Q, given the others, if m solutions u^l of the system (7.72) are known on the set $\Gamma = \Gamma_0 \cup \Gamma_1$, $\Gamma_0 = \{(x,t): x \in R^n,\ t = 0\}$, $\Gamma_1 = \{(x,t): x_n = 0,\ t > 0\}$:

$$u^l|_\Gamma = \psi^l(x,t), \quad l = 1, 2, \dots, m. \tag{7.124}$$

Problem 5 was investigated by S. P. Belinskiĭ in [39] and [40]. We denote by $\mathfrak{M}$ the class of matrices whose elements $s_{ij}(x,t)$ belong to the class $C^\nu(R^n)$ in x, $\nu = [n/2] + 2$, and are jointly continuous on G in x and t; we denote by $\mathfrak{M}(K)$ the class of matrices in $\mathfrak{M}$ having $C(G)$-norm not exceeding K and by $\mathfrak{M}_0$ the class of matrices S whose elements s_{ij} can be represented in the form

$$s_{ij}(x,t) = \sum_{k=1}^{N_{ij}} c_{ijk}(\overline{x}) b_{ijk}(t).$$

Let $A_i \in \mathfrak{M}(K)$, $i = 1, \dots, n$. We denote by $\Omega = \Omega(K)$ the region of the space (x,t) bounded by the closed surface $S = S_0 \cup S_1$, where S_0 is a portion of the plane $t = 0$ and S_1 is a smooth surface based on the boundary of S_0, situated in the half-space $t > 0$, and such that

$$\left(\left[\tau_0 E + \sum_{i=1}^n \tau_i A_i \right] u, u \right) \geq 0$$

on S_1. Here $(\tau_0, \tau_1, \dots, \tau_n)$ denotes the vector of the outer normal to S_1; E is the identity matrix. Let Ω_1 be the projection of Ω onto the plane $x_n = 0$, and let Ψ be the matrix consisting of the columns ψ^l, $l = 1, \dots, m$.

The following two theorems hold.

THEOREM 8. *If* $A_i \in \mathfrak{M}(K)$, $\Psi \in \mathfrak{M}$, $Q \in \mathfrak{M} \cap \mathfrak{M}_0$, *and the functions* F *and* F_t *have derivatives with respect to* x *to order* ν *which are continuous in* G, *then under the condition*

$$\det \Psi(\overline{x}, 0, t) \neq 0, \quad (x,t) \in \Omega_1,$$

the matrix Q *is uniquely determined in* Ω_1 *for known matrices* A_i *by giving* $\Psi(x,t)$ *on* $S_0 \cup \Omega_1$.

THEOREM 9. *If $A_i \in \mathfrak{M}(K)$, $\Psi \in \mathfrak{M}$, $Q \in \mathfrak{M}$, and the vector F satisfies the conditions of Theorem 8, then a matrix $A_k \in \mathfrak{M}(K) \cap \mathfrak{M}_0$ is uniquely determined in Ω_1 for known other matrices of the system (7.72) by giving Ψ on $S_0 \cup \Omega_1$ if for $(x,t) \in \Omega_1$ the following conditions are satisfied:*

$$\begin{aligned} &\det\left(\frac{\partial}{\partial x_k}\Psi(\overline{x},0,t)\right) \neq 0, && k \neq n, \\ &\det\left(\Psi_t + \sum_{i=1}^{n-1} A_i \Psi_{x_i} + Q\Psi - \hat{F}\right) \neq 0, && k = n. \end{aligned}$$

Here $\hat{F}$ denotes the square matrix, all m columns of which coincide with the vector of the right side $F(x,t)$.

The proof of these two theorems is based on using the machinery of energy inequalities for symmetric systems and the Sobolev imbedding theorems [39], [40].

§5. Inverse problems for parabolic equations of second order

In this section we consider some questions pertaining to the general theory of inverse problems, namely, the connection of inverse problems for equations of hyperbolic and parabolic types, the method of descent in inverse problems, and questions related to the investigation of concrete classes of inverse problems for equations of parabolic type.

1. *The connection between solutions of direct problems for equations of hyperbolic and parabolic types and inverse problems.* A connection between solutions of partial differential equations of different types was observed long ago. The usefulness of this connection in the investigation of inverse problems was first pointed out by K. G. Reznitskaya [207]. The essence of this consists in the following. Suppose we consider two problems of determining a uniformly elliptic operator L_q with coefficients depending only on the variable x (in the present case the dimension of the space x plays no role) from two different but at the same time coordinated problems

$$u_{tt} = L_q u + f(x,t), \qquad u|_{t=0} = 0, \qquad u_t|_{t=0} = \varphi(x), \qquad u|_{\mathfrak{M}} = g(x,t), \tag{7.125}$$

$$v_t = L_q v + F(x,t), \qquad v|_{t=0} = \varphi(x), \qquad v|_{\mathfrak{M}} = G(x,t), \tag{7.126}$$

in which $\mathfrak{M}$ is the direct product of some surface S in x-space with the segment $[0,\infty)$ of the t-axis. We suppose that the class of solutions under consideration is such that the Laplace transformation in the variable t can be applied to the functions u, v, f, and F. Then their transforms $\tilde{u}(x,p)$, $\tilde{v}(x,\sigma)$, $\tilde{f}(x,p)$, and

$\tilde{F}(x,\sigma)$ are related by the correspondence

$$p^2\tilde{u}(x,p) = L_q\tilde{u} + \tilde{f}(x,p) + \varphi(x),$$
$$\sigma\tilde{v}(x,\sigma) = L_q\tilde{v} + \tilde{F}(x,\sigma) + \varphi(x).$$

Therefore, if

$$\tilde{f}(x,p) = \tilde{F}(x,p^2), \tag{7.127}$$

then

$$\tilde{u}(x,p) = \tilde{v}(x,p^2). \tag{7.128}$$

In terms of preimages this corresponds to the relation

$$F(x,t) = \frac{1}{2\sqrt{\pi t^3}}\int_0^\infty \tau\exp(-\tau^2/4t)f(x,\tau)\,d\tau, \tag{7.129}$$

$$v(x,t) = \frac{1}{2\sqrt{\pi t^3}}\int_0^\infty \tau\exp(-\tau^2/4t)u(x,\tau)\,d\tau. \tag{7.130}$$

Because of this, the additional data of problems (7.125) and (7.126) intended for finding the operator L_q are connected by the correspondence

$$G(x,t) = \frac{1}{2\sqrt{\pi t^3}}\int_0^\infty \tau\exp(-\tau^2/4t)g(x,\tau)\,d\tau, \quad (x,t)\in\mathfrak{M}. \tag{7.131}$$

Formula (7.131) makes it possible to compute the data of problem (7.126) from the data of problem (7.125). Since for each fixed $x\in S$ the set $\mathfrak{M}$ contains the segment $[0,\infty)$ of the t-axis, formula (7.131) is invertible ($G(x,t)$ up to a factor is the Laplace transform of the function $g(x,\sqrt{z})$ with respect to the argument z with transformation parameter $1/4t$). Hence, (7.131) establishes a one-to-one correspondence between the additional data of problems (7.125) and (7.126). Instead of investigating problem (7.126), we can consider problem (7.125) which has been studied more than problem (7.126). In particular, it is not difficult to paraphrase the results of §3 for the case of an equation of parabolic type.

2. ***The method of descent in inverse problems.*** As is known, the method of descent consists in deriving from a known formula solving some problem in n-dimensional space a formula solving an analogous problem in a space of dimension $n-1$. In inverse problems the idea of using the method of descent was stated by M. V. Klibanov in [126]. We present it for the example of problem (7.126).

Suppose $x\in R^{n-1}$ and $\varphi\equiv\delta(x-x^0)$. We introduce the fundamental solution $H(y,t,y^0)$ of the heat equation:

$$H_t = H_{yy}, \qquad H|_{t=0} = \delta(y-y^0).$$

It is not hard to verify that the product $v \cdot H \equiv w$ satisfies the conditions

$$\begin{aligned} &w_t = L_q w + w_{yy} + F(x,t)H(y,t,y^0), \\ &w|_{t=0} = \delta(x - x^0)\delta(y - y^0), \\ &w|_{\mathfrak{M}'} = G(x,t)H(y,t,y^0) \equiv G_1(x,y,t,y^0). \end{aligned} \tag{7.132}$$

Here $\mathfrak{M}'$ denotes the direct product of $\mathfrak{M}$ with the product of the segments $(-\infty, \infty)$ of the axes y and y^0. If the inverse problem (7.132), consisting in finding the operator L_q, has a unique solution, this means that L_q can be found uniquely from G_1, i.e., in the final analysis from $G(x,t)$; therefore, uniqueness of a solution of problem (7.132) in n-dimensional space implies uniqueness of a solution of problem (7.126) in $(n-1)$-dimensional space. Since we showed above that problems (7.125) and (7.126) are equivalent, all that has been said with regard to problem (7.126) applies in equal measure to (7.125).

We shall present an example of using the method of descent. In §1 we considered the following problem in three-dimensional space: find $q(x)$ if a solution $u_1(x,t,x^0)$ is known for the equation

$$(u_1)_{tt} = \Delta u_1 + q(x)u_0(x,t,x^0) \tag{7.133}$$

with zero initial data on the plane $x_3 = 0$. Here $u_0(x,t,x^0)$ is the fundamental solution of the wave equation with a singularity at the point $(x^0, 0)$. It was indicated in §1 that for fixed x^0 this problem has a unique solution in a class of functions even in the variable x_3. On the basis of this result and the above method of descent, we can assert that in two-dimensional space $x = (x_1, x_2)$ the coefficient $q(x)$ is uniquely determined (in a class of functions even in the variable x_2) by giving the solution of (7.133) for $x_2 = 0$. Indeed, first of all, we can assert that $q(x)$, $x = (x_1, x_2, x_3)$, is uniquely determined (in a class of functions even in x_2) by giving the solution of the Cauchy problem

$$(v_1)_t = \Delta v_1 + q(x)v_0(x,t,x^0), \qquad v_1|_{t=0} = 0, \tag{7.134}$$

for $x_2 = 0$. In (7.134) $v_0(x,t,x^0)$ is the fundamental solution of the heat equation with singularity at the point x^0, 0. But if q does not depend on x_3, the functions v_0 and v_1 for three- and two-dimensional space are related to one another by means of the factor

$$\frac{1}{2\sqrt{\pi t}} \exp\left[-\frac{(x_3 - x_3^0)^2}{4t}\right].$$

Therefore, having data for the two-dimensional equation (7.134), we can always obtain analogous data on the plane $x_2 = 0$ for the three-dimensional equation (7.134). Since $q(x)$ can be found uniquely from three-dimensional data (in a class of functions even in x_2), this implies the corresponding uniqueness theorem for the two-dimensional equation (7.134) and hence for the two-dimensional equation (7.133).

3. ***Inverse problems for equations of parabolic type.*** A number of works have been devoted to the linear inverse problems of determining the right side of a parabolic equation or one of its coefficients in a linearized formulation (see [125]–[130], [156], [157], and [159]). We shall not touch on them here but concentrate our attention on the special class of inverse problems studied by A. D. Iskenderov [116]–[118] and N. Ya. Beznoshchenko [34]–[37]. At a descriptive level this class of inverse problems can be characterized as follows: we consider the problem of finding one or several coefficients of a linear (or quasilinear) parabolic equation which do not depend on one of the spatial variables; to find them, information is given regarding the solution of the Cauchy problem or a boundary value problem of special type: in the formulation of Iskenderov the unknown coefficients occur in the additional condition; in the formulation of Beznoshchenko the additional condition is given on a plane orthogonal to that variable on which the coefficients do not depend. This special prescription of the additional information makes it possible to obtain a closed system of integral equations of the second kind for the unknown coefficients and solution. It is curious to note that the method of Beznoshchenko is based on first obtaining an auxiliary problem in which the unknown coefficient is contained in the additional condition. The rest of the construction is actually based on the method of Iskenderov.

In order to not encumber the exposition of the main idea with generality unnecessary here, we demonstrate the method of Beznoshchenko and Iskenderov with the example of finding the coefficient $q(x_1, x_2, t)$ from the equation

$$u_t = \Delta u + qu \tag{7.135}$$

under the condition

$$u|_{t=0} = \varphi(x) \tag{7.136}$$

on the basis of the given information

$$u|_{x_3=0} = f(x_1, x_2, t). \tag{7.137}$$

We shall assume that the functions q, φ, and f are sufficiently smooth. We note that the consistency condition $\varphi(x_1, x_2, 0) = f(x_1, x_2, 0)$ must be satisfied. We introduce the function $w = u_{x_3x_3}$. It satisfies the conditions

$$w_t = \Delta w + qw, \qquad w|_{t=0} = \varphi_{x_3x_3}, \tag{7.138}$$

$$w|_{x_3=0} = f_t - f_{x_1x_1} - f_{x_2x_2} - q\varphi(x_1, x_2, 0). \tag{7.139}$$

It is not hard to show that problem (7.138), (7.139) is equivalent to problem (7.135)–(7.137) [36]. If $\varphi(x_1, x_2, 0) \neq 0$, then relation (7.139) can be solved for $q(x_1, x_2, t)$:

$$q(x_1, x_2, t) = \frac{1}{\varphi(x_1, x_2, 0)}[f_t - f_{x_1x_1} - f_{x_2x_2} - w|_{x_3=0}]. \qquad (7.139')$$

The equalities (7.138) define w as a nonlinear operator on q. In connection with this (7.139′) can be considered a nonlinear integral equation for q in which the operator $w|_{x_3=0}$ is defined by (7.138). Another approach is possible, however; namely, we replace (7.138) by the integral equation

$$w(x,t) = \int_{R^3} G(x,t,\xi,0)\varphi_{\xi_3\xi_3}(\xi)\,d\xi + \int_0^t d\tau \int_{R^3} G(x,t,\xi,\tau)q(\xi_1,\xi_2,\tau)w(\xi,\tau)\,d\xi. \qquad (7.140)$$

Here $G(x, t, \xi, \tau)$ is the fundamental solution of the equation $w_t = \Delta w$. In (7.139′) in place of $w|_{x_3=0}$ we substitute its expression found from (7.140). We then obtain two nonlinear integral equations of the second kind with a small parameter t. It is clear that for sufficiently small t this system of two integral equations is solvable. We thus obtain an existence and uniqueness theorem for a solution in the small (see [36]). On the basis of these same equations it is also possible to find an estimate of conditional stability of the problem.

More complex problems of determining the coefficients of the leading derivatives of a parabolic equation can be investigated in a similar way.

§6. An abstract inverse problem and questions of its being well-posed

Many formulations of inverse problems for differential equations fit, in the framework of functional analysis, into the following scheme. There is a family of operators A_q acting from a space X to a space Y. This family of operators depends on an element q belonging to some space Q as a parameter. Together with this there is an operator B acting from the space X to a space Z. It is assumed below that X, Y, Z, and Q are normed linear spaces. The problem of solving the equation

$$A_q x = y \qquad (7.141)$$

for given $y \in Y$ and $q \in Q$ is called the direct problem for the operator A_q. The problem inverse to it consists in finding a specific operator A_q belonging to a family of operators A_q (or, equivalently, the parameter $q \in Q$) if regarding it we know that the operator B assigns an element $z \in A$ to a solution of equation (7.141) for fixed y, i.e., if x is a solution of (7.141), then

$$Bx = z. \tag{7.142}$$

In many cases of practical importance the operator B does not have an inverse, while the family of operators A_q possesses nice properties, namely, for each operator A_q, $q \in Q$, there exists a bounded inverse operator A_q^{-1}. In connection with this it is easy to find a solution of the direct problem (7.141), and the inverse problem reduces to solving the operator equation

$$Uq \equiv BA_q^{-1}y = z, \tag{7.143}$$

for $q \in Q$ and fixed y and z.

It is known that many inverse problems for differential equations lead to problems which are classically ill-posed. A. N. Tikhonov suggested altering the concept of a well-posed problem for problems of this type. In application to equation (7.143) Tikhonov's approach to the concept of well-posed problems consists in the following: we consider a set $m \subset Q$ (as a rule, this is some compact set) and its image in the space Z under the mapping effected by operator U (y is fixed). Let $Um = R \subset Z$. The problem of solving (7.143) is called well-posed on the set m if 1) it is known a priori that $z \in R$, 2) a solution of equation (7.143) is unique on the set m, and 3) the solution of this equation is stable with respect to small variations of z on the set R. The set m is here called a set of well-posedness for problem (7.143). Conditions 1)–3) generalize in a natural way the classical conditions of existence, uniqueness, and stability required of a well-posed problem and are physically justified for a large class of inverse problems. For problems well-posed in Tikhonov's sense (conditionally well-posed) there are rather well-developed methods for solving them [113], [145], [185], [266]. In connection with this, it is of primary importance to establish conditional well-posedness of a problem. As is evident from the very definition of conditional well-posedness, the central question in establishing that a problem is well-posed is the question of uniqueness of a solution of the problem.

Investigation of the inverse problem (7.141), (7.142) is impeded by the circumstance that equation (7.143), generally speaking, is nonlinear even in the case where the operators B and A_q are linear. In this section we present two rather general methods of investigating the inverse problem (7.141), (7.142) by means of which, in our view, a large circle of specific inverse problems can be efficiently studied. The content of this section corresponds to [221].

1. *Reduction to the investigation of a two-parameter family of linear equations.* We shall here distinguish a class of inverse problems for which questions of uniqueness and stability of a solution of the nonlinear equation (7.143) can be reduced to the investigation of analogous questions for a family of linear problems. The latter can often be studied much more simply.

We consider the case where the operators A_q and B are linear, and the family of operators A_q admits a special representation. Let m be some set of the space Q.

THEOREM 1. *Suppose B is a linear operator, the operators A_q for $q \in m$ have inverses A_q^{-1}, and the family of operators A_q can be represented in the form*

$$A_q = A + A_q^0, \tag{7.144}$$

where A is a linear operator and A_q^0 is a bilinear operator, i.e., it is linear relative to $x \in X$ and $q \in Q$. For uniqueness of the inverse problem (7.141), (7.142) *on the set $m \subset Q$ for fixed $y \in Y$ and $z \in Z$ it is then sufficient that the two-parameter family of linear operator equations*

$$T_{q_1 q_2} q = 0, \tag{7.145}$$

where q_1 and q_2 are arbitrary elements of the set m and

$$T_{q_1 q_2} q \equiv B A_{q_1}^{-1} A_q^0 A_{q_2}^{-1} y, \tag{7.146}$$

have no nontrivial solutions of the form $q = q_1 - q_2$.

PROOF. Suppose that for fixed y and z a solution of the inverse problem is not unique. Then there exist $q_1, q_2 \in m$, $q_1 \neq q_2$, for which $BA_{q_1}^{-1} y = z$ and $BA_{q_2}^{-1} y = z$. Hence,

$$0 = B(A_{q_1}^{-1} - A_{q_2}^{-1}) y = B A_{q_1}^{-1} (A_{q_2} - A_{q_1}) A_{q_2}^{-1} y = T_{q_1 q_2} q, \tag{7.147}$$

where $q = q_1 - q_2 \neq 0$, and we arrive at a contradiction.

COROLLARY (a sufficient criterion for uniqueness of the inverse problem). *Suppose the family of linear equations* (7.145) *has no nonzero solutions. Then a solution of the inverse problem* (7.141), (7.142) *is unique.*

The next theorem shows that the condition that there be no nontrivial solutions of the family of homogeneous equations (7.145) of the form $q = q_1 - q_2$, where $q_1, q_2 \in m$, is in a certain sense also necessary for uniqueness of the inverse problem (7.141), (7.142) on the set m.

THEOREM 2. *Suppose the operator B and the family of operators A_q satisfy the conditions of Theorem* 1. *Then for uniqueness of a solution of the inverse problem* (7.141), (7.142) *on the set m for fixed $z \in R$ and any fixed*

$y \in Y$ it is necessary that the family of equations (7.145) *have no nontrivial solutions of the form $q = q_1 - q_2$, where $q_1, q_2 \in m$.*

PROOF. Suppose for some $q_1 \neq q_2$ equation (7.145) has a solution $q = q_1 - q_2$. Further, let $BA_{q_1}^{-1}y = z_1$ and $BA_{q_2}^{-1}y = z_2$. Then, using (7.147), we find that $z_1 - z_2 = -T_{q_1 q_2} q = 0$. Hence, for given y also $z = z_1 = z_2$, and the solution of the inverse problem is not unique.

Investigation of stability of the family of nonhomogeneous operator equations corresponding to the operators $T_{q_1 q_2}$ makes it possible to draw certain conclusions also regarding the stability of the inverse problem (7.141), (7.142).

THEOREM 3. *Suppose the operator B and the family of operators A_q satisfy the conditions of Theorem* 1. *Suppose, moreover, that the inverse problem* (7.141), (7.142) *for fixed $y \in Y$ has a unique solution on a set $m \subset Q$ and the family of linear operators $T_{q_1 q_2}$, $q_1, q_2 \in m$, possesses the property that a solution of the operator equations*

$$T_{q_1 q_2} q = z \tag{7.148}$$

is uniformly stable(with regard to distinct $q_1, q_2 \in m$) with respect to small perturbations of the right side. Then solution of the inverse problem (7.141), (7.142) *is stable with respect to small variations of the element z under the condition that they do not take a solution of the problem beyond the confines of the set m.*

We denote by R the set belonging to the space Z which is the image of the set m under the mapping effected by U (see (7.143)). We choose any two elements $z_1, z_2 \in R$. To them there then uniquely correspond elements $q_1, q_2 \in m$. Setting $\tilde{z} = z_1 - z_2$ and $\tilde{q} = q_1 - q_2$, by computations analogous to those above we find that

$$T_{q_1 q_2} \tilde{q} = -\tilde{z}. \tag{7.149}$$

We now choose an arbitrary $\varepsilon > 0$. The uniform stability of a solution of (7.149) then implies that it is possible to find a $\delta > 0$, not depending on q_1 and q_2, such that $\|\tilde{q}\| < \varepsilon$ for $\|\tilde{z}\| < \delta$. But this means stability of the inverse problem (7.141), (7.142) with respect to small changes of z which do not take the solution of the inverse problem beyond the confines of the set m. The set m is thus a set of conditional well-posedness for the inverse problem (7.141), (7.142).

We shall consider some examples of concrete formulations of inverse problems in order to illustrate the problems to which investigation of equations (7.145) can lead.

EXAMPLE 1. Let X be the space of functions $x(s,t)$, $s = (s_1, \ldots, s_n)$, which are twice continuously differentiable with respect to their arguments in the region $D = \{(s,t)\colon |s| < \infty,\ 0 \le t < \infty\}$ and for $t = 0$ satisfy the conditions

$$x(s,0) = 0, \qquad x_t(s,0) = 0; \tag{7.150}$$

let Y be the space of functions $y(s,t)$ continuous in the region D; let Q be the space of functions $q(s)$ continuous in the region $D_0 = \{s\colon |s| < \infty\}$; and let Z be the space of functions $z(s_1, \dots, s_{n-1}, t)$ which are twice continuously differentiable with respect to their arguments in the region $D_1 = \{(s,t)\colon s_n = 0,\ t \ge 0\}$.

We determine the operator A_q from the equality

$$A_q x = x_{tt} - \Delta_s x + q(s)x, \tag{7.151}$$

in which the symbol Δ_s is the Laplace operator in the variables $s_1, \dots, s_n$, and we determine the operator B from the equality

$$Bx = x(s,t)|_{s_n=0}; \tag{7.152}$$

for these A_q and B we consider problem (7.141), (7.142).

For $n = 1$ and under the additional condition $q(-s) = q(s)$ this problem is equivalent to the spectral formulation of the inverse Sturm-Liouville problem studied in detail by V. A. Marchenko, I. M. Gel′fand, B. M. Levitan, M. G. Kreĭn, and other authors [165], [177]. For $n = 2, 3$ this problem in a spectral formulation was first considered by Yu. M. Berezanskiĭ [42]–[44]; later it was also studied in [217] and [223].

We denote by $G_q(s, s^0, t-t^0)$ the Green function corresponding to equation (7.141). Then (7.145) can be written in the form

$$\iint\limits_D G_{q_1}(s, s^0, t-t^0) q(s^0) \left\{ \iint\limits_D G_{q_2}(s^0, \xi, t^0 - \tau) y(\xi, \tau)\, d\xi\, d\tau \right\} ds^0\, dt^0 = 0. \tag{7.153}$$

We recall here that the Green function $G_q(s, s^0, t-t^0)$ is nonzero only in the region $t - t^0 \ge |s - s^0|$. Equation (7.153) then takes the form

$$\int_{|s-s^0| \le t} q(s^0) \rho(s, s^0, t)\, ds^0 = 0, \tag{7.154}$$

where the weight function $\rho(s, s^0, t)$ is computed for fixed q_1, q_2, and y from the formula

$$\rho(s, s^0, t) = \int_0^{t-|s-s^0|} dt^0 \int_{|\xi - s^0| \le t^0} d\xi \int_0^{t^0 - |\xi - s^0|} G_{q_1}(s, s^0, t-t^0) \times G_{q_2}(s^0, \xi, t^0 - \tau) y(\xi, \tau)\, d\tau. \tag{7.155}$$

We thus arrive at the necessity of studying the following problem of integral geometry: it is known that for all spheres of arbitrary radius $0 < t < \infty$ whose centers belong to the plane $s_n = 0$ the integrals of the function q in product with a known weight function ρ are equal to zero; does this imply that $q = 0$?

If the answer to this question is positive for any $q_1, q_2 \in m$, then uniqueness of the solution of the inverse problem (7.141), (7.142) on the set m follows from Theorem 1. In the case $\rho = 1$ problem (7.154) is investigated in [139]; for surfaces of more general structure than spheres it is studied in [217] and [223] (see Chapter VI).

It is interesting to note that in the special case where the function $y(s,t) = \delta(s-s^1,t)$, where $\delta(s-s^1,t)$ is the Dirac delta function (a generalized function) concentrated at the point $(s^1,0)$, $s^1 = (s_1^1,\dots,s_{n-1}^1,0)$, the weight function ρ assumes the especially simple form

$$\rho(s,s^0,t) = \int_{|s^0-s^1|}^{t-|s-s^0|} G_{q_1}(s,s^0,t-t^0)G_{q_2}(s^0,s^1,t^0)\,dt^0, \tag{7.156}$$

and, since $\rho \equiv 0$ for $t^0 < |s^0 - t^1|$, (7.154) reduces to the form

$$\int_{|s^0-s|+|s-s^1|\le t} q(s^0)\rho(s,s^0,t)\,ds^0 = 0, \quad (s,t) \in D_1. \tag{7.154$'$}$$

The problem arising here of determining a function in terms of integrals over ellipsoids of rotation having one focus fixed (at the point s^1) while the other runs through a set of points of the plane $s_n = 0$ is considered in [209].

Thus, investigation of inverse problems for hyperbolic equations is closely related to investigation of problems of integral geometry. In the case where $q(s)$ is a function of a single variable, (7.154) is an ordinary one-dimensional Volterra equation of the first kind whose study is considerably simpler than investigation of problems of integral geometry.

EXAMPLE 2. Let X be the space of functions $x(s)$, $s = (s_1,\dots,s_n)$, which are twice continuously differentiable in a finite region D with smooth boundary Γ and which vanish on Γ:

$$x(s)|_\Gamma = 0; \tag{7.157}$$

let Y be the space of functions $y(s)$ continuous in D; let Q be the space of functions $q(s)$ with support in $D_0 \subset D$ and continuous in D_0; and let Z be the space of functions $z(s)$ defined and continuous in a region $D_1 \subset D$, where $D_1 \cap D_0 = \varnothing$. Further, let

$$A_q x = Lx + q(s)x, \tag{7.158}$$

where L is a given elliptic operator of second order in the variables $s_1,\dots,s_n$, and let B be the projection operator assigning to a function $x(s)$ its values on the set of points $s \in D_1$,

$$Bx = x(s), \quad s \in D_1. \tag{7.159}$$

For problem (7.141), (7.142), the equality (7.145) then reduces by means of the Green function $G_q(s, s^0)$ to the form

$$\int_{D_0} q(s^0)\rho(s^0, s)\, ds^0 = 0, \quad s \in D_1, \tag{7.160}$$

where ρ is defined by

$$\rho(s^0, s) = G_{q_1}(s, s^0) \int_D y(\xi) G_{q_2}(s^0, \xi)\, d\xi. \tag{7.161}$$

The problem thus reduces to investigation of an equation of the type of a Fredholm equation of the first kind. We mention that equations similar to (7.160) were obtained in studying inverse problems for equations of elliptic type (see §1 of this chapter). Certain limit relations equivalent to linearization of the original inverse problem were here used to obtain equations of the type (7.160).

Equation (7.160) is characteristic for the investigation of problems of finding the coefficients of linear differential operators of elliptic and parabolic types.

2. ***The method of linearization in investigating the inverse problem.*** At the beginning of the section it was noted that under certain constraints on the family of operators A_q, problem (7.141), (7.142) reduces to solving the equation

$$Uq = z. \tag{7.162}$$

Here the operator U is nonlinear, generally speaking. In investigating the inverse problem, in many cases it is much simpler to investigate a linearized problem corresponding to the original nonlinear problem. If in a neighborhood of an element q_0 a continuous Fréchet derivative U'_q of the operator U exists and a bounded inverse operator $[U'_{q_0}]^{-1}$ also exists, then, invoking a familiar theorem on the existence of an inverse operator (see, for example, [121]), it can be asserted that the nonlinear problem (7.162) is well-posed in a sufficiently small neighborhood of q_0. However, investigation of specific multidimensional inverse problems for differential equations has shown that the linearized inverse problems are often not well-posed in the classical sense: generally speaking, the inverse operator $[U'_{q_0}]^{-1}$ is not bounded. This makes it impossible to use the theorem on the existence of an inverse operator. In this connection the following question arises: if the linearized inverse problem is conditionally well-posed, then does the original nonlinear problem possess this property? A partial answer to this question is given below. It turns out that it is possible to distinguish a class of compact sets on which conditional well-posedness of the linearized problem implies conditional well-posedness of the nonlinear problem.

DEFINITION. A set $m \subset Q$ is called *locally similar relative to an element* $q_0 \in Q$ if there exists a closed ball $S(q_0, r) \subset Q$ with center at q_0 of radius $r > 0$ such that the set $m_r(q_0) = m \cap S(q_0, r)$ is nonempty and the condition $q \in m_r(q_0)$ implies that the element $\overline{q} = q_0 + (q - q_0)r/\|q - q_0\|$ also belongs to $m_r(q_0)$.

As is evident from this definition, for locally similar sets it is characteristic that together with elements $q \in m$ contained inside a ball $S(q_0, r)$ the set m also contains elements $\overline{q}$ which are obtained by means of similarity projection of elements q on the surface of $S(q_0, r)$ relative to a center of similarity located at q_0. Indeed, $\|\overline{q} - q_0\| = r$. From this it is clear that sets such as a closed sphere with center at q_0 and a line passing through q_0 are sets locally similar relative to q_0. A line not passing through q_0 is an example of a set not possessing the property of similarity relative to q_0.

Below we shall consider compact sets in Q which are locally similar relative to an element q_0. A large class of such sets can be obtained by the following construction: we choose a closed ball $S(q_0, r)$ and a compact set m_0 of elements q lying on the surface of the ball, $\|q - q_0\| = r$, and we construct line segments joining the points q_0 and q for all $q \in m_0$; then the set m obtained by taking the union of all such segments $[q_0, q]$, $q \in m_0$, is a compact set in Q which is locally similar relative to the element q_0. Suppose, for example, that $Q = G(D)$, where D is a closed region of Euclidean space R^n. Then the set of functions $q \in C(D)$ for which the inequalities

$$\|q - q_0\|_C \leq r, \qquad \|\nabla(q - q_0)\|_C \leq K\|q - q_0\|_C, \quad K > 0, \tag{7.163}$$

hold is a compact set in $C(D)$ which is locally similar relative to q_0. Compact, locally similar sets possess the following property: if Q is a Banach space and the linear problem defined by the Fréchet derivative U'_{q_0} of the operator U is conditionally well-posed, then problem (7.162) is also conditionally well-posed. For the proof we first establish an auxiliary lemma.

LEMMA. *Suppose a linear operator A defined on a Banach space Q with range in a normed linear space Z possesses the following properties*: 1) *the operator A is continuous*; 2) *the equation $Aq = 0$ has only the trivial solution $q = 0$. Then on any compact set m in Q which is locally similar relative to an element $q_0 \in Q$*

$$\|A(q - q_0)\| \geq \alpha\|q - q_0\|, \tag{7.164}$$

where α is a positive constant depending only on the set m.

PROOF. For the set m we consider the ball $S(q_0, r)$ appearing in the definition of a locally similar set. Let m_0 be the intersection of m with the surface of this ball. Since the set m_0 is compact while the operator A is continuous and its kernel consists of 0, $0 \notin m_0$, there exists $\alpha > 0$ for which $\|Aq\| = \alpha r$ and $q \in m_0$. By the definition of a locally similar set for $q \in m$ the estimate (7.164) then holds.

REMARK. From the proof of the lemma it is evident that to obtain the estimate (7.164) on a specific compact set locally similar relative to an element q_0, it suffices that the operator A possess properties 1) and 2) only on the closure of the set m.

Inequality (7.164) is obviously equivalent to boundedness of the operator A^{-1} on the image Am of the set m. The next result follows from this in an elementary way.

THEOREM 4. *Suppose at the point q_0 the nonlinear operator U has a Fréchet derivative U'_{q_0} satisfying the conditions of the lemma. Then for any compact set m belonging to the Banach space Q which is locally similar relative to the element q_0, there exists a closed ball $S(q_0, r)$, $r > 0$, such that on the set $m_r(q_0) = m \cap S(q_0, r)$ the inequality*

$$\|Uq - Uq_0\| \geq \gamma \|q - q_0\| \tag{7.165}$$

holds, where γ is a positive constant.

The condition that the operator U have a Fréchet derivative at the point q_0 means that

$$Uq = Uq_0 + U'_{q_0}(q - q_0) + \omega(q, q_0),$$

where $\|\omega(q, q_0)\| = o(\|q - q_0\|)$. By the lemma on the set m we then have

$$\begin{aligned}\|Uq - Uq_0\| &\geq \|U'_{q_0}(q - q_0)\| - \|\omega(q, q_0)\| \\ &\geq \alpha\|q - q_0\| - o(\|q - q_0\|) = \left[\alpha - \frac{o(\|q - q_0\|)}{\|q - q_0\|}\right] \|q - q_0\|.\end{aligned}$$

From this it is clear that by choosing elements $q \in m$ for which $\|q - q_0\|$ is sufficiently small it is possible to arrange, for example, that

$$\|Uq - Uq_0\| \geq \alpha\|q - q_0\|/2.$$

This proves the theorem.

THEOREM 5. *Suppose with regard to the operator $U[Q \to Z]$ the conditions of Theorem 4 are satisfied, and let $Uq_0 = z_0$; then for each compact set m which is locally similar relative to the element q_0 there exists a closed ball $S(q_0, r)$, $r > 0$, such that on the set $m_r(q_0) = m \cap S(q_0, r)$ a solution of the equation*

$$Uq = z_0 \tag{7.166}$$

is unique and stable relative to small variations of z_0 *on the set* $R = Um_r(q_0) \subset Z$.

This theorem is a corollary of Theorem 4. Indeed, uniqueness follows immediately from (7.146). Stability of the solution relative to small variations of z_0 on the set R follows from the same inequality. Let $z \in R$; then, denoting by q its preimage in the set $m_r(q_0)$ (any preimage if there are several), from (7.165) we find that $\|q - q_0\| \leq \|z - z_0\|/\gamma$, which implies stability of the solution in the sense indicated above.

From this theorem it follows that the problem of solving equation (7.166) on the set $m_r(q_0)$ is conditionally well-posed. We further note a corollary of this theorem.

COROLLARY. *If the operator* U *satisfies the conditions of Theorem* 4, *then there exists a set* $M(q_0)$ *which is starlike relative to an element* q_0 *and contains elements* $q \in Q$ *along any direction issuing from* q_0 *on which a solution of equation* (7.166) *is unique.*

The corollary becomes obvious if we observe that any segment $[q_0, q]$ joining a pair of points q_0, q in the space Q is a compact set which is locally similar relative to q_0.

In conclusion we note a connection between the equations to which we have reduced the investigation of the inverse problem (7.141), (7.142) and the equation corresponding to the linearized inverse problem. It is not hard to verify that in the case studied in subsection 1 (a family of operators representable in the form (7.144)) the Fréchet derivative of the operator U (see (7.143)) has the form $U'_{q_0} = T_{q_0 q_0}$, where the operator $T_{q_0 q_0}$ is defined by (7.146). Thus, the equation arising in linearization of problem (7.141), (7.142) belongs to the two-parameter family of linear operator equations (7.148).

In [228] a case is considered where the family of operators A_q is not linear. It is shown that the method of investigating the inverse problem presented in subsection 1 can be generalized to a fairly large class of nonlinear operators A_q which includes practically all cases of interest in applications.

Bibliography*

[1] A. S. ALEKSEEV, *Some inverse problems in wave propagation theory.* I, II, Izv. Akad. Nauk SSSR Ser. Geofiz. **1962**, 1514–1522, 1523–1531; English transl. in Bull. (Izv.) Acad. Sci. USSR Geophys. Ser. **1962**.

[2] ——, *Inverse dynamical problems of seismology*, Methods and Algorithms for the Interpretation of Geophysical Data (M. M. Lavrent′ev, editor), "Nauka," Moscow, 1967, pp. 9–84. (Russian) R. Zh. Mat. **1968** #6Б562.

[3] A. S. ALEKSEEV et al., *A numerical method for solving the three-dimensional inverse kinematic problem of seismology*, Mat. Problemy Geofiz. Vyp. 1 (1969), 179–201. (Russian)

[4] A. S. ALEKSEEV et al., *A numerical method for determining the structure of the earth's upper mantle*, Mat. Problemy Geofiz. Vyp. 2 (1971), 143–165. (Russian) R. Zh. Mat. **1971** #10Б816.

[5] A. S. ALEKSEEV AND A. G. MEGRABOV, *The direct and inverse problems of the scattering of plane waves by inhomogeneous transition layers.* I, Mat. Problemy Geofiz. Vyp. 3 (1972), 8–36. (Russian) R. Zh. Mat. **1973** #5Б442.

[6] ——, *Inverse problems of the scattering of spherical waves by inhomogeneous media*, Mat. Problemy Geofiz. Vyp. 4 (1973), 8–29. (Russian) R. Zh. Mat. **1974** #1Б362.

[7] ——, *Inverse problems for a string with an oblique derivative condition at one end, and inverse problems of plane wave scattering by inhomogeneous layers*, Dokl. Akad. Nauk SSSR **219** (1974), 308–310; English transl. in Soviet Phys. Dokl. **19** (1974/75).

* *Editor's note.* The periodical Mat. Problemy Geofiz., frequently cited herein, is edited by M. M. Lavrent′ev and A. S. Alekseev, and published by the Computing Center, Siberian Branch, Academy of Sciences of the USSR, Novosibirsk.

[8] B. K. AMONOV AND S. P. SHISHAT·SKIĬ, *An a priori estimate of the solution of the Cauchy problem with data on a time-like surface for a second-order parabolic equation, and related uniqueness theorems*, Dokl. Akad. Nauk SSSR **206** (1972), 11–12; English transl. in Soviet Math. Dokl. **13** (1972).

[9] YU. E. ANIKONOV, *On uniqueness of the solution of inverse problems*, Mat. Problemy Geofiz. Vyp. 1 (1969), 26–40. (Russian) R. Zh. Mat. **1970** #12Б589.

[10] ——, *On geometric methods of studying inverse problems*. I, Mat. Problemy Geofiz. Vyp. 2 (1971), 7–53. (Russian) R. Zh. Mat. **1971** #12Б624.

[11] ——, *On a class of operator equations*, Sibirsk. Mat. Zh. **13** (1972), 1383–1386; English transl. in Siberian Math. J. **13** (1972).

[12] ——, *A problem on the definition of a Riemannian metric*, Dokl. Akad. Nauk SSSR **204** (1972), 1287–1288; English transl. in Soviet Math. Dokl. **13** (1972).

[13] ——, *On an application of geometric methods*, Mat. Problemy Geofiz. Vyp. 3 (1972), 63–76. (Russian) R. Zh. Mat. **1973** #5A706.

[14] ——, *On operator equations of the first kind*, Dokl. Akad. Nauk SSSR **207** (1972), 257–258; English transl. in Soviet Math. Dokl. **13** (1972).

[15] ——, *On quasimonotone operators*, Mat. Problemy Geofiz. Vyp. 3 (1972), 86–99. (Russian) R. Zh. Mat. **1973** #5Б896.

[16] ——, *Uniqueness of the solution of integral equations of the first kind*, Mat. Zametki **14** (1973), 493–498; English transl. in Math. Notes **14** (1973).

[17] ——, *A uniqueness theorem for a multidimensional inverse kinematic problem of seismology*, Mat. Problemy Geofiz. Vyp. 5 (1974), part 2, 18–29. (Russian) Zbl. **332** #73093.

[18] ——, *A uniqueness theorem for the solution of the inverse problem for the wave equation*, Mat. Zametki **19** (1976), 211–214; English transl. in Math. Notes **19** (1976).

[19] ——, *A uniqueness theorem for the solution of the inverse problem for a quasilinear hyperbolic equation*, Differentsial′nye Uravneniya **12** (1976), 2265–2266; English transl. in Differential Equations **12** (1976).

[20] ——, *The solvability of a certain problem of integral geometry*, Mat. Sb. **101(143)** (1976), 271–279; English transl. in Math. USSR Sb. **30** (1976).

[21] ——, *A remark on the theory of inverse problems*, in [331], pp. 16-23. (Russian) Zbl. **355** #35078

[22] ——, *Some methods for the study of multidimensional inverse problems for differential equations*, "Nauka," Novosibirsk, 1978. (Russian)

[23] YU. E. ANIKONOV, N. B. PIVOVAROVA AND L. V. SLAVINA, *The three-dimensional velocity field of the Kamchatka focal zone*, Mat. Problemy Geofiz. Vyp. 5 (1974), part 1, 92–117. (Russian) R. Zh. Geol. **1974** #12A474.

[24] YU. T. ANTOKHIN, *An analytic approach to the problem of equations of the first kind*, Dokl. Akad. Nauk SSSR **167** (1966), 727–730; English transl. in Soviet Math. Dokl. 7 (1966).

[25] ——, *Ill-posed problems in Hilbert space and stable methods for solving them*, Differentsial′nye Uravneniya **3** (1967), 1135–1156; English transl. in Differential Equations **3** (1967).

[26] V. YA. ARSENIN, *On methods of solving ill-posed problems*, Moskov. Inzh.-Fiz. Inst., Moscow, 1973. (Russian)

[27] V. YA. ARSENIN AND V. V. IVANOV, *On the solution of certain convolution type integral equations of the first kind by the method of regularization*, Zh. Vychisl. Mat. i Mat. Fiz. **8** (1968), 310–321; English transl. in USSR Comput. Math. and Math. Phys. **8** (1968).

[28] ——, *The effect of regularization of order p*, Zh. Vychisl. Mat. i Mat. Fiz. **8** (1968), 661–663; English transl. in USSR Comput. Math. and Math. Phys. **8** (1968).

[29] N. I. AKHIEZER, *Lectures on the theory of approximation*, 2nd rev. ed., "Nauka," Moscow, 1965; English transl. of 1st ed., Ungar, New York, 1956.

[30] A. B. BAKUSHINSKIĬ, *A numerical method for solving Fredholm integral equations of the first kind*, Zh. Vychisl. Mat. i Mat. Fiz. **5** (1965), 744–749; English transl. in USSR Comput. Math. and Math. Phys. **5** (1965).

[31] ——, *On a general technique for the construction of regularizing algorithms for a linear ill-posed equation in Hilbert space*, Zh. Vychisl. Mat. i Mat. Fiz. **7** (1967), 672–677; English transl. in USSR Comput. Math. and Math. Phys. **7** (1967).

[32] ——, *Selected questions on the approximate solution of ill-posed problems*, Lecture notes, Vychisl. Tsentr, Moskov. Gos. Univ., Moscow, 1968. (Russian)

[33] A. B. BAKUSHINSKIĬ AND V. N. STRAKHOV, *On the solution of certain integral equations of the first kind by the method of successive approximations*, Zh. Vychisl. Mat. i Mat. Fiz. **8** (1968), 181–185; English transl. in USSR Comput. Math. and Math. Phys. **8** (1968).

[34] N. YA. BEZNOSHCHENKO, *On determining a coefficient in a parabolic equation*, Differensial′nye Uravneniya **10** (1974), 24–35; English transl. in Differential Equations **10** (1974).

[35] ——, *On determining the coefficients of the leading derivatives in a parabolic equation*, Differential'nye Uravneniya **11** (1975), 19–26; English transl. in Differential Equations **11** (1975).

[36] ——, *On determining the coefficients of lower-order terms in a parabolic equation*, Sibirsk. Mat. Zh. **16** (1975), 473–482; English transl. in Siberian Math. J. **16** (1975).

[37] ——, *Determining the coefficient of the lowest-order term in a general parabolic equation*, Differentsial'nye Uravneniya **12** (1976), 175–176; English transl. in Differential Equations **12** (1976).

[38] G. YA. BEĬL'KIN, *Stability and uniqueness of the solution of the inverse kinematic problem of seismology in higher dimensions*, Zap. Nauchn. Sem. Leningrad. Otdel. Mat. Inst. Steklov (LOMI) **84** (1979), 3–6; English transl. in J. Soviet Math. **21** (1983), no. 3.

[39] S. P. BELINSKIĬ, *On the inverse problem for linear symmetric t-hyperbolic systems of first-order equations*, Mat. Problemy Geofiz. Vyp. 6 (1975), part 2, 100–109. (Russian) Zbl. **357** #35077.

[40] ——, *On the inverse problem for linear symmetric t-hyperbolic systems with $n+1$ independent variables*, Differensial'nye Uravneniya **12** (1976), 15–23; English transl. in Differential Equations **12** (1976).

[41] ——, *A uniqueness theorem for the inverse problem for a first-order hyperbolic system*, in [331] pp. 24–30. (Russian) Zbl. **355** #35079.

[42] YU. M. BEREZANSKIĬ, *On the uniqueness of determining a Schrödinger equation from its spectral function*, Dokl. Akad. Nauk SSSR **93** (1953), 591–594. (Russian)

[43] ——, *On the inverse problem of spectral analysis for the Schrödinger equation*, Dokl. Akad. Nauk SSSR **105** (1955), 197–200. (Russian)

[44] ——, *The uniqueness theorem in the inverse problem of spectral analysis for the Schrödinger equation*, Trudy Moskov. Mat. Obshch. **7** (1958), 3–62; English transl. in Amer. Math. Soc. Transl. (2) **35** (1964).

[45] S. N. BERNSTEIN, *Extremal properties of polynomials and best approximation of continuous functions of a real variable*, ONTI, Moscow, 1937. (Russian)

[46] I. N. BERNSHTEĬN AND M. L. GERVER, *A problem of integral geometry for a family of geodesics and an inverse kinematic problem of seismology*, Dokl. Akad. Nauk SSSR **243** (1978), 302–305; English transl. in Dokl. Earth Sci. Sections **243** (1978).

[47] A. S. BLAGOVESHCHENSKIĬ, *The inverse problem in the theory of seismic-wave propagation*, Problemy Mat. Fiz., vyp. 1, Izdat. Leningrad. Univ., Leningrad, 1966, pp. 68–81; English transl. in Topics in Math. Phys.,

no. 1, Plenum Press, New York, 1967; erratum, ibid., vyp. 3 (1968), 103=no. 3 (1969).

[48] ——, *A one-dimensional inverse boundary value problem for a second-order hyperbolic equation*, Zap. Nauchn. Sem. Leningrad. Otdel. Mat. Inst. Steklov. (LOMI) **15** (1969), 85–90; English transl. in Sem. Math. V.A. Steklov Math. Inst. Leningrad **15** (1969).

[49] ——, *An inverse problem for the wave equation with unknown origin*, Problemy Mat. Fiz., vyp. 4, Izdat. Leningrad. Univ., Leningrad, 1970, pp. 27–39; English transl. in Topics in Math. Phys., no. 4, Plenum Press, New York, 1971.

[50] ——, *The various formulations of the one-dimensional inverse problem for the telegraph equation*, Problemy Mat. Fiz., vyp. 4, Izdat. Leningrad. Univ., Leningrad, 1970, pp. 40–41; English transl. in Topics in Math. Phys., no. 4, Plenum Press, New York, 1971.

[51] ——, *On a local method of solving a nonstationary inverse problem for a nonhomogeneous string*, Trudy Mat. Inst. Steklov **115** (1971), 28–38; English transl. in Proc. Steklov Inst. Math. **115** (1971).

[52] ——, *An inverse boundary value problem in the theory of propagation of waves in an anisotropic medium*, Trudy Mat. Inst. Steklov **115** (1971), 39–56; English transl. in Proc. Steklov Inst. Math. **115** (1971).

[53] ——, *A quasi-two-dimensional problem for the wave equation*, Trudy Mat. Inst. Steklov **115** (1971), 57–69; English transl. in Proc. Steklov Inst. Math. **115** (1971).

[54] V. F. BONCHKOVSKIĬ, *Internal structure of the earth*, Izdat. Akad. Nauk SSSR, Moscow, 1953. (Russian)

[55] B. M. BUDAK AND A. D. ISKENDEROV, *Difference method for the solution of coefficient boundary value problems*, Dokl. Akad. Nauk SSSR **171** (1966), 1054–1057; English transl. in Soviet Phys. Dokl. **11** (1966/67).

[56] ——, *On a class of boundary value problems with unknown coefficients*, Dokl. Akad. Nauk SSSR **175** (1967), 13–16; English transl. in Soviet Math. Dokl. **8** (1967).

[57] ——, *On a class of inverse boundary value problems with unknown coefficients*, Dokl. Akad. Nauk SSSR **176** (1967), 20–23; English transl. in Soviet Math. Dokl. **8** (1967).

[58] KEITH EDWARD BULLEN, *An introduction to the theory of seismology*, 3rd ed., Cambridge Univ. Press, 1963.

[59] A. L. BUKHGEĬM, *A class of Volterra operator equations of the first kind*, Funktsional. Anal. i Prilozhen. **6** (1972), no. 1, 1–9; English transl. in Functional Anal. Appl. **6** (1972).

[60] ——, *Some problems of integral geometry*, Sibirsk. Mat. Zh. **13** (1972), 34–42; English transl. in Siberian Math. J. **13** (1972).

[61] ——, *On a problem of integral geometry*, Mat. Problemy Geofiz. Vyp. 4 (1973), 69–73. (Russian) R. Zh. Mat. **1974** #1Б523.

[62] ——, *On a class of integral equations of the first kind*, Dokl. Akad. Nauk SSSR **215** (1974), 15–16; English transl. in Soviet Math. Dokl. **15** (1974).

[63] ——, *Normal solvability of some special operator equations of the first kind (a sufficient condition)*, Mat. Problemy Geofiz. Vyp. 6 (1975), part 1, 42–54. (Russian) Zbl. **322** #47031.

[64] ——, *On the analyticity of the solution of special integral equations of the first kind*, Mat. Problemy Geofiz. Vyp. 6 (1975), part 2, 110–119. (Russian) Zbl. **355** #45002.

[65] ——, *Necessary conditions for the stability of a class of integrodifferential equations*, Computational Methods and Programming (V. V. Shaĭdurov, editor), Vychisl. Tsentr Sibirsk. Otdel. Akad. Nauk SSSR, Novosibirsk, 1975, pp. 78–85. (Russian) R. Zh. Mat. **1976** #8Б792.

[66] ——, *Volterra operator equations in scales of Banach spaces*, Dokl. Akad. Nauk SSSR **242** (1978), 272–275; English transl. in Soviet Math. Dokl. **19** (1978).

[67] A. L. Bukhgeĭm and V. G. Yakhno, *Two inverse problems for differential equations*, Dokl. Akad. Nauk SSSR **229** (1976), 785–786; English transl. in Soviet Math. Dokl. **17** (1976).

[68] ——, *On problems of identification of hyperbolic equations*, Preprint No. 24, Vychisl. Tsentr, Sibirsk. Otdel. Akad. Nauk SSSR, Novosibirsk, 1976. (Russian) R. Zh. Mat. **1977** #5Б216.

[69] V. V. Vasin, *On a projection method for solving ill-posed problems*, Izv. Vyssh. Vchebn. Zaved. Mat. **1971**, no. 11 (114), 28–32. (Russian)

[70] ——, *On the problem of computing the values of an unbounded operator in B-spaces*, Izv. Vyssh. Uchebn. Zaved. Mat. **1972**, no. 5 (120), 22–28. (Russian)

[71] V. V. Vasin and V. P. Tanana, *Approximate solution of operator equations of the first kind*, Ural. Gos. Univ. Mat. Zap. **6** (1968), no. 4, 27–37. (Russian)

[72] V. V. Voevodin, *The method of regularization*, Zh. Vychisl. Mat. i Mat. Fiz. **9** (1969), 673–675; English transl. in USSR Comput. Math. and Math. Phys. **9** (1969).

[73] G. N. Watson, *A treatise on the theory of Bessel functions*, 2nd ed., Cambridge Univ. Press, Cambridge, and Macmillan, New York, 1944.

[74] V. S. VLADIMIROV, *Methods of the theory of functions of many complex variables*, "Nauka," Moscow, 1964; English transl., M.I.T. Press, Cambridge, Mass., 1966.

[75] ——, *Equations of mathematical physics*, "Nauka," Moscow, 1966; English transl., Marcel Dekker, New York, 1971.

[76] R. M. GARIPOV, *A nonhyperbolic boundary value problem for the wave equation*, Dokl. Akad. Nauk SSSR **219** (1974), 777–780; English transl. in Soviet Math. Dokl. **15** (1974).

[77] R. M. GARIPOV AND V. B. KARDAKOV, *The Cauchy problem for the wave equation with a nonspatial initial manifold*, Dokl. Akad. Nauk SSSR **213** (1973), 1047–1050; English transl. in Soviet Phys. Dokl. **18** (1973/74).

[78] I. M. GEL'FAND, *Integral geometry and its relation to representation theory*, Uspekhi Mat. Nauk **15** (1960), no. 2 (92), 155–164; English transl. in Russian Math. Surveys **15** (1960).

[79] I. M. GEL'FAND AND S. G. GINDIKIN, *Nonlocal inversion formulas in real integral geometry*, Funktsional. Anal. i Prilozhen. **11** (1977), no. 3, 12–19; English transl. in Functional Anal. Appl. **11** (1977).

[80] I. M. GEL'FAND AND M. I. GRAEV, *Geometry of homogeneous spaces, representations of groups in homogeneous spaces and related questions of integral geometry.* I, Trudy Moskov, Mat. Obshch. **8** (1959), 321–390; English transl. in Amer. Math. Soc. Transl. (2) **37** (1964).

[81] ——, *Integrals over hyperplanes of test functions and generalized functions*, Dokl. Akad. Nauk SSSR **135** (1960), 1307–1310; English transl. in Soviet Math. Dokl. **1** (1960).

[82] ——, *Integral transformations connected with line complexes in a complex affine space*, Dokl. Akad. Nauk SSSR **138** (1961), 1266–1269; English transl. in Soviet Math. Dokl. **2** (1961).

[83] I. M. GEL'FAND, M. I. GRAEV AND N. YA. VILENKIN, *Generalized functions.* Vol. 5: *Integral geometry and representation theory*, Fizmatgiz, Moscow, 1962; English transl., Academic Press, 1966.

[84] I. M. GEL'FAND, M. I. GRAEV AND Z. YA. SHAPIRO, *Integral geometry on a manifold of k-dimensional planes*, Dokl. Akad. Nauk SSSR **168** (1966), 1236–1238; English transl. in Soviet Math. Dokl. **7** (1966).

[85] I. M. GEL'FAND AND S. V. FOMIN, *Calculus of variations*, Fizmatgiz, Moscow, 1961; English transl., Prentice-Hall, Englewood Cliffs, N. J., 1963.

[86] I. M. GEL'FAND AND G. E. SHILOV, *Generalized functions.* Vol. 1: *Operations on them*, Fizmatgiz, Moscow, 1958; English transl., Academic Press, 1964.

[87] M. L. GERVER AND V. M. MARKUSHEVICH, *Investigation of ambiguity in determination of seismic-wave velocities using travel-time curves*, Dokl. Akad. Nauk SSSR **163** (1965), 1377–1380; English transl. in Dokl. Earth Sci. Sections **163** (1964).

[88] ——, *Determining seismic-wave velocities from travel-time curves*, Methods and Programs for the Analysis of Seismic Observations (Vychisl. Seismologiya, vyp. 3; V. I. Keĭlis-Borok, editor), "Nauka," Moscow, 1967, pp. 3–51; English transl. in Computational Seismology, Plenum Press, New York, 1972.

[89] ——, *Characteristic properties of seismic Lodographs*, Dokl. Akad. Nauk SSSR **175** (1967), 334–337; English transl. in Dokl. Earth Sci. Sections **175** (1967) (1968).

[90] V. B. GLASKO et al., *Solution of the inverse problem of gravimetry for a contact surface using the regularization method*, Izv. Akad. Nauk SSSR Phys. Zemli **1973**, no. 2, 30–41; English transl. in Izv. Acad. Sci. USSR Phys. Solid Earth **1973**.

[91] V. B. GLASKO, V. V. KRAVTSOV AND G. N. KRAVTSOVA, *On an inverse problem of gravimetry*, Vestnik Moskov. Univ. Ser. III Phys. Astr. **11** (1970), 174–179; English transl. in Moscow Univ. Phys. Bull. **25** (1970).

[92] A. V. GONCHARSKIĬ, A. S. Leonov and A. G. Yagola, *On a regularizing algorithm for ill-posed problems with an approximately given operator*, Zh. Vychisl. Mat. i Mat. Fiz. **12** (1972), 1592–1594; English transl. in USSR Comput. Math. and Math. Phys. **12** (1972).

[93] A. V. GONCHARSKIĬ, A. M. CHEREPASHCHUK AND A. G. YAGOLA, *Numerical methods for the solution of inverse problems of astrophysics*, "Nauka," Moscow, 1978. (Russian)

[94] A. V. GONCHARSKIĬ AND A. G. YAGOLA, *Uniform approximation of monotone solutions of ill-posed problems*, Dokl. Akad. Nauk SSSR **184** (1969), 771–773; English transl. in Soviet Math. Dokl. **10** (1969).

[95] I. TS. GOKHBERG [ISRAEL GOHBERG] AND M. G. KREĬN, *Theory and applications of Volterra operators in Hilbert space*, "Nauka," Moscow, 1967; English transl., Amer. Math. Soc., Providence, R. I., 1970.

[96] HAROLD JEFFREYS, *The earth: its origin, history and physical structure*, 3rd ed., Cambridge Univ. Press, 1952.

[97] I. N. DOMBROVSKAYA, *On the solution of ill-posed linear equations in Hilbert space*, Ural. Gos. Univ. Mat. Zap. **4** (1963/64), tetrad' 4, 36–40. (Russian)

[98] I. N. DOMBROVSKAYA AND V. K. IVANOV, *On the theory of certain linear equations in abstract spaces*, Sibirsk. Mat. Zh. **6** (1965), 499–508. (Russian)

[99] S. ELUBAEV, *Oman inverse problem for the telegraph equation*, Dokl. Akad. Nauk SSSR **189** (1969), 461–463; English transl. in Soviet Math. Dokl. **10** (1969).

[100] V. G. ZHALNIN AND S. V. USPENSKIĬ, *Reconstruction of functions specified by integrals over a family of second-order surfaces*, Theory of Cubature Formulas and Application of Functional Analysis to Problems of Mathematical Physics, Trudy Sem. S. L. Sobolev. 1976, part 1, Inst. Mat. Sibirsk. Otdel. Akad. Nauk SSSR, Novosibirsk, 1976, pp. 32–40. (Russian)

[101] A. S. ZAPREEV, *A uniqueness theorem for the solution of a plane inverse problem for the Helmholtz equation*, in [331], pp. 46–63. (Russian) Zbl. **355** #35080.

[102] A. S. ZAPREEV AND V. A. TSETSOKHO, *An inverse problem for the Helmholtz equation*, Preprint No. 22, Vychisl. Tsentr Sibirsk. Otdel. Akad. Nauk SSSR, Novosibirsk, 1976. (Russian) R. Zh. Mat. **1977** #8Б499.

[103] A. ZYGMUND, *Trigonometric series*, 2nd rev. ed., Vol. 1, Cambridge Univ. Press, 1959.

[104] V. K. IVANOV, *Integral equations of the first kind and the approximate solution of the inverse problem of potential theory*, Dokl. Akad. Nauk SSSR **142** (1962), 998–1000; English transl. in Soviet Math. Dokl. **3** (1962).

[105] ——, *On linear ill-posed problems*, Dokl. Akad. Nauk SSSR **145** (1962), 270–272; English transl. in Soviet Math. Dokl. **3** (1962).

[106] ——, *On ill-posed problems*, Mat. Sb. **61(103)** (1963), 211–223. (Russian)

[107] ——, *On a type of ill-posed linear equation in topological vector spaces*, Sibirsk. Mat. Zh. **6** (1965), 832–839. (Russian)

[108] ——, *Uniform regularization of unstable problems*, Sibirsk. Mat. Zh. **7** (1966), 546–558; English transl. in Siberian Math. J. **7** (1966).

[109] ——, *On the approximate solution of operator equations of the first kind*, Zh. Vychisl. Mat. i Mat. Fiz. **6** (1966), 1089–1094; English transl. in USSR Comput. Math. and Math. Phys. **6** (1966).

[110] ——, *Ill-posed problems in topological spaces*, Sibirsk. Mat. Zh. **10** (1969), 1065–1074; English transl. in Siberian Math. J. **10** (1969).

[111] ——, *The problem of quasi-inversion for the heat equation in a uniform metric*, Differensial′nye Uravneniya **8** (1972), 652–658; English transl. in Differential Equations **8** (1972).

[112] ——, *Estimating the stability of quasisolutions on noncompact sets*, Izv. Vyssh. Uchebn. Zaved. Mat. **1974**, no. 5 (144), 97–103; English transl. in Soviet Math. (Iz. VUZ) **18** (1974).

[113] V. K. IVANOV, V. V. VASIN AND V. P. TANANA, *Theory of linear ill-posed problems and its applications*, "Nauka," Moscow, 1978. (Russian)

[114] V. V. IVANOV AND V. YU. KUDRINSKIĬ, *Approximate solution of linear operator equations in Hilbert space by the method of least squares.* I, II, Zh. Vychisl. Mat. i Mat. Fiz. **6** (1966), 831–841; **7** (1967), 475–496; English transl. in USSR Comput. Math. and Math. Phys. **6** (1966); **7** (1967).

[115] V. M. ISAKOV, *Uniqueness of the solution of some inverse hyperbolic problems*, Differentsial′nye Uravneniya **10** (1974), 165–167; English transl. in Differential Equations **10** (1974).

[116] A. D. ISKENDEROV, *Inverse boundary value problems with unknown coefficients for certain quasilinear equations*, Dokl. Akad. Nauk SSSR **178** (1968), 999–1002; English transl. in Soviet Math. Dokl. **9** (1968).

[117] ——, *On an inverse problem for elliptic equations*, Inverse Problems for Differential Equations, Vychisl. Tsentr Sibirsk. Otdel Akad. Nauk SSSR, Novosibirsk, 1972, pp. 107–115. (Russian)

[118] ——, *On an inverse problem for quasilinear parabolic equations*, Differentsial′nye Uravneniya **10** (1974), 890–898; English transl. in Differential Equations **10** (1974).

[119] ——, *Inverse problems for determining filtration and thermophysical characteristics*, Nonclassical Methods in Geophysics (Materials All-Union School, 1976; M. M. Lavrent′ev, editor), Vychisl. Tsentr Sibirsk. Otdel Akad. Nauk SSSR, Novosibirsk, 1977, pp. 54–63. R. Zh. Mat. **1978** #9Б416.

[120] FRITZ JOHN, *Plane waves and spherical means applied to partial differential equations*, Interscience, 1955.

[121] L. V. KANTOROVICH AND G. P. AKILOV, *Functional analysis in normed spaces*, Fizmatgiz, Moscow, 1959; English transl., Macmillan, 1964.

[122] S. I. KABINIKHIN, *On the formulation of a two-dimensional inverse problem for the wave equation*, in [331], pp. 64–73. (Russian) Zbl. **356** #35078.

[123] V. M. KAĬSTRENKO, *The Cauchy problem for a second-order hyperbolic equation with data on a time-like surface*, Sibirsk. Mat. Zh. **16** (1975), 395–398; English transl. in Siberian Math. J. **16** (1975).

[124] V. B. KARDAKOV, *A stability estimate for the solution of a nonhyperbolic boundary value problem for the wave equation*, Mat. Problemy Geofiz. Vyp. 6 (1975), part 2, 157–166. (Russian) Zbl. **354** #35010.

[125] M. KLIBANOV, *Uniqueness of the solution of an inverse problem for a parabolic equation*, Mat. Problemy Geofiz. Vyp. 5 (1974), part 2, 59–71. (Russian) Zbl. **323** #35067.

[126] ——, *On the method of descent in solving inverse problems*, Mat. Problemy Geofiz. Vyp. 5 (1974), part 2, 72–77. (Russian) Zbl. **328** #35071.

[127] ——, *On a class of inverse problems for a parabolic equation and for problems of integral geometry*, Dokl. Akad. Nauk SSSR **222** (1975), 29–31; English transl. in Soviet Math. Dokl. **16** (1975).

[128] ——, *Recovery of a function given by integrals over ellipsoids of revolution, and an inverse problem for a parabolic equation*, Mat. Problemy Geofiz. Vyp. 6 (1975), part 1, 97–116. (Russian) Zbl. **324** #53053.

[129] ——, *On a problem of integral geometry and an inverse problem for a parabolic equation*, Sibirsk. Mat. Zh. **17** (1976), 75–84; English transl. in Siberian Math. J. **17** (1976).

[130] ——, *The inverse problem for a parabolic equation, and a problem of integral geometry*, Sibirsk. Mat. Zh. **17** (1976), 564–570; English transl. in Siberian Math. J. **17** (1976).

[131] ——, *On a class of operator equations of the first kind*, Funktsional. Anal. i Prilozhen. **11** (1977), no. 3, 82–83; English transl. in Functional Anal. Appl. **11** (1977).

[132] L. F. KORKINA, *On the regularization of operator equations of the first kind*, Izv. Vyssh. Uchebn. Zaved. Mat. **1969**, no. 8 (87), 26–29. (Russian)

[133] P. O. KOSTELYANETS AND YU. G. RESHETNYAK, *Determination of a completely additive function by its values on half-spaces*, Uspekhi Mat. Nauk **9** (1954), no. 3 (61), 135–140. (Russian)

[134–135] S. G. KREĬN, *On classes in which certain boundary value problems are well-posed*, Dokl. Akad. Nauk SSSR **114** (1957), 1162–1165. (Russian)

[136] ——, *Linear differential equations in Banach space*, "Nauka," Moscow, 1967; English transl., Amer. Math. Soc., Providence, R. I., 1971.

[137] S. G. KREĬN AND YU. I. PETUNIN, *Scales of Banach spaces*, Uspekhi Mat. Nauk **21** (1966), no. 2 (128), 89–168; English transl. in Russian Math. Surveys **21** (1966).

[138] S. G. KREĬN AND O. I. PROZOROVSKAYA, *Approximate methods for solving ill-posed problems*, Zh. Vychisl. Mat. i Mat. Fiz. **3** (1963), 120–130; English transl. in USSR Comput. Math. and Math. Phys. **3** (1963).

[139] R. COURANT, *Methods of mathematical physics*. Vol. II: *Partial differential equations*, Interscience, 1962.

[140] M. M. LAVRENT'EV, *On the inverse problem of potential theory*, Dokl. Akad. Nauk SSSR **106** (1956), 389–390. (Russian)

[141] ——, *Quantitative refinements of interior uniqueness theorems*, Dokl. Akad. Nauk SSSR **110** (1956), 731–734. (Russian)

[142] ——, *On the Cauchy problem for the Laplace equation*, Izv. Akad. Nauk SSSR Ser. Mat. **20** (1956), 819–842. (Russian)

[143] M. A. LAVRENT′EV AND B. V. SHABAT, *Methods of the theory of functions of a complex variable*, 2nd rev. ed., Fizmatgiz, Moscow, 1958; German transl. of 3rd ed., VEB Deutscher Verlag Wiss., Berlin, 1967.

[144] M. M. LAVRENT′EV, *Integral equations of the first kind*, Dokl. Akad. Nauk SSSR **127** (1959), 31–33. (Russian)

[145] ——, *Some ill-posed problems of mathematical physics*, Izdat. Sibirsk. Otdel. Akad. Nauk SSSR, Novosibirsk, 1962; English transl., Springer-Verlag, 1967.

[146] ——, *On a class of nonlinear integral equations*, Sibirsk. Mat. Zh. **4** (1963), 837–844. (Russian)

[147] ——, *On an inverse problem for the wave equation*, Dokl. Akad. Nauk SSSR **157** (1964), 520–521; English transl. in Soviet Math. Dokl. **5** (1964).

[148] ——, *A class of inverse problems for differential equations*, Dokl. Akad. Nauk SSSR **160** (1965), 32–35; English transl. in Soviet Math. Dokl. **6** (1965).

[149] ——, *Inverse problems and special operator equations of the first kind*, Internat. Congr. Math. (Nice, 1970), Papers of Soviet Mathematicians, "Nauka," Moscow, 1972, pp. 130–136. (Russian) Zbl. **261** #35072.

[150] ——, *Inverse problems for differential equations*, Collected Papers from the Symposium "Inverse Problems for Differential Equations," "Nauka," Novosibirsk, 1972, pp. 7–13. (Russian)

[151] ——, *Conditionally well-posed problems for differential equations*, Novosibirsk. Gos. Univ., Novosibirsk, 1973. (Russian)

[152] ——, *Two directions in the theory of ill-posed problems of mathematical physics*, Numerical Methods in Mathematical Physics, Geophysics and Optimal Control (Sympos., Novosibirsk, 1976) (G. I. Marchuk and J. L. Lions, editors), "Nauka," Novosibirsk, 1978, pp. 58–63. (Russian)

[153] M. M. LAVRENT′EV AND A. L. BUKHGEĬM, *On a class of problems of integral geometry*, Dokl. Akad. Nauk SSSR **211** (1973), 38–39; English transl. in Soviet Math. Dokl. **14** (1973).

[154] ——, *On a class of operator equations of the first kind*, Funktsional. Anal. i Prilozhen. **7** (1973), no. 4, 44–53; English transl. in Functional Anal. Appl. **7** (1973).

[155] M. M. LAVRENT′EV AND V. G. VASIL′EV, *On the formulation of some ill-posed problems of mathematical physics*, Sibirsk. Mat. Zh. **7** (1966), 559–576; English transl. in Siberian Math. J. **7** (1966).

[156] M. M. LAVRENT′EV AND M. V. KLIBANOV, *On an integral equation of the first kind, and the inverse problem for a parabolic equation*, Dokl. Akad. Nauk SSSR **221** (1975), 782–783; English transl. in Soviet Math. Dokl. **16** (1975).

[157] ____, *On an inverse problem for a parabolic equation*, Differentsial′nye Uravneniya **11** (1975), 1647–1651; English transl. in Differential Equations **11** (1975).

[158] M. M. LAVRENT′EV AND V. G. ROMANOV, *On three linearized inverse problems for hyperbolic equations*, Dokl. Akad. Nauk SSSR **171** (1966), 1279–1281; English transl. in Soviet Math. Dokl. **7** (1966).

[159] M. M. LAVRENT′EV, V. G. ROMANOV AND V. G. VASIL′EV, *Multidimensional inverse problems for differential equations*, "Nauka," Novosibirsk, 1969; English transl., Lecture Notes in Math., vol. 167, Springer-Verlag, 1970.

[160] E. M. LANDIS, *Some questions in the qualitative theory of elliptic and parabolic equations*, Uspekhi Mat. Nauk **14** (1959), no. 1 (85), 21–85; English transl. in Amer. Math. Soc. Transl. (2) **20** (1962).

[161] ____, *Some questions in the qualitative theory of second-order elliptic equations (case of several independent variables)*, Uspekhi Mat. Nauk **18** (1963), no. 1 (109), 3–62; English transl. in Russian Math. Surveys **18** (1963).

[162] ____, *Behavior of a solution of a parabolic equation on a characteristic*, Mat. Zametki **12** (1972), 257–262; English transl. in Math. Notes **12** (1972).

[163] E. M. LANDIS AND O. A. OLEĬNIK, *Generalized analyticity and related properties of solutions of elliptic and parabolic equations*, Uspekhi Mat. Nauk **29** (1974), no. 2 (176), 190–206; English transl. in Russian Math. Surveys **29** (1974).

[164] R. LATTÈS AND J.-L. LIONS, *Méthode de quasi-réversibilité et applications*, Dunod, Paris, 1967; English transl., North-Holland, 1969.

[165] B. M. LEVITAN, *Generalized shift operators and some of their applications*, Fizmatgiz, Moscow, 1962; English transl., Israel Program for Sci. Transls., Jerusalem, and Davey, New York, 1964.

[166] O. A. LISKOVETS, *Ill-posed problems with a closed noninvertible operator*, Differentsial′nye Uravneniya **3** (1967), 636–646; English transl. in Differential Equations **3** (1967).

[167] ____, *A method of regularization for nonlinear problems with a closed operator*, Sibirsk. Mat. Zh. **12** (1971), 1311–1317; English transl. in Siberian Math. J. **12** (1971).

[168] A. I. MARKUSHEVICH, *Theory of analytic functions*, GITTL, Moscow, 1950; English transl., vols. 1, 2, 3, Prentice-Hall, Englewood Cliffs, N. J., 1965, 1967.

[169] A. G. MEGRABOV, *The direct and inverse problems of the scattering of plane waves by inhomogeneous transition layers*. II (*elliptic case*), Mat. Problemy Geofiz. Vyp. 3 (1972), 113–123. (Russian) R. Zh. Mat. **1973** #5Б423.

[170] ——, *Inverse problems of the scattering of plane waves by inhomogeneous layers with a free or fixed boundary* (*hyperbolic case*), Mat. Problemy Geofiz. Vyp. 4 (1973), 84–102. (Russian) R. Zh. Mat. **1974** #1Б363.

[171] ——, *Inverse problems for an elliptic equation in a strip connected with the problem of the scattering of plane waves by inhomogeneous layers*, Mat. Problemy Geofiz. Vyp. 4 (1973), 103–115. (Russian) R. Zh. Mat. **1974** #1Б364.

[172] ——, *Inverse problems for hyperbolic and elliptic equations with data regarding a countable set of solutions, and inverse problems of the scattering of plane waves*, Mat. Problemy Geofiz. Vyp. 4 (1973), 116–130. (Russian) R. Zh. Mat. **1974** #1Б365.

[173] ——, *A method for reconstructing the density and velocity in an inhomogeneous layer as functions of depth on the basis of a family of plane waves reflected from the layer at various angles*, Mat. Problemy Geofiz. Vyp. 5 (1974), part 2, 78–107. (Russian) R. Zh. Mat. **1975** #4Б636.

[174] ——, *Inverse problems for elliptic equations in the half-plane and strip, and inverse problems on the scattering of plane waves by inhomogeneous layers*, Dokl. Akad. Nauk SSSR **220** (1975), 315–317; English transl. in Soviet Phys. Dokl. **20** (1975).

[175] ——, *Inverse problems for equations of mixed type*, Mat. Problemy Geofiz. Vyp. 6 (1975), part 1, 122–144. (Russian) Zbl. **324** #35066.

[176] ——, *Some inverse problems for an equation of mixed type*, Dokl. Akad. Nauk SSSR **234** (1977), 305–307; English transl. in Soviet Math. Dokl. **18** (1977).

[177] V. A. MARCHENKO, *Sturm-Liouville operators and their applications*, "Naukova Dumka," Kiev, 1977. (Russian)

[178] V. A. MOROZOV, *On the regularization of ill-posed problems and the choice of the regularization parameter*, Zh. Vychisl. Mat. i Mat. Fiz. **6** (1966), 170–175; English transl. in USSR Comput. Math. and Math. Phys. **6** (1966).

[179] ——, *Methods of solving unstable problems*, Mimeographed notes (special course), Vychisl. Tsentr Moskov. Gos. Univ., Moscow, 1967. (Russian)

[180] ——, *Regularizing families of operators*, Vychisl. Metody i Programmirovanie Vyp. 8 (1967), 63–95. (Russian)

[181] ——, *A stable method for computing the values of unbounded operators*, Dokl. Akad. Nauk SSSR **185** (1969), 267–270; English transl. in Soviet Math. Dokl. **10** (1969).

[182] ——, *On pseudosolutions*, Zh. Vychisl. Mat. i Mat. Fiz. **9** (1969), 1387–1391; English transl. in USSR Comput. Math. and Math. Phys. **9** (1969).

[183] ——, *The differentiation problem and some algorithms for approximation by experimental information*, Vychisl. Metody. i Programmirovanie Vyp. 14 (1970), 46–62. (Russian)

[184] ——, *Linear and nonlinear ill-posed problems*, Itogi Nauki: Mat. Anal., vol. 11, VINITI, Moscow, 1973, pp. 129–178; English transl. in J. Soviet Math. **4** (1975), no. 6.

[185] ——, *Regular methods for solving ill-posed problems*, Izdat. Moskov. Gos. Univ., Moscow, 1974; English transl., Springer-Verlag, 1984.

[186] V. A. MOROZOV AND N. N. KIRSANOVA, *On a generalization of the method of regularization*, Vychisl. Metody i Programmirovanie Vyp. 14 (1970), 40–45. (Russian)

[187] R. G. MUKHOMETOV, *On a problem of integral geometry*, Mat. Problemy Geofiz. Vyp. 6 (1975), part 2, 212–242. (Russian) Zbl. **371** #53051.

[188] ——, *The inverse kinematic problem of seismology on the plane*, Mat. Problemy Geofiz. Vyp. 6 (1975), part 2, 243–254. (Russian) Zbl. 365 #73097.

[189] ——, *The problem of reconstructing a two-dimensional Riemannian metric, and integral geometry*, Dokl. Akad. Nauk SSSR **232** (1977), 32–35; English transl. in Soviet Math. Dokl. **18** (1977).

[190] ——, *On the problem of reconstructing an anisotropic Riemannian metric in an n-dimensional region*, Preprint No. 135, Vychisl. Tsentr Sibirsk. Otdel. Akad. Nauk SSSR, Novosibirsk, 1978. (Russian)

[191] R. G. MUKHOMETOV AND V. G. ROMANOV, *On the problem of finding an isotropic Riemannian metric in an n-dimensional space*, Dokl. Akad. Nauk SSSR **243** (1978), 41–44; English transl. in Soviet Math. Dokl. **19** (1978).

[192] G. MYUNTS [Ch. H. Müntz], *Integral equations*. Vol. I, ONTI, Moscow, 1934. (Russian)

[193] L. P. NIZHNIK, *The inverse time-dependent scattering problem for a hyperbolic system of equations*, Linear and Nonlinear Boundary Value Problems (Yu. A. Mitropol′skiĭ and A. A. Berezovskiĭ, editors), Izdanie Inst. Mat. Akad. Nauk Ukrain. SSR, Kiev, 1971, pp. 205–210. (Russian)

[194] ——, *The time-dependent scattering problem for the Dirac equation on a semiaxis*, Boundary Value Problems of Mathematical Physics (Yu. A. Mitropol′skiĭ, editor), Izdanie Inst. Mat. Akad. Nauk Ukrain. SSR, Kiev, 1971, pp. 303–313.

[195] ——, *The inverse problem of time-dependent scattering*, Dokl. Akad. Nauk SSSR **196** (1971), 1016–1019; English transl. in Soviet Math. Dokl. **12** (1971).

[196] ——, *The inverse time-dependent scattering problem for the Dirac equation*, Ukrain. Mat. Zh. **24** (1972), 110–113; English transl. in Ukrainian Math. J. **24** (1972).

[197] ——, *The inverse time-dependent scattering problem*, "Naukova Dumka," Kiev, 1973. (Russian)

[198] P. NOVIKOFF [P. S. Novikov], *Sur le problème inverse du potentiel*, C.R. (Dokl.) Acad. Sci. URSS **18** (1938), 165–168.

[199] I. G. PETROVSKIĬ, *Partial differential equations*, 3rd ed., Fizmatgiz, Moscow, 1961; English transl., Saunders, Philadelphia, Pa., 1967.

[200] ——, *Ordinary differential equations*, 5th ed., "Nauka," Moscow, 1964; English transl., Prentice-Hall, Englewood Cliffs, N. J., 1966.

[201] G. I. PLAKSIN, *On the expression of a function in terms of its integrals over ellipsoids*, Dokl. Akad. Nauk SSSR **166** (1966), 548–550; English transl. in Soviet Math. Dokl. **7** (1966).

[202] ——, *On a problem of Gel′fand*, Dokl. Akad. Nauk SSSR **170** (1966), 783–785; English transl. in Soviet Math. Dokl. **7** (1966).

[203] A. I. PRILEPKO, *Uniqueness of the shape of a body determined from exterior values of the potential*, Dokl. Akad. Nauk SSSR **160** (1965), 40–43; English transl. in Soviet Math. Dokl. **6** (1965).

[204] ——, *Inverse contact problems of generalized magnetic potentials*, Dokl. Akad. Nauk SSSR **181** (1968), 1065–1068; English transl. in Soviet Math. Dokl. **9** (1968).

[205] ——, *Inverse problems of potential theory*, Mat. Zametki **14** (1973), 755–767; English transl. in Math. Notes **14** (1973).

[206] K. G. REZNITSKAYA, *An existence and uniqueness theorem for a one-dimensional linear inverse problem of the theory of heat conduction*, Mat. Problemy Geofiz. Vyp. 4 (1973), 131–134. (Russian) R. Zh. Mat. **1974** #1Б384.

[207] ——, *The connection between solutions of the Cauchy problem for equations of different types and inverse problems*, Mat. Problemy Geofiz. Vyp. 5 (1974), part 1, 55-62. (Russian) Zbl. **328** #35072.

[208] ____, *A uniqueness theorem for some inverse problems for the diffusion equation*, Mat. Problemy Geofiz. Vyp. 6 (1975), part 1, 154–159. (Russian) Zbl. **323** #35073.

[209] V. G. ROMANOV, *The recovery of a function from its integrals over ellipsoids of revolution with one fixed focus*, Dokl. Akad. Nauk SSSR **173** (1967), 766–769; English transl. in Soviet Math. Dokl. **8** (1967).

[210] ____, *Reconstructing a function by means of integrals along a family of curves*, Sibirsk. Mat. Zh. **8** (1967), 1206–1208; English transl. in Siberian Math. J. **8** (1967).

[211] ____, *The one-dimensional inverse problem for the telegraph equation*, Differentsial'nye Uravneniya **4** (1968), 87–101; English transl. in Differential Equations **4** (1968).

[212] ____, *Formulation of the inverse problem for the generalized wave equation*, Dokl. Akad. Nauk SSSR **181** (1968), 554–557; English transl. in Soviet Math. Dokl. **9** (1968).

[213] ____, *A problem of integral geometry and a linear inverse problem for a hyperbolic equation*, Sibirsk. Mat. Zh. **10** (1969), 1364–1374; English transl. in Siberian Math. J. **10** (1969).

[214] ____, *An inverse problem on the propagation of electric oscillations in wires*, Mat. Problemy Geofiz. Vyp. 1 (1969), 92–102. (Russian) R. Zh. Mat. **1970** #12Б502.

[215] ____, *Inverse problems and integral geometry*, Collection of Papers All-Union Sympos. Inverse Problems, Vychisl. Tsentr Sibirsk. Otdel. Akad. Nauk SSSR, Novosibirsk, 1971, pp. 53–63. (Russian)

[216] ____, *A uniqueness theorem for a one-dimensional inverse problem for the wave equation*, Mat. Problemy Geofiz. Vyp. 2 (1971), 100–142. R. Zh. Mat. **1971** #12Б471.

[217] ____, *Some inverse problems for equations of hyperbolic type*, "Nauka," Novosibirsk, 1969; English transl., *Integral geometry and inverse problems for hyperbolic equations*, Springer-Verlag, 1974.

[218] ____, *Uniqueness theorems for a class of inverse problems*, Dokl. Akad. Nauk SSSR **204** (1974), 1075–1076; English transl. in Soviet Phys. Dokl. **17** (1972/73).

[219] ____, *A uniqueness and stability theorem for a nonlinear operator equation*, Dokl. Akad. Nauk SSSR **207** (1972), 1051–1053; English transl. in Soviet Math. Dokl. **13** (1972).

[220] ____, *The problem of determining the one-dimensional propagation speed of signals in a half-space on the basis of the regime of oscillations of a point of this space*, Mat. Problemy Geofiz. Vyp. 3 (1972), 164–186. (Russian) R. Zh. Mat. **1973** #5Б446.

[221] ——, *An abstract inverse problem, and questions of whether it is well-posed*, Funktsional. Anal. i Prilozhen. **7** (1973), no. 3, 67–74; English transl. in Functional Anal. Appl. **7** (1973).

[222] ——, *A one-dimensional inverse problem for the wave equation*, Dokl. Akad. Nauk SSSR **211** (1973), 1083–1084; English transl. in Soviet Phys. Dokl. **18** (1973/74).

[223] ——, *Inverse problems for differential equations*, Novosibirsk. Gos. Univ., Novosibirsk, 1973. (Russian)

[224] ——, *On a uniqueness theorem for a problem of integral geometry on a family of curves*, Mat. Problemy Geofiz. Vyp. 4 (1973), 140–146. (Russian) R. Zh. Mat. **1974** #2Б572.

[225] ——, *On a uniqueness class for a solution of the inverse kinematic problem*, Mat. Problemy Geofiz. Vyp. 4 (1973), 147–164. (Russian) R. Zh. Mat. **1974** #1Б422.

[226] ——, *Uniqueness of the determination of an isotropic Riemannian metric inside a domain by means of the distance between points of the boundary*, Dokl. Akad. Nauk SSSR **218** (1974), 295–297; English transl. in Soviet Math. Dokl. **15** (1974).

[227] ——, *On the uniqueness of a solution of the inverse kinematic problem in a disc in a class of velocities close to constants*, Mat. Problemy Geofiz. Vyp. 5 (1974), part 2, 108–142. (Russian) Zbl. **331** #73096.

[228] ——, *On the question of whether inverse problems for nonlinear equations are well-posed*, Funktsional. Anal. i Prilozhen. **8** (1974), no. 3, 67–70; English transl. in Functional Anal. Appl. **8** (1974).

[229] ——, *On some uniqueness classes for the solution of problems of integral geometry*, Mat. Zametki **16** (1974), 657–668; English transl. in Math. Notes **16** (1974).

[230] ——, *On some uniqueness classes for the solution of Volterra operator equations of the first kind*, Funktsional Anal. i Prilozhen. **9** (1975), no. 1, 81–82; English transl. in Functional Anal. Appl. **9** (1975).

[231] ——, *Volterra operator equations of the first kind. Uniqueness classes*, Some Problems of Numerical and Applied Mathematics (G. I. Marchuk Fiftieth Birthday Vol.; M. M. Lavrent′ev, editor), "Nauka," Novosibirsk, 1975, pp. 123–135. (Russian)

[232] ——, *On an inverse problem for a parabolic equation*, Mat. Zametki **19** (1976), 595–600; English transl. in Math. Notes **19** (1976).

[233] ——, *On an inverse problem for weakly coupled hyperbolic systems of first order*, in [331], pp. 135–148. (Russian) Zbl. **353** #35083.

[234] ____, *On the problem of determining the right-hand side of a hyperbolic system*, Differentsial'nye Uravneniya **13** (1977), 509–515; English transl. in Differential Equations **13** (1977).

[235] ____, *The problem of determining the coefficients of a linear hyperbolic system*, Differentsial'nye Uravneniya **14** (1978), 94–103; English transl. in Differential Equations **14** (1973).

[236] ____, *Integral geometry on the geodesics of an isotropic Riemannian metric*, Dokl. Akad. Nauk SSSR **241** (1978), 290–293; English transl. in Soviet Math. Dokl. **19** (1978).

[237] ____, *Inverse problems for hyperbolic equations and energy inequalities*, Dokl. Akad. Nauk SSSR **242** (1978), 541–544; English transl. in Soviet Math. Dokl. **19** (1978).

[238] ____, *Inverse problems for hyperbolic systems*, Numerical Methods in Mathematical Physics, Geophysics and Optimal Control (Sympos., Novosibirsk, 1976; G. I. Marchuk and J. L. Lions, editors), "Nauka," Novosibirsk, 1978, pp. 75–83; French transl., Étude Numerique des Grands Systèmes (Proc. Sympos., Novosibirsk, 1976), Méthodes Math. de l'Informatique, no. 7, Dunod, Paris, 1978, pp. 100–109.

[239] ____, *Formulation of an inverse problem for first-order symmetric hyperbolic systems*, Mat. Zametki **24** (1978), 231–236; English transl. in Math. Notes **24** (1978).

[240] ____, *Inverse problems for hyperbolic equations and energy inequalities*, Mat. Zametki **24** (1978), 541–544; English transl. in Math. Notes **24** (1978).

[241] ____, *Inverse problems for differential equations. Inverse kinematic problems of seismology*, Novosibirsk. Gos. Univ., Novosibirsk, 1978. (Russian)

[242] V. G. ROMANOV AND S. P. BELINSKIĬ, *On the problem of determining the coefficients of a t-hyperbolic system*, Preprint no. 23, Vychisl. Tsentr Sibirsk. Otdel. Akad. Nauk SSSR, Novosibirsk, 1976, pp. 16–24. (Russian) R. Zh. Mat. **1977** #10Б392.

[243] V. G. ROMANOV AND L. I. SLINYUCHEVA, *An inverse problem for first-order linear hyperbolic systems*, Mat. Problemy Geofiz. Vyp. 3 (1972), 187–215. (Russian) R. Zh. Mat. **1973** #5Б378.

[244] V. G. ROMANOV AND V. G. YAKHNO, *On a linearized formulation of the problem of determining a hyperbolic operator*, Preprint no. 23, Vychisl. Tsentr Sibirsk. Otdel. Akad. Nauk SSSR, Novosibirsk, 1976, pp. 3–15. (Russian) R. Zh. Mat. **1977** #10Б392.

[245] V. O. SERGEEV, *Regularization of a Volterra equation of the first kind*, Dokl. Akad. Nauk SSSR **197** (1971), 531–534; English transl. in Soviet Math. Dokl. **12** (1971).

[246] V. I. SMIRNOV, *A course in higher mathematics.* Vol. IV, 3rd ed., GITTL, Moscow, 1953; English transl., Pergamon Press, Oxford, and Addison-Wesley, Reading, Mass., 1964.

[247] S. L. SOBOLEV, *Applications of functional analysis in mathematical physics*, Izdat. Leningrad. Gos. Univ., Leningrad, 1950; reprint, Izdat. Sibirsk. Otdel. Akad. Nauk SSSR, Novosibirsk, 1962; English transl., Amer. Math. Soc., Providence, R. I., 1963.

[248] ——, *Equations of mathematical physics*, 4th ed., "Nauka," Moscow, 1966; English transl. of 3rd ed., *Partial differential equations of mathematical physics*, Pergamon Press, Oxford, and Addison-Wesley, Reading, Mass., 1964.

[249] V. N. STRAKHOV, *The analytic continuation of two-dimensional potential fields, with applications to the solution of the inverse problem of magnetic and gravitational exploration.* I, II, III, Izv. Akad. Nauk SSSR Ser. Geofiz. **1962**, 307–316, 336–347, 491–507; English transl. in Bull. (Izv.) Acad. Sci. USSR Geophys. Ser. **1962**.

[250] ——, *The theory of approximate solution of linear ill-posed problems in Hilbert space and its utilization in geophysical prospecting.* I, II, Izv. Akad. Nauk SSSR Fiz. Zemli **1969**, no. 8, 30–53; no. 9, 64–96; English transl. in Izv. Acad. Sci. USSR Phys. Solid Earth **1969**.

[251] ——, *Solution of linear ill-posed problems in Hilbert space*, Differentsial′nye Uravneniya **6** (1970), 1490–1495; English transl. in Differential Equations **6** (1970).

[252] ——, *On methods of approximate solution of linear conditionally well-posed problems*, Dokl. Akad. Nauk SSSR **196** (1971), 786–788; English transl. in Soviet Math. Dokl. **12** (1971).

[253] V. P. TANANA, *Ill-posed problems and the geometry of Banach spaces*, Dokl. Akad. Nauk SSSR **193** (1970), 43–45; English transl. in Soviet Math. Dokl. **11** (1970).

[254] ——, *Approximate solution of operator equations of the first kind and geometric properties of Banach spaces*, Izv. Vyssh. Uchebn. Zaved. Mat. **1971**, no. 7 (110), 81–93. (Russian)

[255] ——, *On the structure of classes of uniform regularization in Hilbert space*, Ural. Gos. Univ. Mat. Zap. **9** (1975), no. 2, 125–131. (Russian)

[256] A. F. TIMAN, *Theory of approximation of functions of a real variable*, Fizmatgiz, Moscow, 1960; English transl., Pergamon Press, Oxford, and Macmillan, New York, 1963.

[257] A. N. TIKHONOV, *On the stability of inverse problems*, C. R. (Dokl.) Acad. Sci. URSS **39** (1943), 176–179.

[258] ——, *On the solution of ill-posed problems and a method of regularization*, Dokl. Akad. Nauk SSSR **151** (1963), 501–504; English transl. in Soviet Math. Dokl. **4** (1963).

[259] ——, *On regularization of ill-posed problems*, Dokl. Akad. Nauk SSSR **153** (1963), 49–52; English transl. in Soviet Math. Dokl. **4** (1963).

[260] ——, *Stable methods for the summation of Fourier series*, Dokl. Akad. Nauk SSSR **156** (1965), 268–271; English transl. in Soviet Math. Dokl. **5** (1964), 268–271; errata, ibid. **6** (1965), no. 1, p. 300-d.

[261] ——, *On the solution of nonlinear integral equations of the first kind*, Dokl. Akad. Nauk SSSR **156** (1964), 1296–1299; English transl. in Soviet Math. Dokl. **5** (1964).

[262] ——, *On nonlinear equations of the first kind*, Dokl. Akad. Nauk SSSR **161** (1965), 1023–1026; English transl. in Soviet Math. Dokl. **6** (1965).

[263] ——, *On ill-posed problems of linear algebra and a stable method for their solution*, Dokl. Akad. Nauk SSSR **163** (1965), 591–594; English transl. in Soviet Math. Dokl. **6** (1965).

[264] ——, *On methods for the regularization of optimal control problems*, Dokl. Akad. Nauk SSSR **162** (1965), 763–765; English transl. in Soviet Math. Dokl. **6** (1965).

[265] ——, *Ill-posed problems of optimal planning and stable methods for their solution*, Dokl. Akad. Nauk SSSR **164** (1965), 507–510; English transl. in Soviet Math. Dokl. **6** (1965).

[266] A. N. TIKHONOV AND V. YA. ARSENIN, *Methods of solving ill-posed problems*, "Nauka," Moscow, 1974; English transl., Wiley, 1977.

[267] A. N. TIKHONOV AND V. B. GLASKO, *Approximate solution of Fredholm integral equations of the first kind*, Zh. Vychisl. Mat. i Mat. Fiz. **4** (1964), 564–571; English transl. in USSR Comput. Math. and Math. Phys. **4** (1964).

[268] A. N. TIKHONOV, V. K. IVANOV AND M. M. LAVRENT′EV, *Ill-posed problems*, Partial Differential Equations (Proc. Sympos. Sixtieth Birthday S. L. Sobolev), "Nauka," Moscow, 1970, pp. 224–238; English transl. in Amer. Math. Soc. Transl. (2) **105** (1976).

[269] V. F. TURCHIN, *Solution of a Fredholm integral equation of the first kind in a statistical ensemble of smooth functions*, Zh. Vychisl. Mat. i Mat. Fiz. **7** (1967), 1270–1284; English transl. in USSR Comput. Math. and Math. Phys. **7** (1967).

[270] S. V. USPENSKIĬ, *On the reconstruction of a function defined by integrals over a family of ellipsoids*, Dokl. Akad. Nauk SSSR **202** (1972), 548–550; English transl. in Soviet Math. Dokl. **13** (1972).

[271] ——, *On the reconstruction of a function defined by integrals over a family of ellipsoids*, Sibirsk. Mat. Zh. **13** (1972), 1374–1382; English transl. in Siberian Math. J. **13** (1972).

[272] ——, *On the reconstruction of a function given by integrals over a family of conical surfaces*, Sibirsk. Mat. Zh. **18** (1977), 675–684; English transl. in Siberian Math. J. **18** (1977).

[273] S. V. USPENSKIĬ AND S. B. SADYKOVA, *On some problems in integral geometry*, Dokl. Akad. Nauk SSSR **222** (1975), 295–298; English transl. in Soviet Math. Dokl. **16** (1975).

[274] ——, *On some problems in integral geometry*, Sibirsk. Mat. Zh. **17** (1976), 414–425; English transl. in Siberian Math. J. **17** (1976).

[275] L. D. FADDEEV, *The inverse problem in quantum scattering theory. II*, Itogi Nauki: Sovremennye Problemy Mat., vol. 3, VINITI, Moscow, 1974, pp. 93–180; English transl. in J. Soviet Math. **5** (1976), no. 3.

[276] L. S. FRANK AND L. A. CHUDOV, *Difference methods for solving an ill-posed Cauchy problem*, Numerical Methods in Gas Dynamics, Vychisl. Metody i Programmirovanie Vyp. 4 (1965), 3–27. (Russian)

[277] V. M. FRIDMAN, *The method of successive approximations for a Fredholm integral equation of the first kind*, Uspekhi Mat. Nauk **11** (1956), no. 1 (67), 233–234. (Russian)

[278] B. A. FUKS, *Introduction to the theory of analytic functions of several complex variables*, Fizmatgiz, Moscow, 1962; English transl., Amer. Math. Soc., Providence, R. I., 1963.

[279] V. N. SUDAKOV AND L. A. KHALFIN, *A statistical approach to the well-posed property of problems in mathematical physics*, Dokl. Akad. Nauk SSSR **157** (1964), 1058–1060; English transl. in Soviet Math. Dokl. **5** (1964).

[280] A. A. KHACHATUROV, *Determination of the value of the measure of a region of n-dimensional Euclidean space from its values for all half-spaces*, Uspekhi Mat. Nauk **9** (1954), no. 3 (61), 205–212. (Russian)

[281] A. YA. KHINCHIN, *Continued fractions*, 3rd ed., Fizmatgiz, Moscow, 1961; English transls., Noordhoff, 1963, and Univ. of Chicago Press, Chicago, Ill., 1964.

[282] YU. I. KHUDAK, *Regularization of solutions of integral equations of the first kind*, Zh. Vychisl. Mat. i Mat. Fiz. **6** (1966), 766–769; English transl. in USSR Comput. Math. and Math. Phys. **6** (1966).

[283] L. A. CHUDOV, *Difference schemes and ill-posed problems for partial differential equations*, Vychisl. Metody i Programmirovanie Vyp. 8 (1967), 34–62. (Russian)

[284] B. V. SHABAT, *Introduction to complex analysis*, "Nauka," Moscow, 1969. (Russian)

[285] S. P. SHISHAT·SKIĬ, *On a method for the approximate solution of an ill-posed Cauchy problem for an evolution equation*, Mat. Problemy Geofiz. Vyp. 3 (1972), 216–228. (Russian) R. Zh. Mat. **1973** #5Б717.

[286] ——, *A priori estimates in the problem of extending a wave field from a cylindrical time-like surface*, Dokl. Akad. Nauk SSSR **213** (1973), 49–50; English transl. in Soviet Math. Dokl. **14** (1973).

[287] ——, *Uniqueness of a solution of the Cauchy problem with data on a segment of the time axis for a degenerate parabolic equation*, Differentsial′nye Uravneniya **11** (1975), 1453–1459; English transl. in Differential Equations **11** (1975).

[288] ——, *On determining a function harmonic in a plane region on the basis of its values on three parallel segments*, Questions of the Well-Posedness of Problems of Mathematical Physics (M. M. Lavrent′ev, editor), Vychisl. Tsentr Sibirsk. Otdel. Akad. Nauk SSSR, Novosibirsk, 1977, pp. 143–149. (Russian) R. Zh. Mat. **1978** #10Б172.

[289] ——, *Uniqueness and stability of some Cauchy problems for second-order equations*, Numerical Methods in Mathematical Physics, Geophysics and Optimal Control (Sympos., Novosibirsk, 1976; G. I. Marchuk and J. L. Lions, editors), "Nauka," Novosibirsk, 1978, pp. 209–215; French transl., Étude Numerique des Grands Systèmes (Proc. Sympos., Novosibirsk, 1976), Méthodes Math. de l'Informatique, no. 7, Dunod, Paris, 1978, pp. 92–99.

[290] S. P. SHISHAT·SKIĬ AND È. R. ATAMANOV, *Uniqueness and an estimate of stability in a problem for a pseudoparabolic equation*, Preprint No. 105, Vychisl. Tsentr Sibirsk. Otdel. Akad. Nauk SSSR, Novosibirsk, 1977. (Russian) R. Zh. Mat. **1978** #11Б1341.

[291] S. P. SHISHAT·SKIĬ AND K. S. FAYAZOV, *The Carleman operator for an evolution equation of elliptic type*, Preprint No. 81, Vychisl. Tsentr Sibirsk. Otdel Akad. Nauk SSSR, Novosibirsk, 1977. (Russian) MR **58** #22897.

[292] L. È. ÈL′SGOL′TS, *Differential equations and calculus of variations*, "Nauka," Moscow, 1965; English transl., "Mir," Moscow, 1970.

[293] V. G. YAKHNO, *An inverse problem for the nonlinear equation of a vibrating string*, Mat. Problemy Geofiz. Vyp. 4 (1973), 179–192. (Russian) R. Zh. Mat. **1974** #1Б340.

[294] ____, *A uniqueness theorem for a quasilinear hyperbolic equation*, Mat. Problemy Geofiz. Vyp. 5 (1974), part 1, 73–91. (Russian) (Zbl. **323** #35071.

[295] ____, *An existence and uniqueness theorem for a one-dimensional inverse problem for a second-order quasilinear hyperbolic equation*, Mat. Problemy Geofiz. Vyp. 5 (1974), part 2, 173–183. (Russian) Zbl. **323** #35068.

[296] ____, *Uniqueness theorem for an inverse problem for a hyperbolic equation*, Differentsial'nye Uravneniya **13** (1977), 544–551; English transl. in Differential Equations **13** (1977).

[297] RÉMY ARCANGELI, *Pseudo-solution de l'équation* $Ax = y$, C.R. Acad. Sci. Paris Sér. A-B **263** (1966), A282–A285.

[298] N. ARONSZAJN, *A unique continuation theorem for solutions of elliptic partial differential equations or inequalities of second order*, J. Math. Pures Appl. (9) **36** (1957), 235–249.

[299] TORSTEN CARLEMAN, *Les fonctions quasi analytiques*, Gauthier-Villars, Paris, 1926.

[300] HEINZ OTTO CORDES, *Über die eindeutige Bestimmtheit der Lösungen elliptischer Differentialgleichungen durch Anfangsvorgaben*, Nachr. Akad. Wiss. Göttingen Math.-Phys. Kl. II a: Math.-Phys.-Chem. Abt. **1956**, 239–258.

[301] JIM DOUGLAS, JR., *A numerical method for analytic continuation*, Boundary Problems in Differential Equations (Proc. Sympos., Madison, Wisc., 1959; R. E. Langer, editor), Univ. of Wisconsin Press, Madison, Wisc., 1960, pp. 179–189.

[302] JIM DOUGLAS, JR., AND T. M. GALLIE, JR., *An approximate solution of an improper boundary value problem*, Duke Math. J. **26** (1959), 339–347.

[303] NELSON DUNFORD, *Spectral operators*, Pacific J. Math. **4** (1954), 321–354.

[304] S. R. FOGUEL, *The relations between a spectral operator and its scalar part*, Pacific J. Math. **8** (1958), 51–65.

[305] DAVID W. FOX AND CARLO PUCCI, *The Dirichlet problem for the wave equation*, Ann. Mat. Pura Appl. (4) **46** (1958), 155–182.

[306] JACQUES HADAMARD, *Sur les problèmes aux dérivées partielles et leur signification physique*, Princeton Univ. Bull. **13** (1902), 49–52; reprinted in his *Oeuvres*. Vol. III, Centre Nat. Recherche Sci., Paris, 1968, pp. 1099–1105.

[307] ____, *Le problème de Cauchy et les équations aux dérivées partielles linéaires hyperboliques*, Hermann, Paris, 1932.

[308] FRITZ JOHN, *Bestimmung einer Funktion aus ihren Integralen über gewisse Mannigfaltigkeiten*, Math. Ann. **109** (1933/34), 488–520.

[309] ——, *Abhängigkeiten zwischen den Flächenintegralen einer stetigen Funktionen*, Math. Ann. **111** (1935), 541–559.

[310] ——, *On linear partial differential equations with analytic coefficients*, Comm. Pure Appl. Math. **2** (1949), 209–253.

[311] ——, *Numerical solution of the equation of heat conduction for preceding times*, Ann. Mat. Pura Appl. (4) **40** (1955), 129–142.

[312] ——, *A note on "improper" problems in partial differential equations*, Comm. Pure Appl. Math. **8** (1955), 591–594.

[313] ——, *Differential equations with approximate and improper data*, Lectures, Courant Inst. Math. Sci., New York Univ., New York, 1955.

[314] ——, *Numerical solution of problems which are not well posed in the sense of Hadamard*, Sympos. Numer. Treatment of Partial Differential Equations with Real Characteristics (Rome, 1959), Birkhäuser, 1959, pp. 103–116.

[315] ——, *Continuous dependence on data for solutions of partial differential equations with a prescribed bound*, Comm. Pure Appl. Math. **13** (1960), 551–585.

[316] KYÛYA MASUDA, *A unique continuation theorem for solutions of wave equations with variable coefficients*, J. Math. Anal. Appl. **21** (1968), 369–376.

[317] SIGERU MIZOHATA, *Unicité du prolongement des solutions pour quelques opérateurs différentiels paraboliques*, Mem. Coll. Sci. Univ. Kyoto Ser. A Math. **31** (1958), 219–239.

[318] A. C. MURRAY AND M. H. PROTTER, *Asymptotic behavior and the Cauchy problem for ultrahyperbolic operators*, Indiana Univ. Math. J. **24** (1974), 115–130.

[319] D. J. NEWMAN, *Numerical method for solution of an elliptic Cauchy problem*, J. Math. and Phys. **39** (1960), 72–75.

[320] R. G. NEWTON, *Inverse problems in physics*, SIAM Rev. **12** (1970), 346–356.

[321] M. H. PROTTER, *Properties of solutions of parabolic equations and inequalities*, Canad. J. Math. **13** (1961), 331–345.

[322] CARLO PUCCI, *Sui problemi di Cauchy non "ben posti"*, Atti Accad. Naz. Lincei Rend. Cl. Sci. Fis. Mat. Nat. (8) **18** (1955), 473–477.

[323] ——, *Discussione del problema di Cauchy per le equazioni di tipo ellittico*, Ann. Mat. Pura Appl. (4) **46** (1958), 131–153.

[324] ——, *On the improperly posed Cauchy problems for parabolic equations*, Sympos. Numer. Treatment of Partial Differential Equations with Real Characteristics (Rome, 1959), Birkhäuser, 1959, pp. 140–144.

[325] JOHANN RADON, *Über die Bestimmung von Funktionen durch ihre Integralwerte längs gewisser Mannigfaltigkeiten*, Ber. Verh. Sächs. Ges. Wiss. Leipzig Math. Phys. Kl. **1917**, 262–277.

[326] WILLIAM RUNDELL AND MICHAEL STECHER, *Remarks concerning the supports of solutions of pseudoparabolic equations*, Proc. Amer. Math. Soc. **63** (1977), 77–81.

[327] OTTO NEALL STRAND AND ED R. WESTWATER, *Statistical estimation of the numerical solution of a Fredholm integral equation of the first kind*, J. Assoc. Comput. Mach. **15** (1968), 100–114.

[328] ROBERT S. STRICHARTZ, *The stationary observer problem for* $\Box u = Mu$ *and related equations*, J. Differential Equations **9** (1971), 205–223.

[329] R. J. KNOPS (editor), *Symposium on non-well-posed problems and logarithmic convexity* (*Edinburgh, 1972*), Lecture Notes in Math., vol. 316, Springer-Verlag, 1973.

[330] S. TWOMEY, *On the numerical solution of Fredholm integral equations of the first kind by the inversion of the linear system produced by quadrature*, J. Assoc. Comput. Mach. **10** (1963), 97–101.

[331] M. M. LAVRENT'EV AND A. S. ALEKSEEV (editors), *Ill-posed mathematical problems and problems of geophysics*, Vychisl. Tsentr Sibirsk. Otdel. Akad. Nauk SSSR, Novosibirsk, 1976. (Russian)